All Creatures

Property of Library
Cape Fear Comm College
Wilmington, N. C.

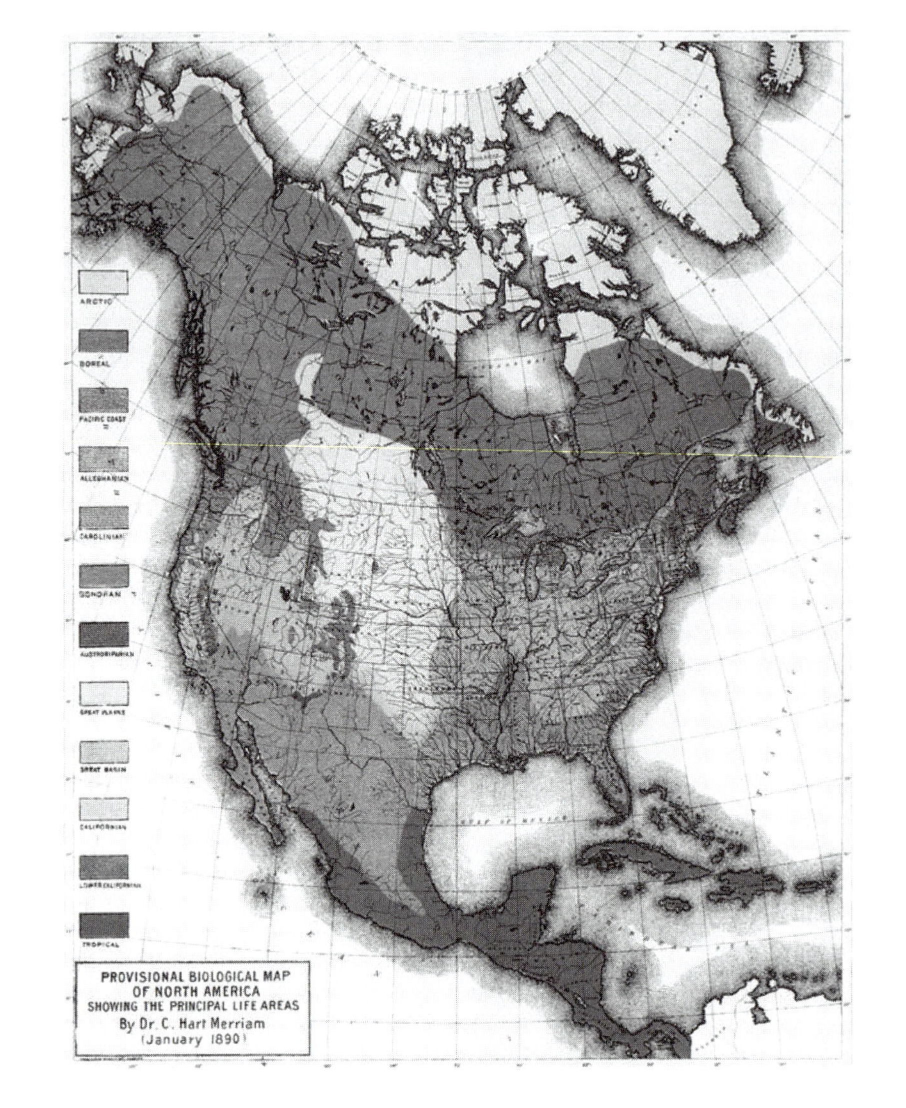
ARCTIC
BOREAL
PACIFIC COAST
ALLEGHANIAN
CAROLINIAN
SONORAN
AUSTRORIPARIAN
GREAT PLAINS
GREAT BASIN
CALIFORNIAN
LOWER CALIFORNIAN
TROPICAL
PROVISIONAL BIOLOGICAL MAP
OF NORTH AMERICA
SHOWING THE PRINCIPAL LIFE AREAS
By Dr. C. Hart Merriam
(January 1890)

All Creatures

Naturalists, Collectors, and Biodiversity, 1850–1950

Robert E. Kohler

PRINCETON UNIVERSITY PRESS
PRINCETON AND OXFORD

Copyright © 2006 by Princeton University Press
Published by Princeton University Press, 41 William Street,
Princeton, New Jersey 08540
In the United Kingdom: Princeton University Press,
3 Market Place, Woodstock, Oxfordshire OX20 1SY
All Rights Reserved

Library of Congress Cataloging-in-Publication Data

Kohler, Robert E.
All creatures : naturalists, collectors, and
biodiversity, 1850–1950 / Robert E. Kohler.
p. cm.
Includes bibliographical references and index.
ISBN-13: 978-0-691-12539-8 (cl : alk. paper)
ISBN-10: 0-691-12539-2 (cl : alk. paper)
1. Natural history—History. 2. Biological specimens—
Collectors and collecting—History. I. Title.

QH15.K642006
508—dc22 2005055095

British Library Cataloging-in-Publication Data is available

This book has been composed in Sabon

Printed on acid-free paper. ∞

pup.princeton.edu

Printed in the United States of America

1 2 3 4 5 6 7 8 9 10

For my students

Contents

Illustrations

Figures

Table

Preface

BOOKS OFTEN GROW out of small but nagging puzzles that open up, when picked at, into unexpected worlds of human experience. This one began when, as I was reading in the history of field biology, I began to wonder how exactly we know about all those species that naturalists—and all the rest of us as well—take for granted. Who were the modern Noahs that went out and discovered and sorted the animals and plants? And how did they go about their work, and where, and when? It seemed important to find answers to these questions: after all, our own future as a species is likely to depend on how well we understand our fellow creatures.

Although naturalists have been finding and naming for centuries, it turns out that the full discovery of biodiversity is in fact quite modern. It was only in the early twentieth century that our knowledge of even the major animal groups became essentially complete. I also discovered that natural history expeditioning was a much bigger business in our own time than I or any historian had imagined. From the 1880s into the 1930s, museums and government surveys dispatched dozens or scores of expeditions a year to collect specimens and map the biogeography of species. It was like coming upon a large town in what was supposed to be an empty place. It made me want to chart its streets and byways.

I call this a discovery, yet this huge scientific effort was hidden in plain sight. Expeditions were well publicized at the time and created abundant paper trails in museums' annual reports and archives. We historians did not see this evidence before because we didn't look, or because we already knew that in the early twentieth century the natural history sciences were being outgrown by the aggressively expansive laboratory sciences. They were outgrown, that is true; but the natural history sciences were not for that reason in decline. In fact, systematics and biogeography were in a period of vigorous growth and intellectual progress, especially in museums and public agencies—overshadowed perhaps, at least in universities, but not shaded out. Recovering this lost history became for me a compelling project.

A second reason to take up this project was that the history turned out to be interestingly complex. The practices of modern systematics and biogeography, I found, were more deeply rooted in environmental, social, and cultural developments than one might have expected. Access to the places where creatures lived, for example, depended on the conjunction of environmental and cultural developments. The natural environments of North America (my particular area) were in an unusual transition between the 1880s and 1930s: readily accessible for the first time, by rail and increasingly by road, yet still relatively wild, with their floras and faunas still intact though increasingly squeezed. The same was true in the other parts of the world where American naturalists made their expeditions, where prairies and rain forests were being opened up to trade and extraction but were not yet densely settled.

If these "inner frontiers" were made physically accessible by steam transport, it was changes in society and culture that gave naturalists intellectual and aesthetic reasons to learn about them. Attitudes to nature were changing, as "wilderness" became less threatening and more inviting and new customs arose that brought people into intimate contact with nature. Outdoor recreation especially—camping, mountain walking, rural cottaging, amateur naturalizing—encouraged a naturalistic and, for some, a scientific interest in nature.

Evidence for this new interest is again abundant, and most vividly in the public enthusiasm for museum dioramas, which continue to charm museum-goers to this day. In fact, it was the public taste for dioramas that impelled civic museums to undertake collecting expeditions on a vast scale. I found it necessary to delve into the history of the evolving conventions of nature art—essays, animal sculpture, dioramas—as well as the new forms of outdoor recreation that gave Americans a taste for such art forms and made some willing to pay for scientific expeditions.

This book is a history, too, of how science is organized and reorganized socially: of the transformation of museums from passive repositories of stuff to active collectors and of museum curators from stay-at-home housekeepers to field naturalists. It is a history of the modern scientific expedition as a kind of complex scientific instrument, as exacting and intricate in its design and operation as any ever devised. It is a history of scientific travel; of collecting as an exacting scientific practice; and of the mixed emotions that attend the pursuit of a science that is at once rigorous and recreational.

I have striven to make this book accessible and engaging not just to professional historians of science but to scientific practitioners and to anyone interested in why sciences of all kinds flourish in modern society. I hope specialist historians will also learn from the book, but I wrote it more for historical generalists and curious general readers. I have focused on individual naturalists and their practices and experiences more than on the arcane knowledge they produced. I have tried to give readers a vivid sense of what it was like for naturalists and others to participate in discovering the full guest list of our celestial ark.

It is never easy to strike a balance between professional historians' demands for rigorous argument and general readers' desire for a bold and engaging story. But the subject invites and rewards attempts to try. Enjoy.

Finally, there are intellectual debts to acknowledge. First and foremost to the many archivists and librarians who have made researching this book a deeply pleasurable voyage of discovery. They are a noble race! May they thrive and multiply in the age of infojunk and Internet. To John Pollack, Christopher Jones, and Matthew Hersch goes my thanks for help with digital images. Sally Kohlstedt, Jennifer Light, and Jeremy Vetter read a penultimate draft of the manuscript and provided valuable reactions and suggestions. But above all, my thanks go to Frances Kohler for applying (yet again) her incomparable editorial skills to this book.

All Creatures

CHAPTER ONE

Nature

WE HUMANS are one in a million: to be exact, one species among 1,392,485, according to a recent tally by the zoologist Edward O. Wilson. Those are the ones we know: estimates of the total number of living species range from five to thirty million and up, depending on how one reckons.[1] A substantial majority of Earth's species are insects: something like 751,000 by Wilson's tally. Plants account for another 248,428, the vast majority being flowering plants (which coevolved with insect pollinators). Among the vertebrates, bony fishes are the largest group, with 18,150 species, leaving aside the 63 species of jawless fishes and the 843 cartilaginous fishes (lampreys, sharks). Amphibia and reptiles account for 4,184 and 6,300 more species; birds for 9,040, and mammals for 4,000, give or take. Not to mention invertebrates other than insects: tunicates and cephalochordata (1,273), molluscs (roughly 50,000), and arthropods (12,161). And single-cell organisms: algae (26,900), fungi (46,983), protozoa and microbes (36,560).[2] Of our fellow vertebrates we have an inventory that is nearly complete—over 90 percent, it is estimated. On the plants and invertebrates, however, we may only have made a start. We earthlings sail through the void on an ark that is impressively biodiverse.

Biodiversity is a lively issue these days, mainly because of the number of species that are going extinct, either by natural causes, or because we space-hungry humans are destroying their habitats. Wilson estimates that perhaps 17,500 species (mostly insects) go extinct each year in tropical forests, and that we humans have accelerated the historical rate of extinction by a factor of one thousand to ten thousand.[3] Biologists and conservationists are concerned that vast numbers of species may be forced into extinction ahead of schedule (extinction is the ultimate fate of all species) before they can be found and classified. There is concern, too, that in our ignorance we may be destroying species vital to the fabric of ecosystems on which we depend for our own survival.

Systematic biology, or taxonomy, is reputed to be a humdrum, cataloging science—a reputation entirely undeserved, let it be said.[4] We depend on those few among us who collect, describe, name, and classify our fellow passengers on the global ark. But how exactly do we find, collect, identify, and order those millions of species? That is my subject here: not the biology or the ethics of biodiversity, but its practices and their history. Though people have always named plants and animals, the science of species inventory is relatively new, beginning with the big bang of Carl von Linné's invention of the (Linnaean) binomial system of naming in the mid-eighteenth century.[5] And though much has been written on theories of species, relatively little is known of the practical work that produced the empirical base for theorizing. When and how were those inventories created and made robust? Who organized and paid for collecting expeditions, collected and prepared specimens in the field, compiled lists, built museums and herbaria, and kept vast collections in good physical and conceptual order? Of these practical activities we do not as yet know much. This book is a step toward acquiring such knowledge.

The history of our knowledge of biodiversity is first and foremost a history of collecting and collections. Remarkably little has been written about the craft and social history of scientific collecting: it remains a "black box," as the historian Martin Rudwick observed a few years ago, an activity that has "barely been described by historians, let alone analyzed adequately."[6] There are now signs of a growing interest in the history of collecting science, but it is perhaps understandable why this black box is only now being opened. Although collecting is a widespread and varied obsession, modern scientific collecting is sober and businesslike, not irregular or idiosyncratic. It is done en masse and methodically, because modern taxonomy requires large and comprehensive collections. Scientific collecting is exacting and quantitative science, as methodical and organized as taking stock of galaxies, subatomic particles, or genes. Modern specimen collections are quite unlike the romantic "cabinets of curiosities" of earlier centuries. Modern herbaria consist of cases filled with hundreds of thousands of large folios of pressed plants in paper. Museum study collections are rooms of metal boxes, each with trays of animal skins and skulls in neat rows neatly labeled—all seemingly humdrum and unromantic.

Yet the scientific visions that inspire collectors to go afield, and the varied activities that go into making large collections, are anything but humdrum. Collecting is an activity that has engaged diverse sorts of people—unlike laboratory science, which is restricted to a relatively few approved types.[7] The botanist Edgar Anderson once did an experiment, in which he took a manila folder at random from an herbarium case (a Southwestern grass, it turned out to be), to discover the kinds of people who had collected the specimens. It was an amazingly diverse lot: a botanist on the Mexican Boundary Survey of the early 1850s; an immigrant intellectual German who had come to America in 1848 to escape political persecution; the wife of a mining engineer stationed in a remote mountain range, who dealt with the isolation by studying the local flora; a Boston gentleman, who made collecting trips to New Mexico for thirty years; a Los Alamos scientist and amateur botanist; university professors of botany; and college students who bought a second auto and spent a summer holiday collecting. "Though they have sometimes been contemptuously referred to as 'taxonomic hay' by other biologists," Anderson concluded, "herbarium specimens can be quite romantic in their own dry way."[8]

Anderson's experiment is easily replicated: page through museums' accession lists, and you will see hundreds of names of people who contributed specimens to scientific collections, from a few odd skins to tens of thousands. Read taxonomists' checklists—which give for each species the name of the naturalist who first described it, and when—and you will glimpse a living community of collectors and naturalists stretching back 250 years, in which amateurs have the same honor and dignity as the most eminent professionals. Species collectors are as diverse as the species they collect, and no other community of scientists preserves such a deep sense of its collective identity and past. Taxonomists' elaborate system of keeping track of names, which anchors each species to the name historically first given to it and to the actual specimen first described—the "type" specimen—keeps the past forever present. All sciences have their heroes and founding myths, but taxonomy is about the only one with a living memory of all past contributors, famous and obscure.

Scientific collecting was (and is) also an unusually complex and varied kind of work. Collecting expeditions are more complex socially than anything one might find, say, in a biochemistry or gene-sequencing lab. They require a great deal of book knowledge, but also practical skills of woodcraft and logistics, as well as firsthand experience of animal habits and habitats. Modern natural history is an exacting science whose practitioners must also cope and improvise in difficult field conditions. Collecting expeditions afford an experience of nature that mixes scientific and recreational culture in a way that lab sciences never do. Collecting parties usually travel light and depend on local inhabitants for information and support, making survey collecting a diversely social experience. And because of that diversity, the identity of scientific collectors has been less fixed than that of laboratory workers. In the black box of modern expeditionary collecting, there is much of interest.

We know nature through work, the environmental historian Richard White has observed, whether it is poling canoes against the current of a great river (his particular case), or building dams across it to tap its energy, or hauling fish out of it, or diverting its waters for irrigated farming—or, historians may add, studying its hydrology and natural history.[9] So too is our scientific knowledge of nature acquired through the work of mounting expeditions; observing plant and animal life; and collecting, preparing and sorting specimens. Historians have only recently begun to address the work of field science.[10] And of all the field sciences, natural history survey is an exceptionally inviting subject—because the work of systematic, scientific collecting is so varied.

One is also struck, paging through scientific inventories of species, by the lumpiness of the history of their discovery. Species have accumulated steadily, but more rapidly in certain periods than in others. The first such period of discovery was the Linnaean: roughly the second half of the eighteenth century. Then, after a pause of a few decades in the early nineteenth century, another period of rapid discovery set in from the 1830s to the 1850s, which I shall call "Humboldtian," after the encyclopedic author of *Cosmos*, Alexander von Humboldt.[11] Following another pause, the pace of finding and naming again quickened from the 1880s into the 1920s, by which time a substantial proportion of vertebrate species had been found and named. Since the mid-twentieth century the pace of discovery of new vertebrate species has been a fitful trickle (though lists of invertebrates grow ever longer).

These cycles of collecting and naming vary a good deal from one group of animals to another, depending on their accessibility and interest to us. Those that are large, fierce, freakish, beautiful, edible, lovable, or dangerous were inventoried early on. These include birds, carnivores, primates, and large game. Inconspicuous or insignificant creatures, or those that do not appeal—because they are slimy, cold-blooded, annoying, nocturnal, or just very good at avoiding our notice—were not fully inventoried until the surveys of the late nineteeth and early twentieth century or even later. These groups include rodents, bats, insectivores, amphibians, and reptiles.

Birds—those visible, audible, and beloved objects of watchers and collectors—were so well inventoried in the Linnaean and Humboldtian periods that the discoveries of the later survey phase show up as mere blips on a declining curve of discovery.[12] (Fig. 1–1.) In contrast, discoveries of mammalian species display the most pronounced cyclic pattern, with marked activity in the first two phases, but the most productive collecting in the survey period.[13] (Fig. 1–1.) The pattern for North American mammals is even more pronounced, with discoveries more concentrated in the 1890s, and the earlier peak shifted from the 1830s and 1840s to the 1850s and 1860s. Different groups of mammals show some variation in this basic pattern. Most carnivore species were described in the eighteenth century, and most of the rest in the 1820s and 1830s—we humans have taken a keen interest in our closest competitors. Rodents, in contrast, were hardly known to Linnaean describers and not fully known to science until the age of survey, when it first became apparent just how prolific of species this group has been—it would appear that the Creator loves rodents as well as He does beetles. (Fig. 1–1.) Insectivores display the same strikingly lumpy pattern of discovery; as do also Chiroptera (bats), though with a stronger period of discovery in the mid-nineteenth century and a less striking peak in the early twentieth. Discoveries of North American reptiles and amphibians also display this periodicity, though less markedly: relatively few were described before 1800, most in the 1850s, with small peaks in the age of survey and after.[14] (Data on world species of these groups is either absent or harder to tabulate.)

These distinctive periods in the pace of collecting and describing suggest that the process of discovery was not random and individualistic, but that individual efforts were synchronized by larger cultural, economic, and social trends. This is not a novel thought. It is a commonplace (and doubtless true, as well) that early modern naturalists were inspired by the flood of new knowledge that was

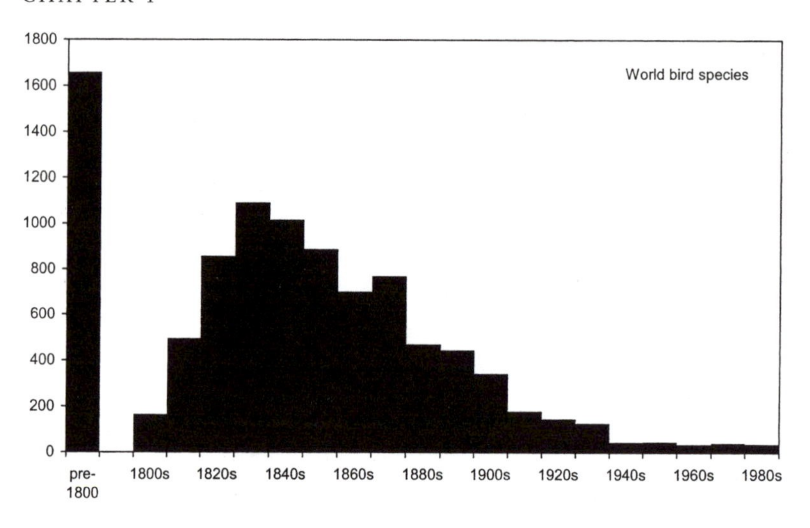

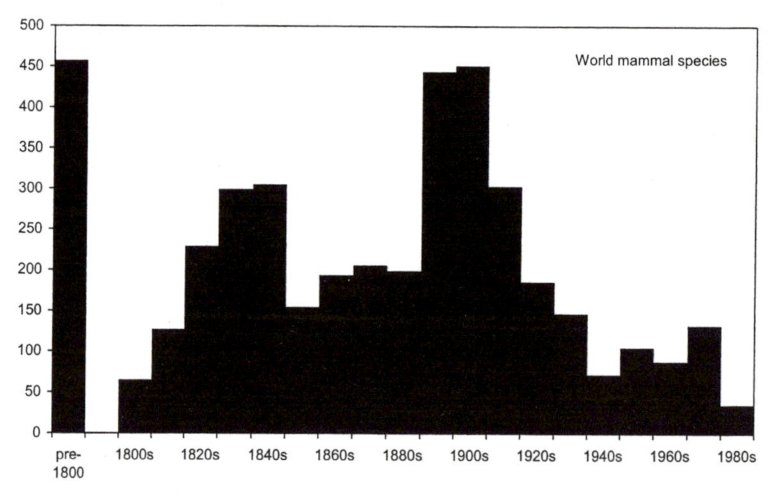

a by-product of the expanding global reach of European trade and conquest.[15] And we now also know that Linnaean taxonomy grew out of the widespread interest in Enlightenment Europe in state-sponsored agricultural improvement, including schemes for acclimatizating exotic species to northern countries.[16]

It is also clear that the early-nineteenth-century flowering of collecting and naming resulted from the greater affordability of transoceanic steam travel and from European imperial expansion and settlement, especially in the rich tropical environments of the

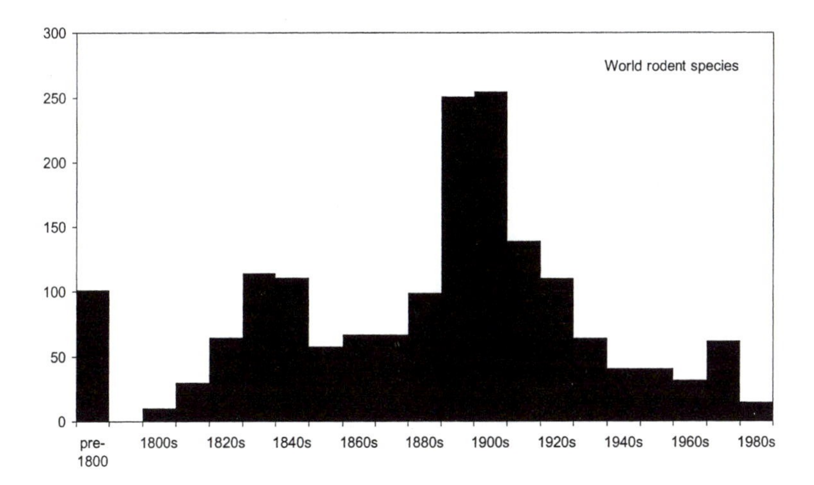

Fig. 1–1. First descriptions of bird (1a), mammal (1b), and rodent (1c) species, by decade, showing peaks of collecting and discovery. Sources: Charles G. Sibley and Burt L. Monroe, Jr., *Distribution and Taxonomy of Birds of the World* (New Haven: Yale University Press, 1990); and James H. Honacki, Kenneth E. Kinman, and James W. Koeppl, *Mammal Species of the World: A Taxonomic and Geographic Reference* (Lawrence, Kans.: Allen Press, 1982). Graphics courtesy of Jack Kohler.

southern hemisphere.[17] In North America, naturalists like John James Audubon followed the military frontier into the species-rich environments of the southeastern United States. And the western boundary and transport surveys of the 1850s took naturalists like Spencer Baird into the faunally diverse and virtually unworked areas of the American West.[18] No one has tried to map the historical geography of taxonomic knowledge onto that of imperial expansion and settlement, but I would expect a close correlation. If trade has followed flags, so also have naturalists and collectors. Access was crucial: wherever improved transportation technology and colonial infrastructure afforded ready access to places previously expensive or dangerous to reach, there the pace of discovery of new species will soon pick up.

The third of these cycles of collecting—I have without fanfare been calling it "survey" collecting—is the least well known and the most surprising. We do not think of the late nineteenth and early twentieth centuries as being a great age of discovery in natural history; but they were. One need only peruse the annual reports of national and civic museums to appreciate the enormous enthusiasm for expeditions and collecting. In the United States alone dozens or scores of collecting expeditions were dispatched each year to the far corners of the world between 1880 and 1930: hundreds in all, or thousands—perhaps as many as in the previous two hundred years of scientific expeditioning.[19] They certainly produced as much knowledge of the world's biodiversity as any of the earlier episodes of organized collecting.

It was in the age of survey that scientists became fully aware of the world's biodiversity. In places that were explored but not intensively worked, like the American West or much of South America, faunas and floras that had seemed closed books were reopened and vastly expanded. In its first two years of operation in the western states, the U.S. Biological Survey turned up seventy-one new vertebrate species—an abundance that some zoologists found hard to credit.[20] Inventories of vertebrate animals became so complete that subsequent discoveries of new species became media events. Why, then, has this phase in the discovery of biodiversity remained the least well known?

One reason is that collecting expeditions were mostly small and unpretentious, unlike the grand voyages of imperial exploration. Scientific collecting in the age of survey was accomplished mostly by small parties (three to half a dozen) whose purpose was to send back not exotica and accounts of heroic adventure and discovery, but rather crates of specimens. It is the dramatic explorations of the earlier periods that have caught the eye, because they were designed to catch the eye—of investors, princes, publishers, readers, chroniclers.[21] It is no accident that the heroic voyaging of eighteenth- and early-nineteenth-century explorers—Cook, Vancouver, Lapérouse, Humboldt, Bougainville, Murchison—is well documented and remembered. Or that historians have dwelt on the feats of American explorers from Lewis and Clark to later ventures like the Harriman Alaska Expedition, or the adventures of polar explorers, rather than on the more numerous but less flashy modern discoverers of biodiversity.[22] Still, this imbalance needs to be set right, and I hope this book will help do that.

My version of this history is of necessity an exploration, not a survey, because the empirical foundations for a more comprehensive treatment have yet to be laid. I deal not at all with the earlier episodes of collecting and inventory, only with the survey period of the late nineteenth and early twentieth centuries. Nor do I offer a global and comparative history of natural history surveys, but treat only those organized by American institutions. Many of these surveys took place within the United States and its North American neighbors, in part because state and national governments supported only surveys of their own political territories. Civic museums, however, did have a global reach, and I have made use of historical evidence from expeditions to South America and elsewhere, because the evidence is rich and pertinent. My subject is American natural history collectors and collections—not collections of American animals.

Globalism and cross-cultural comparison are in vogue these days among historians, but these ideals should not deter us from studies of a particular time and place. Comparative methods are for subjects that have a well-developed history, which this one decidedly does not have. Besides, survey and collecting, even more than most scientific practices, are specific to particular natural and cultural environments. They can be properly understood only by intensive study in their particular contexts. For example, histories of Linnaean or Humboldtian collecting will necessarily focus on western Europe and its empires; here American work is the sideshow. Transcultural and transtemporal comparisons will be highly rewarding, but for now they are beyond our empirical reach. We have not yet entered a survey period in the history of natural history.

Another limitation of this history is that it treats mainly vertebrate zoology and some botany, but insects and other invertebrates hardly at all. (The "all" creatures of the title should not be taken too literally.) This is not an arbitrary limitation: survey collecting in my period, especially by museums, concentrated on vertebrate animals, because scientific fieldwork piggybacked on collecting for exhibits of vertebrate animals. (Insects, plants, and mollusks did not have quite the same potential for eye-catching displays.) In addition, invertebrates are discouragingly numerous for comprehensive survey inventories, and they remained the province of amateur specialists long after vertebrate animals became the objects of organized survey. Invertebrates have recently become the object of systematic inventory, but in ways quite different from earlier surveys.

Like any scientific (or any cultural) practice, natural history survey had its particular period and life cycle. It arose out of a particular set of environmental, cultural, and scientific circumstances; ran its course; then gave way to new and different ways of studying nature's diversity. It was especially well developed in the United States, though not exclusively there. My aim is to describe what natural history survey was in its heyday, the reasons it flourished where it did, and how it worked in practice.

Readers should approach what follows not as the last word on the subject, but as a reconnaissance. I hope it may inspire further study, and one day perhaps the full, comparative survey that the subject deserves.

Natural History Survey

But what exactly was natural history survey, and how did it differ from other, earlier modes of collecting? It was organized, systematic, and sustained, in contrast to sporadic individual efforts; at least, that was the ideal. In practice, natural history surveys varied widely. Some were little more than summer projects of college professors and their students, who set out to inventory the plants and animals of their state or region. The Nebraska Botanical Survey organized by Charles Bessey and Roscoe Pound was the best of this type, but there were many lesser ones. Other surveys were more public and institutionalized, like the U.S. Biological Survey founded by C. Hart Merriam, which was in animal biogeography what the U.S. Geological Survey was in topographical and geological mapping. Research museums like the Museum of Vertebrate Zoology at Berkeley or the University of Michigan's museum undertook systematic survey collecting of a region or state. Finally, large civic museums like the American Museum of Natural History in New York or the Field Museum in Chicago organized systematic collecting of animals and plants not just in North America but in South and Central America, Oceania, and other parts of the world. But whether it was the private passion of one man or an official government project, and whether it lasted months or decades, natural history survey was organized, planned, and long-term. It aimed at a comprehensive, total inventory. Survey expeditions produced vast public collections, in the millions of specimens, all prepared and arranged according to standardized procedures (again, the ideal).

It is useful to contrast survey with exploration, though the categories overlap. Explorers liked places little known to people of the West: places that were hard to reach and where the infrastructure of modern travel and communication did not yet extend. Exploration was an activity of world frontiers, where agricultural and commercial societies were pushing into regions of hunting-gathering or swidden farming. Often exploration was prologue to war, trade, or settlement, and voyages of exploration served mainly commercial, military, or political ends, and only incidentally scientific ones. Scientists attached to exploration parties were frequently guests and hangers-on, more tolerated than encouraged, and their collecting was catch-as-catch-can. Of course, we should not underrate the role of pure (or impure) curiosity: Europeans are a famously curious people and attracted to the exotic, and voyages of exploration were often voyages of discovery as well.[23] But scientific discovery was more a by-product of exploration than its intended purpose.

Survey collecting expeditions, in contrast, were primarily scientific ventures, dispatched to map and inventory the world's flora and fauna. Although economic and nationalistic rationales were often deployed to get funding for surveys, their aim in practice was to gather facts about the earth and its natural history. The trend was toward science. In the United States, for example, the Mexican Boundary Survey and railway surveys of the 1850s were primarily military and political and only incidentally—though very productively—scientific. The great western surveys of the 1870s were officially economic and imperial but three of the four were led by naturalists pursuing knowledge of the region's natural history. A decade later the U.S. Biological Survey began officially as a project in economic biology but soon became openly what it was always intended to be: an all-out faunal survey. Survey expeditions in their heyday were generally organized by universities or civic museums, and staffed by curators and biologists who could set itineraries and schedules to suit their own aims, not someone else's. Whereas exploratory collecting was catch-as-catch-can, survey collecting was methodical and guided by scientific agendas. Whereas amateur and commercial collectors valued novelties most highly, survey collectors put equal value on full and exact knowledge of known and common species and on detailed mapping of the ranges of all the

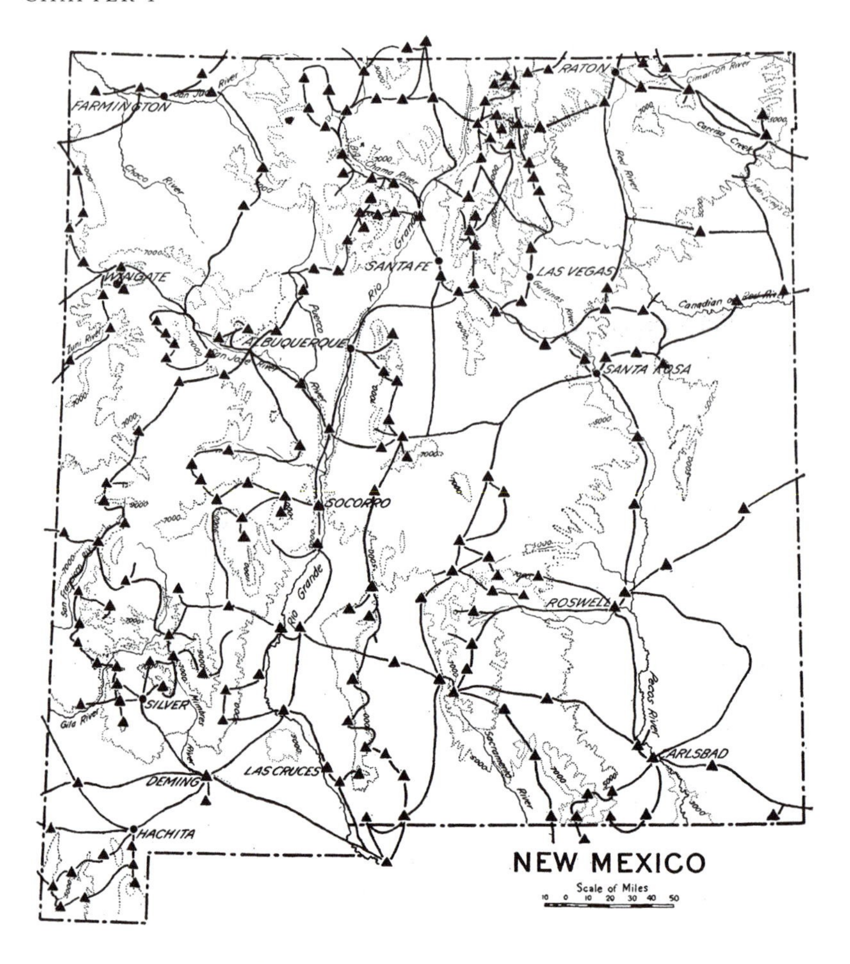

Fig. 1–2. Routes and collecting stations of U.S. Biological Survey parties in New Mexico, showing the crisscross pattern typical of intensive survey (prepared by Vernon Bailey). In Florence M. Bailey, *Birds of New Mexico* (Albuquerque: New Mexico Department of Fish and Game, 1928), p. 14.

species within biogeographic regions. Explorations typically skimmed the cream of nature's biodiversity; surveys were as thorough and complete as time and hard work could make them. Explorer-collectors were opportunists; survey collectors made plans and executed them as best they could.

Most important, survey collecting was both extensive and intensive. Field parties surveyed whole regions but paused regularly at collecting sites or "stations," where they worked intensively for days or weeks. And they returned repeatedly to a region until its biodiversity was thoroughly inventoried and its biogeography mapped in detail. Explorations, in contrast, were extensive but generally not intensive. Their itineraries marked out linear projections across territory for the purpose of reconnaissance, not complete inventory; they meant only to sample. Survey collecting also differed from the practices that later became the dominant mode of field biology from the mid-twentieth century. These practices were intensive but not extensive and addressed some particular theoretical issue (say, in population genetics or ecosystem ecology) rather than making an inventory. It was the combination of both extensive coverage and intensive local work that gave natural history survey its distinctive character. It was a transitional, mixed practice, combining the extensive element of exploration with the intensive features of modern, empirical science. This combination of extensive and intensive practices is a point on which much will hang.

I want to say that natural history survey was an "exact science" but cannot, because that term has come to refer more narrowly to the sciences that can be reduced to mathematics, like mechanics or astronomy. So I can only say that natural history survey employed exact methods, and that it was an "exact*ing*" science. The word gets the idea across without claiming too much: survey was methodical, systematic, and disciplined. Although the term "exacting science" is not an actors' category, it is not an anachronism either. The terms "exact" and "exact method" were widely applied to all the empirical sciences in my period, including sciences of the field, and exactness seems to have been understood as the hallmark of modern science.

Exploration and survey were also carried on in different kinds of places. Although unexplored regions were tempting to survey collectors, because they were likely to harbor undiscovered species, most expeditions were dispatched to places that were already known and partly settled by Europeans; places that were still wild but also accessible. It is one thing for explorers to transit uninhabited or hostile territory, but quite another for field parties to carry out intensive survey collecting beyond the infrastructure of settled society.[24] The logistics of receiving regular shipments of supplies and returning bulky shipments of specimens kept collecting parties close to railroad and telegraph nets. If exploration was an activity of frontiers, biological survey was distinctly one of "inner frontiers"—a concept I will explain shortly.

Survey and exploration also gave participants different experiences of place. Explorers were always just passing through; survey collectors lingered and revisited. Explorers treasured the exotic; survey teams aspired to make places familiar, to know them as well as if they lived there (though they could never quite do that). Survey collecting lost none of its cultural worth, as exploration did, by being carried on in semidomesticated places, because adventure was not one of its purposes, as it was of exploration. It was the display of courage and resourcefullness that gave authenticity to explorers' efforts to know the world. For survey naturalists it was the other way around: if a survey turned into an adventure, it was a sign that something had gone wrong in its planning or performance. Survey knowledge was local or "residential" knowledge made systematic and scientific.

Finally, survey collecting was carried on in a particular scientific context. It was a practice of the downward slope of the discovery curve—recall the graphs of the rates of species discovery. It was collecting in a situation where many or even most species were already known and described. It was not filling in the very last gaps (which requires more serendipity than system), but making maps and inventories essentially complete—say, 90 percent. Survey collecting was a mode of fieldwork suited to a situation in which a complete inventory was a realistic aim, but where the species remaining to be discovered and described were not excessively rare. That is one reason why survey collecting was both intensive and extensive: for complete stocktaking, collectors could leave no corner unexamined and no stone unturned.

Collecting in other periods was quite different: for the Linnaeans it was often a matter of mining existing collections and texts. All those known species ready for proper scientific names and pigeon holes—it was hardly essential to go afield at all (though many naturalists did, or ran networks of local collectors). Collecting on the upward trend of discovery, in a Humboldtian mode, required fieldwork but was more extensive than intensive. In places newly opened to naturalists, collectors did not have to be systematic and exacting to reap rich rewards of new species. Opportunistic cream-skimming—collecting whatever was most easily collected—was more cost effective than trying to get every last, hard-to-get species.

(Much collecting in this period was done by commercial collectors, for whom time was money.) And after the survey period, when inventories of species were substantially complete, collecting tended to be intensive, local, and focused on solving some particular problem—"project" collecting, we might call it. When only a few percent of species remained to be collected and described, the rewards hardly justified full-scale survey. Natural history survey was poised between cream skimming and gleaning. Earlier it was not necessary; later it did not pay.

Little has been done to classify and compare different kinds of collecting practice, but one thing is sure: that they will differ depending on the degree to which inventories are already complete. Collecting is not one activity but a diverse family of practices—and a subject ripe for historical survey.

Why did natural history flourish when it did in North America and elsewhere? What kind of society would invest in comprehensive collecting and inventory, and what kind of scientists invested careers in survey fieldwork? How were expeditions justified and financed, and how did they operate in the field? These are the themes of the story I want to tell, and it turns out to be a more complex story than one might imagine. To tell it requires a mix of historical genres: environmental history of "inner frontiers," to begin with; then a social history of nature-going; and some cultural history of artistic representations of nature. Institutional history is also required: of the museums that organized collecting expeditions, and of the patrons who paid for them, and of changes in curators' professional role that allowed them to go afield as collectors. Methods of science studies and history of science are required to deal with field practices, and methods of intellectual history to explain how taxonomic categories changed with collecting practices.

The practices of survey collecting depended on a whole set of environmental and social conditions. One was a continent (North America) and a world that had only recently been knit together by cheap and rapid steam transport, making nature physically accessible. Another was a culture whose interest in nature was becoming more naturalistic than sentimental, and that made survey science seem familiar and (so to speak) natural. Third, scientific institutions evolved (universities, museums, government agencies) that supported field collecting that was both far-ranging and exact. And finally there was a taxonomic science that for the first time had data that were more or less complete and could turn new facts from the field reliably into scientific discoveries and careers. The practices of survey science were specific to these circumstances, as those of Enlightenment and Victorian natural history were to the institutions, cultures, and events of European exploration and expansion. None of these elements—an abundant nature, the cultural incentives to scientific study, and institutional rewards for discovery—by itself accounts for the popularity of natural history survey: but taken together, they do. That in a nutshell is the message of this book.

As history, natural history survey is doubtless a harder sell than either exploration or experimental field science. On the one side it lacks the romantic appeal of exploration, with its grand voyages, its cast of flamboyant, self-fashioning characters, and its aura of heroism and imperial power. On the other side, survey lacks (or seems to lack) the grand theories and exact experimental methods that we now accept as exemplary of modern science. Few survey naturalists and collectors have brand-name recognition, and many are obscure figures who passed inconspicuously across the stage of history. The creatures that fired their interest were likewise often rather unromantic: weedy plants, lizards and snakes, small nocturnal rodents that few of us could find even if we wanted to and could not tell one from another if we did. We associate survey work with "mere" collecting, observing, record keeping, and pigeonholing: practices often taken to be a lower kind of science—what scientists do before they are able to do proper experiments or theoretical modeling.

Yet it is these very qualities that make natural history survey a particularly rewarding historical subject. Its varied cast of characters, its intimate relation with the culture of vacationing and outdoor recreation, its combination of high science and craft practices, the ambiguous careers and scientific identity of its practitioners—all pique historical interest. Besides, as the distant descendents of Adam and Eve, the first people to survey and name, we should know how we have become acquainted with our fellow passengers on our biodiverse celestial ark.

Let us begin then with nature: the ambiguous, liminal nature of the inner frontiers.

Inner Frontiers

In 1893 the historian Frederick Jackson Turner famously declared that the frontier of Euro-American settlement had become history. He took his cue from the U.S. Census Bureau, which in 1890 announced that the westward-moving frontier of settlement was so broken up by scattered pockets of settlement that it was no longer a line at all. The frontier had ceased to be a significant feature of North America's human geography.[25] With the end of the Indian wars in the 1870s, no part of North America except the taiga and the polar north was occupied by migratory hunter-gatherer societies. Every part of the American West was inhabited by Euro-American agriculturalists: the Hispano-American settlements of the Southwest; the expanding homesteading areas of the high plains; the Mormon irrigation settlements in the Great Basin; the coast and valley farms and orchards of Oregon and California; the mining and logging camps of the western mountains.[26] Farther east, older settled areas of New England and Appalachia were being unsettled, as hill farmers sold up and moved west to homestead. The North American landscape was becoming a mosaic of intermingled settled and uninhabited areas—a continent not of one linear frontier but of varied inner frontiers.

Historians ever since Turner wrote his famous essay have been much excercised over the significance of the "closing" of the free-land frontier—in modern terms, the displacement of mobile hunter-gatherer societies by intensive agricultural ones. Although Turner's views of the formative influence of the frontier on American society and politics have long been discarded, his ghost still haunts American history. Environmental historians still begin with Turner, if only to signal that they are untainted with the academic heresy of geographical "determinism."[27] But perhaps Turner's ghost is hard to exorcise because he was on to something. How people settle, inhabit, and experience places does profoundly shape their culture and politics.[28] The human geography of America did change in fundamental ways a century ago, with the transition from a two-zone to a mosaic landscape. However, it was not the end of geographical influence, as Turner believed, but a transition from one kind of geographical influence to another: from the sharp gradient of a settlement frontier to the shaded gradients and ambiguous landscapes of inner frontiers. The experience of these inner frontiers was crucial to Americans' evolving conceptions of nature and culture, I believe—and also to field naturalists' changing ideas and practices.

Here is the argument. For a period of some four or five decades, from the 1870s to the 1920s, the landscape of North America afforded an unusual intimacy between settled and natural areas. Densely inhabited and wild areas were jumbled together. Areas of relatively undisturbed nature, with much of its original flora and fauna intact (except for large game animals and predators), were accessible to people who lived in towns and cities, with their cultural and educational institutions. It was this combination of wildness and accessibility that defined the inner frontiers. No longer were densely settled and cultivated landscapes separated from relatively unsettled ones by a demographic and cultural line, as had been the case for more than two centuries. Nor was the landscape—yet—so extensively occupied that natural areas were reduced to small relicts and artificially preserved parks and refuges, as was increasingly the case in much of North America after the mid-twentieth century. To experience nature in the inner frontiers one did not cross a line where the infrastructure of settled society stopped and travel became suddenly much slower, more expensive, and hazardous. Nor was it—yet—a problem for biologists to discover places to study nature au naturel, rather than as isolated and degraded preserves.

In the intermingled landscapes of the inner frontiers, the boundaries between wild and settled were unusually extensive and permeable. Opportunities for working and knowing nature and rethinking its cultural meanings were unusually rich, especially in natural science but also in other outdoor pursuits, like hunting and camping. Some of the best evidence of what life was like in the inner frontiers comes from popular sport and recreational magazines, and naturalists' reports.

What made the inner frontiers so extensive in this period was the inflationary way that the western half of America was settled after the end of the Civil War. This was the age of railroad booms and homestead and timber acts, which constituted one of the greatest giveaways of state lands in modern history. It was comparable in scale to the medieval forest clearances in Central Europe, and much faster. The politics of sectionalism and manifest destiny; the expansive dynamism of an intercontinental market economy; and a commercial culture that put the highest value on owning property and turning natural resources as quickly as possible into money—these elements combined to produce a period of rapid, unregulated, and ragged settlement and cycles of land exploitation and abandonment.[29]

The historical geographer Carville Earle has argued that American frontier expansion was always cyclical, with periods of inflationary geographical expansion alternating with periods of relative stability. These cycles of restlessness and land speculation were driven by macroeconomic business cycles, Earle argues, and came to an end in the 1840s, when population growth in the most densely settled areas outstripped that in newly opened frontier areas.[30] However, the expansion of the late nineteenth century in the West appears to be another of these inflationary pulses—it was simply the first to take place in an urbanizing world.[31] Earle's schema seems quite consistent with the concept of inner frontiers.

It was not population pressure in eastern urban areas that drove the inflationary westward-driving phase of settlement, but economic and cultural forces: nationalist zeal and the quest for quick speculative profits. The social machinery of rectangular survey and wholesale land grants to states, railroads, projectors, and homesteaders were designed to encourage rapid occupation.[32] This period of inflationary settlement was characterized by homestead bubbles in marginal dry lands, boom-and-bust mining and timbering, and a transportation infrastructure built out well in advance of economic or demographic demand. Some areas were settled too rapidly, while others were too hastily abandoned. The result was a mosaic landscape of dense settlement mingling with lightly occupied or abandoned areas.

Some of these inner frontiers were places unsuited to intensive human uses—too dry, too wet, too sandy, too rugged. The Lake Michigan dune area east of Chicago was such a place—a lightly inhabited island left behind in a sea of cornfields and expanding suburbs.[33] The sand region of central Illinois was another such place. Here a tributary of the Illinois River had piled up a large inner delta of sandy glacial outwash, which was too dry and infertile for cornbelt farming, and too forbidding (with local names like "Devil's Neck" and "Devil's Hole") even for the vacationers and nature worshipers who flocked to the lakeshore dunes. Such areas remained inner frontiers in the middle of one of the most intensively cultivated regions in the world.[34] Farther west on the semiarid high plains were the large badland areas of the Dakotas and other places, bizarrely etched and rapidly eroding areas in which few plants can grow and/or humans live—wild islands in a sea of productive ranchland. Farther east there were the sandy pine barrens of southern New Jersey, a mazy wilderness in an area of intensive truck farming, where outsiders were almost certain to lose their way.[35]

Another island of nature in the agricultural heartland was the so-called driftless area centered in southwest Wisconsin. Once an island between two lobes of the continental ice sheet, this area is a preglacial landscape of wooded ridges and valleys. Less fertile than the sheets of glacial till that surround it and too hilly for machine agriculture, this area remained a refuge for plants and animals, as it was surrounded by a rising sea of agriculture—a kind of human glaciation.[36] The cold and rugged uplands of the Adirondack and Catskill mountains likewise remained islands of wild and semiwild country—tourist islands, eventually—in a landscape webbed by towns, canals, and railroads.[37]

Actual islands are relatively rare in the interior of North America, but they too can afford protection against human change. Isle Royale in Lake Superior, for example, was protected by its isolation and rugged terrain from clear-cut logging and the spectacular fires that scorched the northern cut-overs. The copper-mining boom of the 1840s was short-lived and relatively undestructive; and relicts of mining and lake shipping, plus a growing tourist industry, provided access and ready-made facilities for visiting naturalists.[38]

Great swamps like Virginia's Great Dismal Swamp, or the Everglades and Okefenokee in Florida, or much of the Gulf coast constituted another kind of inner frontier: islands of wild and inaccessible nature close to coastal cities.[39] These were created when rising postglacial sea levels flooded estuaries and low peninsulas. But the greatest areas of marsh and swamp were in the glaciated regions of the upper Mississippi Valley and Great Lakes, where they covered much of northern Indiana, northeast Ohio, southeast Michigan, and parts of northeast Illinois. This vast wetland was created in the most recent advance and retreat of the Laurentide ice sheets, which bulldozed established drainage systems, then covered them

Fig. 1–3. "Mountain Resorts of New York" (1884), showing the pockets of wild nature within an enveloping transportation web, characteristic of inner frontiers. In Walton Van Loan, *Van Loan's Catskill Mountain Guide* (New York: Walton Van Loan, 1888).

with moraines and layers of till. Stream and river systems were slow to reform because in this flattened terrain back-cutting erosion was very slow. Differences of just a few feet of elevation could send waters to the Atlantic or the Gulf of Mexico. Bypassed in the early westward migrations, these wetlands remained largely intact in the late nineteenth and early twentieth centuries. Likewise the glaciated regions of northern Iowa, Minnesota, and the Dakotas, with their marshy sloughs and countless prairie pothole ponds. Much of northern Minnesota, with its thousands of interconnected lakes, is a low indefinite watershed—three, in fact: north to Hudson Bay, east to Lake Superior, and south to the Mississippi.

These coastal and interior wetlands were potentially rich agricultural areas, but expensive to drain; in effect, humans had to do in a few decades what it took nature ten or twenty millennia to achieve. There may originally have been 125 million acres of swamp and seasonally flooded land in the United States—an area the size of France—of which an estimated 80 million acres remained undrained in 1915.[40] Some had simply been bypassed for more accessible land farther west; in others, speculators "preempted" huge tracts and set them aside unused for decades until rising land prices made drainage projects economical. Thanks to the quirks of American land and tax laws, large areas were not drained and plowed until the 1910s; meanwhile they remained pockets of undisturbed nature teeming with wildlife—a vast inner frontier to entice naturalists and collectors.

In the Central Valley of California in the 1910s, where large-scale reclamation projects were rapidly transforming delta and desert into irrigated monoculture, Joseph Grinnell could still find patches and strips of "waste" land that harbored their original rosters of small mammals.[41] In the lake and swamp district of northern Indiana, despite decades of draining and road building, the zoologist Carl Eigenmann could still report in 1895 that the whole region "gives one the impression that it has changed but little since the ice left it." It was hardly pristine: dams and ditches had turned swamps into ponds and streams; but enough was left to attract sportsmen and naturalists.[42] The Huron River valley in southeastern Michigan, with its kettle ponds, moraines, bogs, and swamps, was an outdoor ecological preserve for biologists at the University of Michigan, and just minutes away by urban trolley.[43]

Fig. 1–4. Wetland drainage operations active circa 1920, indicating immense area of wetlands close to densely settled areas, especially in the upper Middle West. In "Irrigation and Drainage," vol. VII of *The Fourteenth Census of the United States, Made in 1920* (Washington: Government Printing Office, 1922), p. 348.

The largest remaining areas of wetland in 1920 were in the Southeast and Gulf Coast states, especially Florida; but significant portions of the marshes of the North Central states still survived, though the more accessible parts in Ohio and Indiana had been drained.[44] A map of active drainage projects circa 1920 vividly reveals both the vast size of the original wetlands and the monumental efforts that were made in the 1910s to turn these inner frontiers into productive agricultural landscapes.[45]

Large river floodplains were also extensive inner frontiers before reclamation enclosed and destroyed them in the 1910s and 1920s. The Mississippi Valley was one of the great flyways of North America, providing continuous and relatively undisturbed habitat for animals and huge flocks of migrating birds. The Illinois River, before it was leveed in the 1910s, was an important commercial fishery and a popular center for sport fishing and fowling, just a few hours train ride from Chicago. Its floodplain forests had been timbered but only selectively, leaving much intact, and floodplain fields produced alternating crops of fish and corn as they flooded and dried out in an annual cycle.[46] River towns with a well-developed infrastructure of boat building and repair gave sportsmen and collectors ready access to a wilderness of sloughs, swamps, and oxbow and fluctuating lakes—an inner frontier. So did other major deltas and floodplains, for example, the delta of the Sacramento and San Joaquin rivers in California, where in 1918 the mammalogist Joseph Grinnell worked "the frontiers of reclamation" and tried to stay one step ahead of the dredgers and dikers.[47]

The roster of inner frontiers could be extended: for example, to the deserts of the Southwest and Southern California; the canyon country of the Colorado Plateau; and the caprock canyons of west Texas, where southern and northern faunas mingled.[48] The well-watered mountain massifs rising like islands out of southwestern deserts remained, because of their desert isolation, remote and untouched even as large cities grew nearby. The San Francisco Mountains of Arizona, for example, were a popular resort for naturalists decades before becoming a tourist mecca. Scouting Arizona's deserts for a site for a field station in 1903, botanists Daniel MacDougal and Frederick Coville were attracted to these peaks: "charming situations in the mountains of the desert, remote from civilization, rich and remarkable in their flora, furnished with an abundance of pure, never-failing water, and altogether delightful

Fig. 1–5. The Illinois River at Havana, a sport and commercial fishing area in the middle of one of the most intensely farmed regions of the world. A kind of inner frontier. Courtesy of the University of Illinois Archives, Urbana-Champaign, Illinois, Record Series 39/2/20, photo file ADA 12–2.

in their surroundings . . . treasure spots for the camping naturalist." (They settled for a more accessible and convenient site on the outskirts of Tucson.)[49] The Panamint and other mountains arising out of the California deserts were similar places, and the rugged high country of the Sierra Nevada and Rocky mountains remained quite wild, yet accessible to visitors via old trails and stage roads. Farther north, dense coastal rain forests were more forbidding islands of nature (some interior parts of the Olympic Mountains and Vancouver Island were virtually impassible), but at the heart of an urbanizing region.

The deciduous forests of the southern Appalachian Mountains were another inner frontier, preserved in part by the rugged, unglaciated terrain, and in part by local land-use customs. Hardscrabble hill farmers occupied mainly the bottom lands of valleys and their lower slopes, using the wooded mountain slopes lightly for foraging pigs and livestock, hunting, and selective timbering, and for operating illegal moonshine stills. The result was a forest area lightly inhabited and used, but well supplied with trails. Residents notorious for their suspicion of outsiders further increased the region's isolation. It was biotically very rich—for a million years and some seventeen ice ages it had been a refuge for northern species forced south by cold—yet remained incompletely known into the 1920s and 1930s.[50]

An inventory of remnant old-growth forests in 1920 showed substantial areas in the Pacific Northwest and western Montana; in coastal Florida, especially the panhandle; and in the lower Mississippi River Valley. Smaller areas remained in the mountains of central Appalachia, in Northern Michigan and Minnesota, and in the mountains of the Northwest. Eastern old-growth forests were mere vestiges of the vast forests that covered much of the region in 1850, but in places like the Adirondacks or White Mountains they constituted islands of wild nature in a largely humanized landscape.[51]

Other inner frontiers were created by quirks of capitalist economics and landholding laws that gave access to wild places without (for a time) completely transforming them. There was no more powerful force creating inner frontiers than railroads and interurban trolleys. The peculiar economics of the highly capital-intensive railroads made it rational for railroad companies to build lines into areas well before there were any people there to use them. It was better to lay down track and take losses up front than to leave potentially profitable areas open to competing roads. This "logic of capital," as William Cronon has termed it, was especially marked in the roads fanning out into the western prairies from Chicago. The western transcontinental railroads, which became vast landholders as a result of huge grants of public land, also became powerful promoters of rapid settlement in the agriculturally marginal areas of the prairie-plains transition. To stimulate settlement—or to simulate it—railroads plopped down standardized "towns" at regular intervals, hoping that they would become centers for the collection of freight, and built dense networks of feeder lines across thinly settled countryside.[52] This premature infrastructure gave people from cities and towns easy access to areas that had very recently been reachable only by stage-coach or on horseback and were still relatively wild, but now just hours away from cities to the east and west.

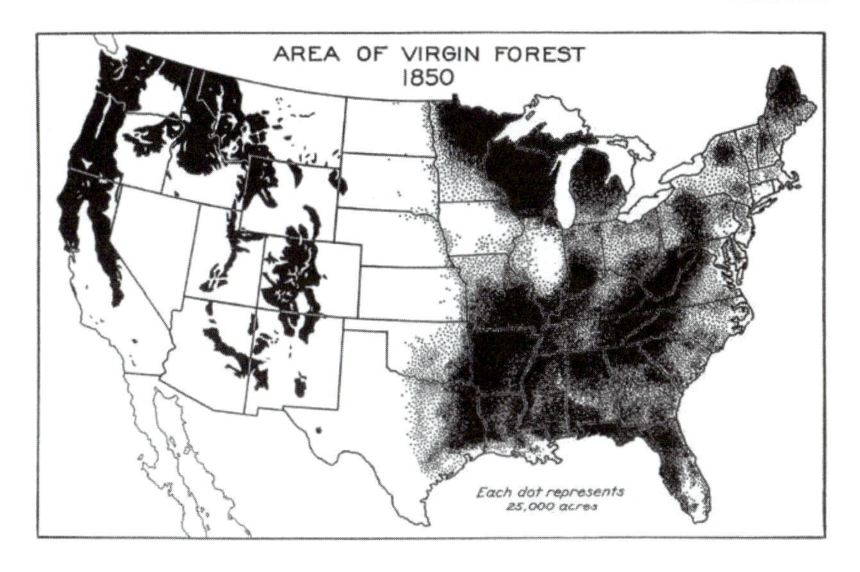

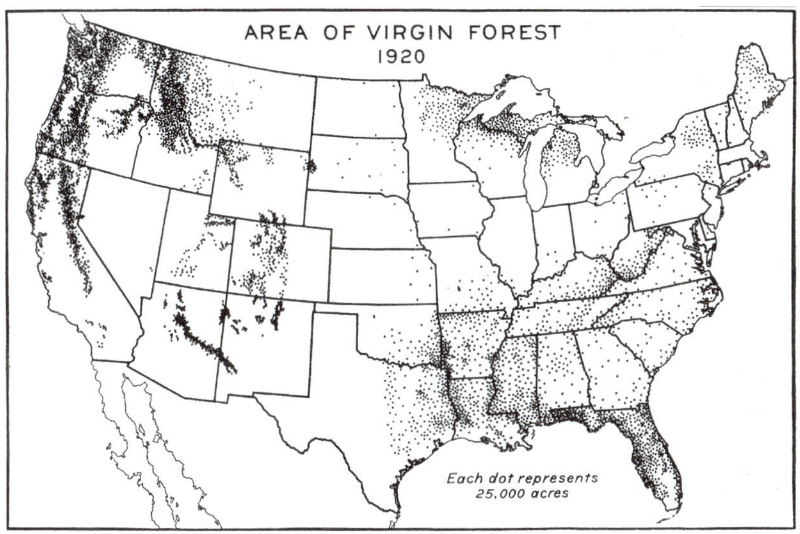

Fig. 1–6. Old growth forests remaining in 1850 and 1920—an indicator of vestigial inner frontiers. In William Greeley, "Relation of geography to timber supply," *Economic Geography* 1 (1925): 1–11, pp. 4, 5.

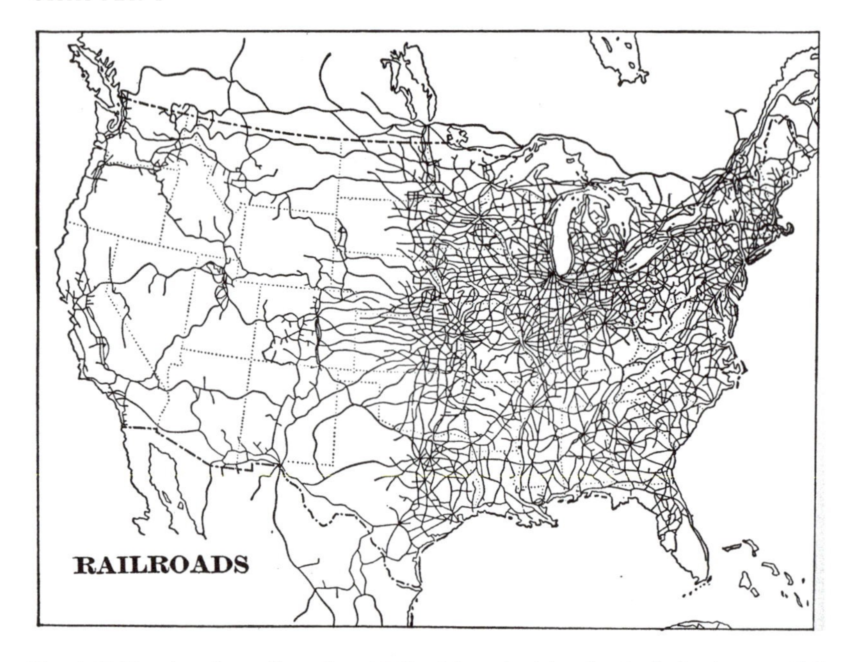

Fig. 1–7. Trunk railway lines, late 1890s. Note the islands of relative inaccessibility in the West, Southeast, New England, and the upper Midwest. In Jacques W. Redway and Russell Hinman, *Natural Advanced Geography* (New York: American Book Co., 1901), p. 60. This map was evidently adapted from "Gray's New Trunk Railway Map of the United States" (New York: Charles P. Gray, 1898).

In 1887, for example, a local botanist in Hastings, in south-central Nebraska, found a stretch of original prairie just five miles from town, where the only signs of human presence were an old wagon track and abandoned sod huts. This biotic island was created by the hop-skip geography of prairie settlement. The earliest settlers took homestead claims in the Platte River valley, believing that to be the best land (because it was wooded); the second wave, arriving on the transcontinental railroad, settled along the rail line, where land was cheap and well advertised. In between river and railroad, strips of untouched prairie remained within a few miles of a city of twelve thousand.[53]

A similar process played out in the regions of boom-and-bust logging and mining. In the pineries of the upper Great Lakes, which were cut from the early 1870s to the mid-1890s, networks of local railroads built to haul out lumber remained in place when the pines were gone and the loggers moved on, to haul in sportsmen and naturalists.[54] Here too railroads were active promoters of settlement and of the recreational industry that replaced logging as an economic mainstay after the pineries were stripped. Natural areas that just a few years before had been accessible only by logging parties or sportsmen with Indian guides could be reached comfortably in less than a day and at reasonable cost from urban centers. And as the outer fringes of the railroad frontier pushed into ever more remote areas of North America, naturalists were never far behind. When a line was built in 1931 to the edge of the arctic tundra at Churchill, on Hudson Bay, parties of naturalists arrived soon after, to see a biome that until then would have required a full-fledged expedition to visit.[55]

Railroads encouraged tourists and naturalists by offering cheap excursion fares, building hotels and roads in prime recreational areas, lobbying for national and state parks, and providing information on transportation and lodging in tourist regions.[56] They also on occasion provided information to amateur and commercial collectors, publishing lists of rare plants and directions to find them—much to the dismay of botanists, who often found only the holes where rare species had been dug out.[57]

Areas of logging, mining, and commercial fishing remained suprisingly natural even as they were cut down, dug up, and fished out. The archaeologist Harlan Smith recalled his boyhood summers at Bay Port, Michigan, on the western shore of Lake Huron in the late 1870s and early 1880s, when extractive logging and fishing were in full swing. The signs of human depredation were everywhere: "The beach was strewn with drift wood from the mills perhaps ten feet wide and continuous. The heads of dead sturgeon were so close together on the beach that you were seldom out of smelling distance." Yet the wildlife was abundant—turtles, snakes, even bears. The real change came when the place became a summer resort. Revisiting the place in 1910, Smith found it, in effect, gone: "the country had so completely changed in appearance and was so much fenced up that it simply made me sick," he recalled. The beach was so built up that he could not land a boat.[58] The initial assaults of loggers and commercial fishers had created a landscape that was nothing like the aboriginal landscape of Indian hunter-gatherers; but it was still biodiverse and could recover quickly in a natural succession. After all, die-offs and clearances by fires and blights are natural occurences. Suburbs are not.

The history of the Illinois River illustrates the same point. In 1900 the City of Chicago opened the Chicago Drainage Canal, which redirected the city's raw sewage away from Lake Michigan and into the upper reaches of the Illinois River, down which a putrid tide oozed slowly south to the Mississippi. Although the initial impact of this assault damaged river wildlife, a decade later an ecological survey showed that the river had largely recovered. Indeed, the influx of nutrients seemed to have increased the river's fish populations. The greater danger, the ecologist Stephen Forbes warned, was the headlong draining and leveeing of the river's floodplain (from 6,700 acres in 1899 to 124,205 acres in 1914) that was already destroying the rich river fisheries by draining seasonal breeding grounds and turning the river into a barren and unhealthful drainage ditch.[59] Boom-and-bust exploitation of natural resources created the inner frontiers; permanent settlement and landscape engineering brought their era to a close.

Twilight Zones

Closer to town, access to rural countryside was also eased by the massive building of suburban and interurban electric trolleys from the 1870s on. Here too the logic of capital prevailed, as trolley companies extended their routes into the deep countryside well before there were large numbers of suburbanites there to ride them into town. It was cheaper to lay down track and run cars where no one yet lived than to buy out competitors later. Companies also banked on the fact that cheap transport would draw commuters to their catchment areas and eventually give them a profitable monopoly—which it did. The result was an inflationary expansion of suburban zones. By 1870 cities were surrounded by suburban fringes fourteen times the size of the inner walking city. By 1900 this zone had doubled again in area and extended five to fifteen miles out. This process was further accelerated between the wars by the advent of the family automobile and paved roads.[60]

Eventually, of course, exurban fringes became suburbs, and suburbs became city neighborhoods, obliterating the natural environments that had attracted people to the fringes in the first place. But for a brief period the logic of capital gave town and city dwellers ready access to what was still deep countryside. In just a few hours

Fig. 1–8. A "twilight zone" on the exurban fringe of Colorado Springs, Colorado, showing a relict patch of bunchgrass prairie. In Frederic E. Clements, "The relict method in dynamic ecology," *Journal of Ecology* 22 (1934): 39–68, p. 42.

on interurban trolleys, city dwellers could ride to collecting areas that previously could be reached only laboriously by horse and carriage on unimproved roads.

The "twilight zone between town and country," as the ecologist Frederic Clements once called it, was a good place to find vestiges of presettlement floras—sometimes the only place. In corn or wheat belts, mechanized agriculture had so thoroughly destroyed the original savanna and prairie vegetation that suburbs and vacant lots were the last refugia of the region's native plants. The ecologist Arthur Vestal was amazed to find a patch of original California prairie in the vacant lot just a few doors from his home in Stanford.[61] Twilight zones were hardly picturesque, to be sure, but they afforded an "intriguing" experience of nature as both wild and humanized.[62]

The nature writer John Burroughs, who worked as a government clerk in Washington in the 1860s and early 1870s, found the city and its environs a naturalist's paradise. Unlike northern commercial cities, the capital city had not yet sprawled, "and Nature, wild and unkempt, comes up to its very threshold, and even in many places crosses it."[63] Diverse species of birds inhabited the grounds of the government buildings, and the ravines of Rock Creek, between Washington and Georgetown, were semiwild areas full of wildlife. "There is, perhaps, not another city in the Union that has on its very threshold so much natural beauty and grandeur, such as men seek for in remote forests and mountains," Burroughs rhapsodized. "There are passages . . . as wild and savage, and apparently as remote from civilization, as anything one meets with in the mountain sources of the Hudson or the Delaware."[64]

In the mid-1880s the ornithologist Frank Chapman pursued his avocation while commuting to a job as a bank clerk in New York City; his commuter stop was right in the middle of one of the best birding areas in the region, at least in migration seasons. The experience was as interesting and exciting as any in his subsequent career.[65] Naturalists were delighted when the American Association for the Advancement of Science decided to hold its 1886 meeting in New York City, because of the "wealth of attractive places for collecting near at hand," especially in the "New Jersey collecting grounds." A decade later the biologist Henry Linville found abundant animal life for his students in the Fort Lee Woods, just across the Hudson River from Manhattan, and in the wilds of Staten Island.[66] Hard to believe now, winding in heavy traffic through the endless subdivisions, but so it was, and not all that long ago.

And in Chicago too: the city made by railroads and the paradigm of rapid urban sprawl.[67] In the early 1900s an ornithologist friend of Joseph Grinnell's, Frank S. Daggett, found excellent bird collecting in the suburbs and surrounding country, a few hours away by trolley. Or he would take a train thirty or forty miles out and float back to town down the Des Plaines River in his folding boat, observing and collecting. In the migration season Daggett did not have to go farther than the local city parks, so abundant was the bird life. "These are great migration times here nowadays," Daggett wrote Grinnell in spring 1904, "the parks are loaded with birds . . . and people shooting them with opera glasses."[68] Harry Swarth, another Grinnell protégé, who worked at the Field Museum, would leave his house at about nine, ride an hour and a half on the street car, and return by four-thirty with twenty-five birds or more. Fox Lake, which was a few hours away by railroad, was a center of sport hunting and also a fine place to collect small mammals; the woods were full of shrews, mice, chipmunks, and gophers. Swarth spent three months collecting on the estate of the museum's president just forty miles south of the city.[69]

In the early nineteen-teens collectors found rare animals and even new species in the Sacramento delta, not too far from San Francisco. And Joseph Grinnell collected successfully on the fringes of Central Valley towns—so long as he stayed outside the patrolling range of resident cats.[70] Florida at the turn of the century was a naturalist's and collector's paradise—a vast island of almost untouched nature readily accessible from major towns. Here is Frank Chapman's description of a farmstead where he collected, just five miles from downtown Gainesville:

> It is a beautiful place, surrounded by ground of almost every character, pines, scrub-oak, low hummocks, and the old clearings, and is thus an excellent place in which to trap. There is probably no wilder region in the vicinity. . . . [H]ere the family have their home, a magnificent large log house, and a mile further on, their orange grove, the finest in this region.[71]

Not all inner frontiers were unintended. National forests were set aside for future use or recreation. Amounting to just a few million acres in 1890, mostly in the Sierra Nevada and Cascades, national forests expanded by 1910 to some 200 million acres, mostly in the western mountains. A further 40 million were reserved in the northern Great Lakes, Ozarks, central Appalachians, and New England.[72] In these eastern regions underused or abandoned areas reverted to public ownership through default and purchase. The Adirondack National Park was created in this way in a region still partly forested and wild, and the Green Mountain National Forest (in the late 1920s and 1930s) in a state that just a generation earlier had been largely open land.[73] In effect a national commons, national parks and forests were generally accessible (eastern forests were laced with walking trails), but off the beaten tracks they remained quite wild.[74] Private game parks were another kind of nature preserve, numbering at least 500 in 1910 and some as large as 60,000 acres, though not always accessible to scientific collectors.[75]

Rural cemeteries and school yards were other places where vestiges of nature could be found, as were Indian reservations, if they were not overgrazed. Also railroad rights-of-way, because they

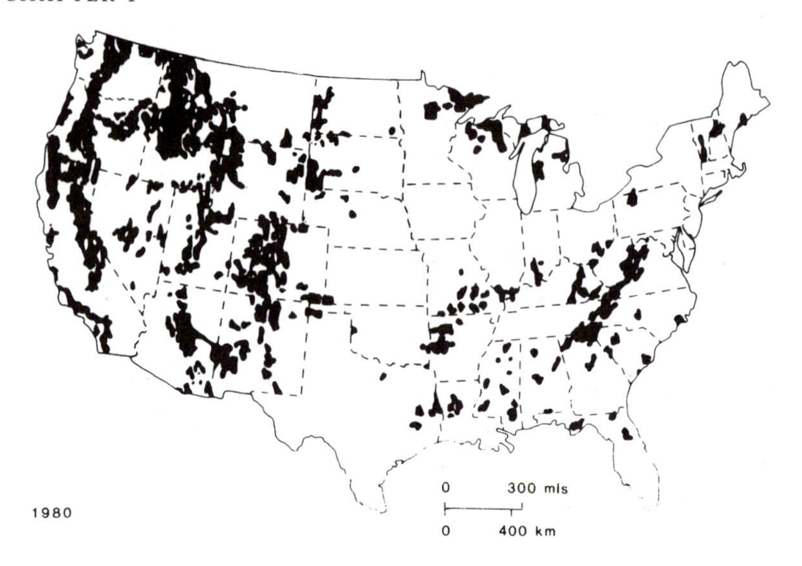

Fig. 1–9. National Forests in 1980. In Michael Williams, *Americans and Their Forests: A Historical Geography* (New York: Cambridge University Press, 1989), p. 408.

were fenced and regularly burned, as the original prairies had been by Indian firing, but with coal-fired locomotives now providing the sparks. Town parks and unkempt rural roadsides were other refugia in the exurban "twilight zone." Roadsides were ready made for ecological survey transects.[76] Such vestigial bits of nature might seem poor places for systematic collecting, but for a few decades there were enough of them around rural towns to serve the purpose. In 1916 a group of ecologists selected Charleston, Illinois, for the site of a survey of invertebrate fauna and found all they wanted in railroad rights of way, an old-growth woodlot (clear-cut just a few years later), and bluff and floodplain forest.[77] Of course, these vestiges of nature were a vanishing resource. Exurban frontiers were relentlessly on the move, engulfing fields and filling in vacant lots. But between 1880 and World War I this twilight zone was a prominent feature of America's inner frontiers.

The opening up of underpriced farmland in the West also created inner frontiers by drawing population from settled areas farther east. Watching suburbs engulfing farms and forests has made us unused to thinking of abandonment and depopulation as common events in history (except in inner cities); but they are a recurring feature of the periodic waves of aboriginal dispossession and westward movement. For example, this cycle played out in South Carolina in the 1820s and 1830s, as leading planter families moved west en masse to newly opened lands in Georgia and Alabama.[78] Nor was abandonment a phenomenon only of places people left. Settlers in frontier areas often bought land that was poor for farming because it was cheap, and because in the feeding frenzy of land rushes they wrongly assumed that the price of any land must rise, and then learned the truth the hard way. So in many areas settlement and abandonment were concurrent processes.[79]

The late nineteenth century was another time of mass migration and abandonment, driven this time by western homesteading after the end of the Plains Indian wars. Homesteading had long been the favored way to get public lands into private hands quickly. It had always been speculative and volatile, making large tracts of land available at prices well below their actual value, spreading the virus of land fever. And as homesteading expanded into the western prairies and semiarid plains, it stimulated a large-scale abandonment of the poorer agricultural regions of the Northeast.

New England (though never depopulated) was the region where wholesale abandonment was most marked, from the 1870s or even earlier. The original settlers of interior New England had preferred to situate their farms on hills than in river valleys, owing to the easier transport on high ground. But as the thin upland soils gave out, people moved into valley towns, leaving the hillsides to revert to woodland. Then as cheap land beckoned in the west, farmers moved out altogether. The state of Vermont, once wholly cleared, reverted to forest within a matter of decades.[80] A study of farm abandonment in New York State in the 1920s showed that the acreage of farmland in 1924 was just 81 percent of what it had been in 1880; and the pace of abandonment was accelerating (60,000 acres per year between 1880 and 1910; 140,000 in the 1910s; and 270,000 in 1920–25). Near cities, farms became suburbs, but in the Catskill and Adirondack mountain areas and the hilly country of south-central New York they became forest.[81]

A Depression-era survey reckoned that 2.3 million acres of cleared land reverted to forest or was suburbanized in New England between 1880 and 1910, plus 0.77 million in the Mid-Atlantic states. Another 13 million acres were abandoned by 1930. The greatest percentage declines in cleared land were in New England

and New York, eastern Michigan, eastern Virginia, and the interior Southeast.[82] The shift from wheat to dairy farming, first in New England and New York, then in Wisconsin, produced a mosaic landscape of pasture and second-growth forest.[83]

A few decades later massive abandonment also occurred in the western plains, as the homestead frontier pushed into the semiarid zone west of the hundredth meridian, where cultivation gave way to ranching. In this climatically volatile region, a thirty-year cycle of wet and drought first lured farmers into marginal areas, then drove them out. One such cycle coincided with the inflationary phase of homesteading in the 1880s; another produced the great plow-up of the southern plains in the 1920s and the subsequent dust-bowl out-migration. Another is presently under way in the northern high plains.[84] Traveling the American west in 1913 on an ecological survey, Frederic Clements counted hundreds of abandoned farms—“dead homes”—and ghost towns in central Oregon, southern Idaho, and South Dakota. Areas a hundred miles across were almost entirely depopulated, and it was even worse in the southern plains. (The “live homes” were even more distressing than the dead ones, Clements felt: bare one-room shacks in bleak and hopeless situations.)[85] Cheap-land frontiers thus became inner frontiers: object lessons in human hubris and favored sites for natural history surveys.

The same pattern of settlement and abandonment occurred in logging and mining areas in the forests of New England, the upper Great Lakes, and the Rocky Mountains, as well as in river fisheries. It was not homesteading that drove the boom-and-bust cycles of resource use, but the legal system that sold rights to cut or mine public land for a pittance. Boom-and-bust harvesting of resources reached a crescendo after the Civil War, when market competition caused loggers, miners, and fishermen to treat natural resources as placer deposits, to use or lose, rather than as bases for sustainable livelihoods. The western public lands resulted from the way that the United States had grown politically, by huge purchases or expropriations; and Americans’ devotion to states’ rights and laissez-faire political economy made these lands a virtually unregulated commons. The result, as is usual where there is no customary system of usufruct rights and mutual obligations, was a commons tragedy.[86] Clear-cut logging and abandonment from the Great Lakes pineries to the rain forests of the Pacific Northwest; mining booms and busts in the Rockies and Sierra Nevada; fishery cycles of bonanza and collapse in the Sacramento and Columbia rivers—all left behind stump lands and depleted soils, ghost towns and valleys choked with mining debris, and half-emptied and impoverished coastal villages.[87]

Logging and mining towns were built and briefly inhabited by people who came not to stay and make a life, but to make their fortunes as quickly as possible and leave. Marginal homesteads were occupied by farmers who expected to "improve" the land, sell at a profit, and move on within a few years. The extractive industries of the late nineteenth century (including tillage agriculture) depended on resources that had accumulated over centuries and millennia and afforded once-only chances to make a financial killing. Placer gold deposts are the classic example, but it is equally true of white pine forests, loess soils, and mountain forests, which follow (respectively) large fires, melting continental ice sheets, and rains of volcanic ash. Tillable high plains are likewise a temporary gift of recurring climate cycles.

Americans' boom-and-bust practices of settlement and land use thus produced a patchwork landscape of densely and lightly settled areas. Whether skipped over, newly settled, or recently abandoned, these inner frontiers were served by a transportation infrastructure more than adequate to the needs of recreationers and visiting naturalists. In such places, town dwellers could experience nature without setting aside urban perceptions. And out of this meeting of culture and nature came new conceptions of their relation. Only rarely has so much semiwild nature been so accessible as between 1870 and 1920; nor is it likely to recur, barring some catastrophic change in global climate or the natural history of human epidemic diseases. As the cheap-land frontier shaped American culture before 1870, so too did the inner frontiers of the 1880s to 1920s, in ways we are only beginning to understand.

Impressions

What was it like to visit the inner frontiers? What did they look and feel like? Ecological journals are full of photographs of such places; but photographs, especially those made artlessly as a scientific record, quite fail to evoke the experience of places so far from town culture yet so near. To capture the meaning of such liminal places, literary documents—word pictures—are more evocative and revealing.

For example, the stories told by hunters and fishermen in sporting journals offer vivid glimpses of what life was like along the railroads that were pushing into the northern plains and north woods in the 1870s and 1880s. Sportsmen were among the first urbanites to take these lines into the inner frontiers, and they possessed literary genres—the fish story and sporting travelogue—designed to evoke their experiences for armchair readers. Sportsmen are famous for stretching the truth, and the credibility of their exploits depended on their ability to give detailed and vivid evocations of natural places and phenomena (more on that in chapter 2). So they make engaging witnesses.

In 1877 the angler-writer "Will O' the Wisp" reported on a trip on the Wisconsin Central Railroad's newly completed line to Lake Winnebago and Ashland, in the North Woods. Winnebago, he discovered, offered good fishing but it had already been claimed by "the noble army of tourists." The country northwest to Stevens Point was beautiful agricultural country, but ruffed grouse had been exterminated by cultivation, and prairie chicken (which thrived in the patchwork landscape of woodlots and fields) had yet to appear. Farther north, conifer forests were being cut, but with Indian guides, the fishing and hunting was good, though not quite what was claimed by tourist promoters. "Will O' the Wisp" found the country unimpressive but provided with good accommodations and not overrun by hunters, and it was an easy trip from Chicago.[88] "Will Wildwood" took the same trip and reported a landscape altered by logging and railroads, yet still attractive to vacationers and sportsmen. It was stripped of its white pine, but islands of hardwood remained among the boreal conifer, and the woods were dotted with lumbering hamlets and camps.[89]

About the same time, another sportsman-writer, "Ingomar," wrote of a trip on the Northern Pacific Railroad into northern Minnesota and Dakota Territory. Before the railroad, Ingomar recalled, this region was virtually inaccessible, and hunting and fishing "was the hardest of hard work; nothing but trails to travel on, without a house or white man's habitation from Lake Superior to the Mississippi River." Now, Pullman cars took sportsmen and their wives and families to hamlets that afforded good hotels, teams and guides, and reasonably priced weekly stage service into the deep boreal forest along the Canadian border, which was "one vast game preserve."[90]

Such reports, even discounting sports writers' hyperbole, give some sense of what it was like to gain access to vast stretches of nature that just a few years before had been the domain of Indian hunters and the most well-heeled and adventurous sportsmen. There was a sense of a new world of nature suddenly opening up; of a world that had before been experienced virtually, through the memoirs of big-game hunters and heroic naturalists, now accessible to middling sorts. There was a sense of the immediacy of nature, and its diversity and richness; and of the intimacy of being so easily in the middle of it all. What had once been accessible through books and travelogues became immediate, lived experience for a generation or two of sportsmen and naturalists.

Many naturalists who came of age in this period later recalled their childhood experiences of twilight zones. Francis Sumner, for example: he grew up on a small orchard farm in the hilly fringe of Oakland, California, and spent his days (he was schooled at home) roaming the hills and ponds; making pets or specimens of birds, eggs, toads, and small mammals; poking sticks in ant hills; and observing the hidden life of nature. (Returning thirty years later he found the place a tawdry, jerry-built development.) As a preteen living in the inexpensive outskirts of Colorado Springs, then a village of 2,500, Sumner hiked in the canyons of the nearby Front Range and collected fossil ammonites and Indian artifacts. As a teenager in Minneapolis, again on the sparsely settled outskirts of the city (the best his impecunious family could afford), he roamed the woods and lakes with shotgun ready, observing the wildlife and feeling "the thrill of real exploration," and later trying his hand at taxidermy. It was not science that intrigued him then—it never occurred to him to consult books of natural history—but nature and the pleasures of discovering it for himself.[91]

The ecologist Henry Gleason recalled his boyish rambles in the outskirts of Decatur, Illinois: "We were not interested in any phase of science or nature study. We merely absorbed nature-lore without books or teachers. We were the only boys who knew where to find the big 'river-bottom' hickory nut or the coffee tree. We knew where the ground robin nested and we picked up crinoid stems from the gravel banks."[92] The political cartoonist and conservationist Jay "Ding" Darling had similarly vivid memories of the vast wetland areas of the Dakotas:

> The potholes and marshes of eastern South Dakota were the chief points of interest when I was a kid. . . . As soon as school was out and I was old enough to be turned loose on my own, I spent all my summers riding our old family horse out of Sioux City across the Big Sioux and into South Dakota and herding anybody's cattle who would give me a job. . . . [T]he lakes and marshes of eastern South Dakota . . . still form my pleasantest recollections. It was the disappearance of all that wonderful endowment which stirred the first instincts I can remember of Conservation.[93]

The naturalist Paul Errington counted himself lucky to have been just old enough to experience, as a boy in the late 1910s, "the last of the primitive abundance of the Dakota prairies, and the irreparable changes since then are most disheartening to a naturalist of my taste."[94] This was the quintessential experience of the inner frontiers: a sense of abundance and endless novelty, yet readily accessible—not a resource to be exploited, or a wilderness to be fought and tamed, but a place to be discovered and understood.

A fixture of biologists' reminiscences and obituaries, such stories are a literary trope: a foretelling of the future scientist in the curious youth. But they also bear witness to experiences of actual places and of a particular time in the environmental history of North America. They are stylized recollections of what it felt like to inhabit the twilight zone, where wild (or wildish) nature was experienced through the ideals of town culture. So it is no surprise that the trope of the boy-naturalist especially appealed to the naturalists whose careers happened to intersect with the inner frontiers. It was an element out of which scientific identities were constructed.

Abandoned regions also had a special feeling. Vestiges of a human presence in vacant places were melancholy reminders of nature's power to outlast us and to reclaim lost terrain. Joseph Grinnell found himself in such a place in 1918 in the coastal hamlet of Morro, halfway between San Francisco and Los Angeles. It was an old dilapidated town, nothing like the usual resort towns, inhabited by Swiss and Portuguese who grew beans and cattle and lived amid some of the most abundant bird life Grinnell had ever seen. "One refreshing feature of the country," Grinnell noted, was "the total lack of no-shooting signs. . . . Beach, marsh, sand dunes, tule swamp, brush lands—all are open ground."[95] (Most collecting areas were private property, and collectors were trespassers.) A few years later in Baja California Grinnell found a landscape once densely inhabited and worked and now depopulated and forgotten, but again about to change:

> When you look on the map you will see various place-names, but for the most part these now apply only to ruins or merely to situations with no living person near. This is a country of the past; but there are signs of re-awakening, or rather, invasion. I saw an oil-well drilling outfit in operation near Cape Colnett! Also there is a marble quarry in action below San Quentin; and old mines are being prospected vigorously.[96]

Ecologists surveying the semi-arid plains in the early 1900s were struck by the broad wagon "roads" or parallel ruts that had been abandoned a generation earlier when the railroads came through and now stood out vividly on the landscape as bands of yellow sunflowers.[97] Roads to a vanished Oz.

The north woods were full of such ghostly places. The University of Michigan's field station at Douglas Lake was set in a landscape that had been cut and burned. "Gaunt black stubs twenty to fifty feet high were a prominent feature of the landscape, and huge pine stubs showed that a magnificent pine forest had been removed some thirty years earlier."[98] Ulysses Cox, working at Lake Vermillion, in Minnesota's north woods, improvised camp tables and a darkroom from remnants of a shack abandoned by one of the many homesteaders who had finally despaired of making a living in such a poor place and had moved on.[99]

The hill country of the Northeast is famously endowed with signs of abandonment. John Burroughs encountered many such places in his rambles. Near the village of Highlands in the Hudson Valley, for example, he found

> a rocky piece of ground, long ago cleared, but now fast relapsing into the wilderness and freedom of nature, and marked by those half-cultivated, half-wild features which birds and boys love. It is bounded on two sides by the village and highway . . . and threaded in all directions by paths and byways, along which soldiers [it was 1863], laborers, and truant schoolboys are passing at all hours of the day.[100]

In the mid-1920s geographers studying abandonment in New York State described a poignant twilight zone: "Disintegrating houses and barns were seen in all directions. Many buildings had disappeared entirely, and nothing but the foundations marked the site. In even more cases the former site could be located only by an old orchard or a clump of lilac bushes." In one typical area, houses were gone on two-fifths of the farms, three-fifths of cleared land was idle, and residents lived by logging second-growth forest.[101] Stone walls and cellar holes are a familiar and still haunting sight in the forests of the Green Mountains. These are landscapes full of ghosts and poised between cycles of human coming and going.

The notion of an inner frontier bears a family resemblance to Leo Marx's conception of "middle landscape." An intellectual historian, Marx invoked the idea to explicate the knot of political and social ideals that shaped American culture in the early nineteenth century: the hope that universal ownership of land would make Americans virtuous; that America could industrialize without losing its agrarian character; and that the free-land frontier would guarantee a democratic future. These ideals, Marx found, were expressed less powerfully in formal political philosophies than in literary and especially artistic representations, and most strikingly in George Inness's painting *Lackawanna Valley* of 1855. Commissioned by the Lackawanna Railroad Company (and accepted with some misgivings by the painter of romantic landscapes), this work depicts a pastoral landscape, evidently not far from a frontier (stump-land is foregrounded), and with railroad, steam locomotive, and workshops in peaceful harmony with rural nature. In this image Inness captured the ideal of an agrarian democracy based on new land, steam-powered machines, and nature. As Marx suggested, Turner's frontier thesis was another expression of this ideal.[102]

For Marx this middle landscape was not a real but a symbolic place, a cultural and moral order embodied in an imagined landscape. Marx was dealing not with physical but "moral geography," as he put it, because the ideals of social and environmental harmony embodied in Inness's painting were not descriptive of American society of that or any period. In the end machines would dominate nature, and property would corrupt democratic ideals, so middle landscapes could only be places of the mind—false dreams.[103]

But is the middle landscape a purely imagined place? To be sure, the scene Inness conjured up is not a particular place, but it may depict an actual *kind* of place. I have in mind interior New England in the early industrial period, when machine manufacture de-

Fig. 1–10. George Inness, *The Lackawanna Valley* (circa 1855). An artistic representation of a "middle landscape," in which nature and industry intermingle. Courtesy National Gallery of Art, Washington, D.C.

pended on water power, and factories were located wherever streams and rivers were unused, which was increasingly in the deep countryside. Steam power would soon reverse this rural dispersal; but for a few decades, there existed a middle landscape or inner frontier in which machines and nature were in fleeting intimacy. Is it fanciful to think that the experience of such places fostered an optimistic view of industrialization like Inness's vision in *Lackawanna Valley*, or Ralph Waldo Emerson's vision of a working nature?[104] Likewise with the inner frontiers created a generation or two later by the forced expansion of railways, suburbs, and homesteading: they too, I think, shaped Americans' perceptions of their natural world.

The point is simply that landscapes embody cultural values, as landscape historians have repeatedly demonstrated.[105] And partly humanized landscapes may be especially potent in evoking such sentiments, because they visibly display both our own and nature's power at the same time. If water mills in rural outbacks could foster an optimistic view of industrialization, so too might the landscapes transformed by steam technology inspire an image like Inness's *Lackawanna Valley*, with engines in the place of vanished water mills. Moral landscapes may be real in experience as well as in memory.

Historians who have used and extended Marx's concept have applied it quite convincingly to particular places. Thomas Schlerith characterizes the campground headquarters of the Chautauqua movement—a quasi-religious intellectual revival—as a "middle landscape of the middle class."[106] And the historian Susanna Zetzel wittily likens Frederick Law Olmsted's Central Park to a "garden in the machine":

> [L]ike the railroad paintings and literary pastorals, it occupies a middle ground between the two extremes of pastorial ideal and urbanism. . . . From the point of view of the "frontier" the park represented civilization; from the point of view of the "city" it represented rural values. It was, in short, a pastoral "middle landscape."[107]

Olmsted's naturalistic urban parks were not, as many scholars have asserted, a nostalgic and phony "wilderness" into which stressed-out bourgeoisie could escape from urban turmoil. In fact, Olmsted meant parks to encourage people of all sorts to socialize, in a way that neither artificial promenades nor rural countrysides could do. Artfully naturalistic settings, with winding walkways and greenswards, were designed to strengthen, not escape, urban folkways. They were an inner-urban twilight zone in which nature served as symbolic guarantee that urbanity was consistent with older rural values. College campuses, whose design was powerfully influenced by Olmsted's ideas (especially land-grant college campuses), were another such place.[108] Here parklike surroundings reassured collegians that intellect and knowledge—including knowledge of nature—had a place in a striving commercial and industrial society.

In the same way, I would argue, the mosaic landscapes of the inner frontiers encouraged Americans to see nature neither as a commodity to be used up, nor as a wilderness to be left alone, but as a place of cultural and scientific interest, to be surveyed, collected, conserved, and understood. My version of the Turner thesis.

Conclusion

The period of the inner frontiers was a brief moment in the environmental history of North America, spanning just a little more than half a century, and the same forces that created them worked to destroy them. Demographic pressure and a commercial society's hunger for property and profit drowned islands of nature in a rising tide of development. Tourist and vacationing economies in time did as much to alter as to preserve semiwild outbacks. And then there is automobility: nothing was as destructive of inner frontiers as the family auto and state-subsidized systems of improved roads, as countrysides were engulfed in waves of suburban sprawl.

But during the time that the inner frontiers were a dominant feature of the landscape, they made the natural world accessible in a way it had not been before. And among those who took advantage of that accessibility were collectors and survey biologists, the Noahs of nature's diversity. The surveys of the late nineteenth and early twentieth century were made possible by the new ease of getting to places that had not hitherto been intensively surveyed and collected. As the infrastructure of global trade and imperialism had opened new areas of the world to European collectors in the early nineteenth century, so did the infrastructure of national expansion and settlement open new areas to American naturalists in the age of inner frontiers. Survey was a mode of scientific practice especially suited to such places, and place and practice arose and declined together.

But physical accessibility is only half the story. Places may be easy to get to, but if people do not have reasons to go there, places will remain unvisited and unknown. Of course, naturalists and collectors always have a compelling reason to visit places they have not yet seen: to find new things. But survey collecting involved more than trips by a few professional collectors. Expeditions required money, organization, institutional support; they depended on public interest. And comprehensive collections could not exist without large museums and herbaria to house them. Survey naturalists operated in a complex web of social relations, and many sorts of people required reasons of their own to take an interest in sending groups of collectors into the inner frontiers.

It was the inner frontiers' intimate mix of town culture and semiwild nature that got Americans to take a scientific interest in nature's biodiversity. Accessibility brought people of various sorts into contact with deep nature, enabling them to experience nature as a part of town culture: the culture of schools, colleges, and amateur natural history societies. The closeness of inner frontiers to towns thus encouraged a more naturalistic interest in the natural world. Where the cheap-land frontier encouraged a view of nature as a commodity and exploitable resource, the inner frontiers encouraged Americans to see nature as a resource for intellectual work. In the one kind of place, nature seemed boundless for the taking; in the other, visiting naturalists were never unaware that this resource was fleeting. It was this new (or newly prevalent) way of experiencing nature that inspired patrons, museum builders, and collectors to survey America's and the whole world's flora and fauna in an organized and comprehensive way.

CHAPTER TWO

Culture

IT IS WELL KNOWN that a profound change occurred in Americans' relationship to their natural environments in the last third of the nineteenth century. Americans rediscovered nature—or reinvented it. Even as they continued to subdue and exploit the continent's abundant natural endowments, Americans began also to conceptualize nature as a place of intellectual and spiritual interest and recreative power. Older conceptions of "nature" and "wilderness" were transformed to make them congruent with the changing landscapes that Americans actually inhabited and used.[1]

Much has been written on the evolving literary and philosophical ideas of "nature" and "wilderness," but rather less on the actual experiences and social customs that nurtured these conceptions. What social activities brought Americans in significant numbers into actual or virtual contact with nature, and how did participants give these activities new cultural meanings? And how exactly did these pursuits—many of them recreational—encourage natural science? I will argue that active experience of nature enlarged the pool of potential recruits to field science and created constituencies to support large-scale natural-history surveys. But more than these simple connections, personal experience of nature as recreation encouraged a view of nature that made a scientific interest familiar and acceptable.

A common feature of many varieties of nature-going in our period is a new naturalism: a pervasive interest in nature as it is rather than in its moral or economic meanings, as moral symbol or "resource." This naturalism in popular understandings of nature resembled the naturalism of science, though for most people it was not scientific per se. Most Americans no more desired to see nature through the abstractions of science than they did through the moral abstractions of an older sentimental culture. They experienced nature not as scientists but as foragers, vacationers, or cultural consumers; science was of interest to the extent that it was congruent with these everyday practices. But the naturalistic attitude that sustained these practices eventually also supported surveys of the world's flora and fauna. Cultural and scientific practices were elements of a single social complex; and as more and more Americans had direct experience of their natural environments, the more likely it was that some of them would wish to understand and perhaps even to support or take part in natural history surveys. A social history of survey science must thus begin with the customs that structured everyday experiences of the inner frontiers.

What were these activities that took Americans into nature? There were the practical economic activities of farming, logging, mining, road and railroad building, and so on, which had created the inner frontiers. These activities required accurate knowledge of nature; however, they were less fraught with cultural meanings than some others, especially those cultural activities that drew middle-class town and city dwellers into the inner frontiers. Most of these activities were recreational, as one might expect, given the growing importance of recreation and tourism in rural economies. Sport hunting and fishing were two such activities, widely enjoyed by all classes; also camping, which became popular first in the northeastern United States, then everywhere; and rural vacationing generally, along with activities associated with outdoor vacationing, like bird-watching and amateur naturalizing. Related to these activities, though less strenuous, were the "walking mania" that swept the eastern seaboard like an epidemic in the 1870s, and the fashion for country driving with horse and buggy in the 1880s. Bicycle touring on country roads became the rage in the 1890s, as did golf a few years later and automobile touring just after the turn of the century.[2] These outdoor activities drew their appeal in part from the natural settings in which they were enjoyed. And some at least could inspire a scientific interest in those settings.

Other indicators of a new interest in nature for its own sake, rather than for the profit it might yield, are the Progressive era campaigns for nature conservation that started in the 1890s with efforts to save the last American bison from extinction and with the Audubon Society campaign to close down the feather trade, which was turning millions of birds into women's hats. Americans' well-known and continuing devotion to birdwatching arose in the early 1900s in part as a result of these movements and grew even more popular and scientific in the 1920s with communal projects of bird banding and migration census.[3] Although many of these outdoor pursuits were especially valued by those in middle-class occupations, they were not limited to social or cultural elites. "Fresh-air" charities gave working-class children a brief exposure to rural nature from the 1880s, and the huge popularity of boy and girl scouting movements after 1900 make it clear enough that Americans' desire to experience nature transcended distinctions of class and sex.

Supporting these direct experiences of nature were the virtual activities of reading nature essays and viewing the dioramas that have captivated generations of museum-goers. The new, or reinvented, nature essay appeared in the mid-1860s and became hugely popular in the 1890s. At about the same time the familiar museum exhibits of stuffed animals in rows were replaced by realistic exhibits of animal groups in strikingly naturalistic settings. A similar trend can be seen in secondary- and elementary-school textbooks, in which natural history began to leaven the more abstract, indoor subjects, and in the "nature study" movement that swept through primary education in the early 1900s.[4] Though these literary and artistic representations afforded only secondhand experiences of nature, they were meant to encourage readers and viewers to see nature firsthand. In effect they were cultural guides for inexperienced nature-goers of what to see and how to feel. Literary and artistic representations made nature accessible to people who might otherwise have had no incentive to visit the inner frontiers.

In short, in the late nineteenth century Americans (and others elsewhere) were inventing customs of nature-going that were less concerned with harvesting resources and more encouraging of an intellectual interest in how nature worked. It may be true, as Richard White suggests, that farmers, miners, loggers, and commercial fishermen know nature in an especially intense and immediate way.[5] But for the history of biological survey and collecting, the more relevant actors are the middle-class people who attached values of learning and self-improvement to their leisure customs. Thus reconceived, these activities encouraged an intellectual and naturalistic interest in nature. It was not the demands of livelihood that impelled their practitioners to know nature accurately, as with practical vocations, but it was something oddly like work. As we will see, middle-class vacationers experienced outdoor recreation as a kind of work, and it was that cultural association that connected outdoor recreation strongly to science.

We will look first at the recreational customs that made nature aesthetically, morally, and intellectually accessible to middle-class nature-goers, then inquire how these activities inspired an intellectual interest in the natural world. If it was the infrastructure of the inner frontiers that made them physically accessible, it was the customs of middle-class vacationing that made them intellectually accessible, by giving large numbers of Americans the imaginative means to experience nature in a naturalistic way.

Nature-Going

The various back-to-nature movements of the late nineteenth century all display a similar pattern of development. Their intellectual roots can be traced to midcentury moralists like Henry Thoreau and Ralph Waldo Emerson, but it was in the 1870s that naturalistic ideas began to be enacted in social habits and behavior. Taken up by a few vanguard groups of social reformers in the 1870s, these activities grew slowly in the 1880s, gained visibility in the early 1890s as organized social movements, and burgeoned as popular activities in the 1900s and 1910s, when they were routinized, commercialized, and transformed into services for mass consumption.

Take recreational camping. The camping habit began in a small way in the 1850s among members of the Boston-Cambridge intelligentsia—artists, professors, students. It then emerged into the public consciousness in the early 1870s in part as a result of William H. "Adirondack" Murray's famous 1869 travel book on the Adirondack Mountains and his sensationally popular lecture tour. (Something like half a million people heard him speak!)[6] New York magazines and newspapers began to report temporary tent encampments in the mid-1870s, and the eminent travel writer John Bachelder included a few pages on the new fashion of outdoor camping in the 1874 edition of his popular guidebook (obviously a last-minute addition). He had come across campers here and there in his travels and was surprised that more people did not use "this charming way to diversify . . . family trips." It was, he thought, "beyond all question the most delightful and healthful way to spend one's summer vacation."[7]

Encouraged by how-to books and media coverage, more and more Americans acquired the camping habit. It was cheaper than resort hotels and less formal, yet it need not involve roughing it in some distant outback. It was an activity of twilight zones and inner frontiers. Hotels began in the 1870s and 1880s to set up tent sites on their grounds, as did enterprising farmers, who would lease land to families to build semipermanent encampments. Some families camped for a few weeks on their own property or nearby. Others created permanent camp buildings in wooded areas like the Adirondacks—or leased a site and built wooden platforms for tents—or hired a horse-drawn wagon and camping outfit and moved from place to place. Groups of families got together in summer colonies, to which they returned year after year.[8]

We lack statistics on the growth of the camping habit (it was too unorganized and insufficiently serious to attract the notice of social statisticians, it seems). But there can be little question that recreational camping was widely popular by 1900, with a well-developed infrastructure and customs that cut across class and gender lines. It was from the start a family affair. Women were encouraged to participate, and it was often wives and children who spent summers in rural resorts or lakeside cottages, with husbands commuting on weekends.[9] And although camping was mainly an activity for the middle class, the fact that it could be done cheaply and with little advance preparation also made it accessible to the working class.

The development of "organized camping"—commercial summer camps for boys and girls—displays a similar pattern of growth: a few scattered experiments before 1890, mainly by idealistic New England clergy and educators who liked its potential for moral improvement; sporadic but steady growth in the 1890s (1.8 new camps per year on average); then rapid growth in the 1900s, when some 10 new camps opened their gates each year. By 1910 summer camping for the kids was "the thing to do." The number of private camps alone grew from 20 in 1900 to 150 in 1910 (one third of those for girls), 180 in 1915, 463 in 1920, and 720 in 1924, plus equal numbers of YMCA camps and camps for Boy Scouts and Girl Scouts. Commerce and youth movements combined to implant the custom of recreational camping firmly and broadly in American society.[10]

The cultural geography of camping seems to map fairly closely on the inner frontiers. The lake districts of New England and the Middle West, the Adirondacks and southern Appalachians, and the boreal forests from Maine to Minnesota were all popular camp areas—as were the high desert and mountains of the Southwest and the second-growth forests that were reclaiming lands cleared by loggers and homesteaders.[11]

Recreational walking was another activity that drew Americans to rural nature. The most elaborate form of recreational walking was the pedestrian tour made by groups who would walk rural roads for a few weeks, leaving the heavy lifting to a baggage wagon in the rear, and spending nights pleasantly in country inns. It was a delightful way to vacation, one can imagine, in the days before automobiles, trucks, and paved roads. From small beginnings in the 1840s this custom blossomed in the 1860s and became quite the thing to do in the 1870s, before fading in the 1880s. Less strenuous than camping or mountain climbing, the pedestrian tour was a vacation activity of the flat rural landscapes and river valleys of the Northeast. It was also popular in New England colleges, where it fed on ideals of physical culture dating back to the 1830s. Between the mid-1860s and early 1880s it was the custom at Dartmouth College for incoming seniors to "do the mountains," and parties of seven to fifteen would roam the roads through the White Mountains and visit standard tourist attractions like Crawford's Notch and Mount Washington. Occupational groups with regular summer leisure—teachers, ministers, artists, writers—were also early pedestrian tourists.[12]

Walking clubs are an indicator of the organized phase of pedestrian touring. The first such club may have been the one active between 1863 and 1866 at Williams College, and it had many imitators in the 1870s. Examples are the Pemigewasset Perambulators of Boston (1866); the Oatmeal Crusaders, who in 1875 memorialized a tour in a thirty-nine-page poem; the "Woburn High School Bummers" (four students who strolled 224 miles in the Massachusetts hills in 1867); and the Wlwascackbwgmsaesfssjhe Pedestrian Association (1881), an organization whose history (if any) is as enigmatic as its name. Walking clubs were popular in cities, especially New York, where members of the Fresh Air Club would take fifty-mile hikes on weekends in all weather, earning a reputation as "fresh-air cranks." This was the "walking mania" that caught the eye of city journalists in the 1870s.[13]

A democratized version of the carriage touring that was popular with the eastern gentry in the 1830s and 1840s, pedestrian touring was both a self-conscious imitation of an outmoded form of elite culture and a harbinger of the emerging age of organized sports. Chroniclers' mock-heroic tone and the funny names of walking clubs suggest that even in its heyday pedestrian touring seemed somewhat archaic and was fated to be short-lived. What killed it is not entirely clear, but it was probably organized sports. When student newspapers began to complain that walking inhibited the growth of collegiate sports, it was a sure sign that the fad would soon end. A student article at Colby College in 1883 referred to "old-time pedestrians." Likewise in cities like New York, athletic clubs and organized sports quickly put pedestrian touring in the shade. Leisurely walks through rural countrysides seemed a poor substitute for the excitement and intense bonding of competitive team sports.[14]

However, one form of walking did survive into the age of sport: mountain climbing—no doubt because of all the forms of perambulation it was the most strenuous and sportlike. Like walking tours, mountain climbing was a revival in a new form of a somewhat older cultural mode. The highest peaks of the Presidential and other northeastern ranges became popular tourist attractions in the 1840s and 1850s—Mount Washington in the White Mountains, Mount Marcy in the Adirondacks, Maine's Mount Katahdin, and others. However, these earlier tourists did not climb but ascended peaks on horseback along bridle trails and spent a comfortable night at summit houses like Mount Washington's Tip-Top House. They ascended not for the purpose of strenuous recreation but for the panoramic views and, guided by the writings of John Ruskin and transcendentalists like Thoreau, for the aesthetic experience of mountain sublimity.[15]

Cut short by the Civil War, the custom of mountain ascents remained in eclipse for some fifteen years (for reasons that remain unclear), and when it did revive in the late 1870s and 1880s, it was in the more strenuous mode of alpine sport. Mountain bridal paths and summit houses gave way to walking trails—some laid out to be arduous—and camping out. The aesthetics of the sublime gave way to the values of strenous exertion and competitive alpine sport.

Fig. 2–1. Summit House and Tip-Top House, Mount Washington, New Hampshire. Note the tourists arriving on horseback. In Peter B. Bulkley, "Identifying the White Mountain tourist, 1853–1854: origin, occupation, and wealth as a definition of the early hotel trade," *Historical New Hampshire* 35 (1980): 106–62, p. 106.

The highest peaks remained the most popular, but the wooded hills near eastern cities were also resorts for weekend walking: for example, Boston's Blue Hills and Middlesex Fells, New York's Hudson Highlands, and New Haven's traprock outcrops. The mountain clubs founded in the 1870s and 1880s are an indicator of the new mode of nature-going (the Appalachian Mountain Club was formed in 1875), as are also the proliferating climbing guidebooks and extensive trail-making of the 1880s and 1890s.[16] Unlike walking tours, mountain climbing was sufficiently strenuous and competitive to survive in the age of organized sport, and it remained an important way for middle-class Americans to experience nature close up.

Sport hunting and fishing also took Americans in large numbers to the inner frontiers. Hunting and fishing are ancient pursuits of peasants and aristocrats, for the different reasons of subsistence and display. But as a middle-class recreation, sport hunting and angling had an early nineteenth-century origin, roughly coincident with rural tourism.[17] Hunting was and still is a broadly popular activity. Though great white hunters were the most visible, because of avid media interest in their grandiose lodges and safaris, they were never typical. Wherever there were fields and woods to harbor deer and small game, people went afield to hunt, and in the inner frontiers that was about everywhere.

One need only turn the pages of Charles Hallock's popular sportsmen's guide to appreciate how well developed the infrastructure of hunting and angling already was by 1877: rail and wagon roads, hotels, boarding farms, and local guides and outfitters catered to sportsmen in most rural areas. Railroads encouraged hunters and sportsmen by offering attractive schedules and rates and by publishing guides to prime hunting areas along their routes. Hallock's guidebook also reveals just how abundant and accessible game was in the inner frontiers of the 1870s and 1880s. Although big game animals—bear, moose, elk—survived only in remote forests, small game and water fowl were abundant in places within easy reach of the largest cities. And the growing network of railroads gave townsfolk ready access to areas previously accessible only to dedicated or commercial hunters.[18] The second-growth succession of abandoned stumplands and agricultural fields also created prime habitat for edge game like deer. And while clearing prairie groves and plowing prairie destroyed the habitats of some game animals (for example, passenger pigeons, which needed oak groves and mast to live and without these went extinct), it created population explosions of other game animals, like prairie chicken and muskrat. These ecological transformations and the rapid expansion of branch-line railroads into deep rural areas made the period from the mid-1860s to the late 1880s a brief golden age for sport hunters: a kind of biological gold rush, fleeting but intense.[19]

The hunter-writer "Ouiskonsin" described the scene along the new railroads in southwestern Minnesota in 1880: "The number of shooters to be found all along this route is simply amazing," he exclaimed. "Almost every railway station has its quota of hunters."[20] Sports clubs proliferated in the 1870s and 1880s (there were at least thirty-six in New York State by 1876). During cisco runs in Lake Geneva, Wisconsin, thousands of local people from the surrounding country gathered to harvest the little fish: the shores were lined with hundreds of tents and camp fires. Hunters and anglers descended on every lake and stream in prime places like the Adirondacks, and when they departed in the fall, the local people who had served them went afield for their own pleasure. Farmers practiced a kind of game gardening, feeding captive quail in the winter, turning them loose in the spring to multiply, and harvesting them in the fall—if tourist hunters and poachers did not get them first. Chicago's game market in the late nineteenth century offered an almost incredible abundance and biodiversity.[21]

It is impossible to know precisely how many Americans hunted, fished, and trapped, but in 1905 the number of hunters was estimated from hunting licenses at 2.5 to 3 million and growing fast; and since much hunting was unofficial, this figure is certainly a lower bound. The proliferation of game regulations and wardens in the Midwest and West in the late 1880s and 1890s suggests an increasing number of hunters chasing a diminishing supply of fish and game.[22] There can be no doubt that sport hunting and fishing gave many Americans an intimate and ongoing experience of nature. In rural towns the opening days of hunting season were—and still are—unofficial local holidays: all business stops.

It is clear enough from this brief survey that for many Americans nature was not just property and exploitable wealth, but also a source of bodily health and intellectual and emotional enrichment. Country people had always known nature through the work of agriculture and resource harvesting, but in the late nineteenth century townsfolk in large numbers also experienced nature as a place of recreation. Urban and rural folkways combined to give nature new cultural meanings. The customs of camping, walking tours, mountain climbing, and sport hunting and fishing were not separate developments but elements of one cultural repertoire. And underlying that repertoire was a new and distinctively middle-class understanding of work and leisure and a newly constituted custom of middle-class vacationing.

Middle-Class Vacation: From Leisure to Recreation

Leisure is as old as humankind, but the vacation is a relatively recent social invention, just about 150 years old. It was also an American invention. "Vacation" (noun) is an American usage, according to the Oxford English Dictionary, and dates from the 1870s, though its practices appeared a few decades earlier. It is also, scholars agree, a distinctively middle-class conception: that is, it was an activity not just for the upper-middle and grand bourgeois classes but also for the middling and petit bourgeois.[23] (A rare surviving hotel guest register, of the Tip-Top House on Mount Washington for 1853–54, records visits not just by well-to-do merchants and proprietors, but also by professionals, clerks and salespeople, artisans, and farmers.)[24] As with the various forms of outdoor recreation, the idea of a regular annual vacation germinated in the 1850s and expanded in the 1870s and 1880s, becoming, along with work, a hallmark of middle-class identity. Taking time off from business had long been an informal, catch-as-catch-can practice, and only gradually became one that was expected and taken for granted. William James wrote in 1873 that clerks and young white-collar workers were coming to expect an annual two-week holiday.[25] John Bachelder's 1874 guide to tourist resorts reveals a well-developed infrastructure of touring and vacationing. Virtually every town in the Northeast and Middle Atlantic states seems to have had its hotels and boarding cottages, and a similar infrastructure was growing up in the Great Lakes states.[26] Just twenty years later Melvil Dewey (inventor of the Dewey decimal system) stated that vacations were recognized "not as a luxury, but as a necessity for those who aim to do a large amount of high-grade work."[27]

We have no economic data on vacationing, but every qualitative indicator confirms that it became constitutive of middle-class life between 1870 and 1900. Americans had never enjoyed as many state and religious holidays as Europeans, and contractual paid vacations were not common for the middle class until the twentieth century (for blue-collar workers not until the 1940s).[28] However, these official indicators are misleading, because most vacationing was informal and improvised—playing hooky was an honored custom well before it became a social entitlement. The editor of one popular sporting magazine advised businessmen to arrange their business so they could take two or three days off every few weeks "for a run to the country" and a spot of hunting or fishing.[29] The tourist-writer Edward Hungerford made a rough estimate of the extent of the custom from New England hotel and railroad receipts. In 1891 three smaller railroads alone (not including the big

Fig. 2–2. Alton Bay, New Hampshire, an early New England resort town. In John B. Bachelder, *Popular Resorts and How to Reach Them* (Boston: The author, 1874), p. 18.

Pennsylvania and New York Central) sold 1.5 million round-trip tickets, mostly to long-distance summer visitors. Hotels from some of the more important resort areas—but excluding the very large resorts of Saratoga, Asbury Park, and Atlantic City—served 444,000 summer visitors. Add to that people who camped, boarded in farms, and rented cottages—again as many as the numbers who stayed in hotels, Hungerford estimated—and the huge traffic of river and coastal day-tripping, and it becomes clear just how big a business summer vacationing had become.[30]

Hungerford was also struck by the social diversity of vacationers. The wealthy were in evidence but were far outnumbered by people of the middle class, especially of the "learned class": college and school teachers; professional and amateur artists; amateur naturalists and collectors. Also professional men, small tradesmen, seamstresses and barbers, and beneficiaries of fresh-air charity. On a tour of the Lake Champlain area, Hungerford found boarding farms and camp grounds packed with factory workers who had been laid off for a week by a shutdown.[31] A zealous promoter of middle-class vacationing, Hungerford may have seen what he wished to see, but even allowing for that, it is clear that middle-class Americans already had the vacationing habit. And not just the resort vacationing of earlier decades: Americans were also vacationing out of doors, in the natural world of the inner frontiers.

Crucial to vacationing, for middle-class Victorians, was its integral connection to work. Vacationing was good less for its own sake than as an essential complement to work, especially white-collar and professional work—"mental" work. This moral dimension of middle-class vacationing was vital, and to understand it we need to see how leisure customs of other classes were appropriated by the middle class and reconstrued as serious and purposeful activity. In a phrase, they refashioned "leisure" into "recreation."[32]

"Recreation" and "leisure" are not words that contemporaries used to describe the cultural change going on, but they neatly sum up the transformation of values that opened leisure to the middle classes. Leisure before the Civil War was mainly limited to the well-to-do, who summered on country estates or frequented grand resort hotels, both of which were simply alternative venues for carrying on the formal social life of commercial or plantation gentries. Leisure, for the commercial and evangelical Protestant middle class, was anti-work: for, as the proverb says, idle hands make the devil's handiwork. However, the sanctions against leisure were less theological than social. Leisure for the middle classes was tainted with aristocracy and what middle-class critics saw as aristocratic vices: extravagant social display, ignorance, idle pleasure-seeking, and dissipation. Likewise on the other end of the social scale, the leisure of the laboring classes was associated with disreputable public-house entertainments (boozing, rough sports) and irregular employment: that is, idleness imposed by necessity or irregular habits. Because leisure was seen to be a defining social character of the aristocratic and laboring classes, it was disapproved of by a middle class then greatly concerned to define its own distinctive and (in their eyes) superior social identity.[33] As the historian Peter Bailey put it: "[I]n a work-oriented value system, leisure represented the irresponsible preoccupations of a parasitic ruling class or the reckless carousing of an irrational working class."[34] Participation in leisure activities—outdoor sports, hunting, camping out, vacationing—could be perceived as evidence of laziness or vice and was best avoided. That was the moral logic.

For leisure to become a defining characteristic of middle-class culture it had to acquire moral qualities that distinguished it from mere idleness and entertainment. It had to acquire purpose, moral seriousness, and spiritual or economic benefits, so that middle-class vacationers could not be mistaken for fast men or business failures. Recreation and vacationing had to be made into a complement to work or even into an exaggerated, particularly strenuous kind of work, as the historian Daniel Rodgers has observed: "The cult of strenuousity and the recreation movement grew together, minimizing the distinction between usefulness and sport, toil and recreation, the work ethic and the spirit of play."[35]

The moral work of refashioning leisure into recreation began in the United States (and in Britain) in the 1850s, when urban reform clergy like Henry Ward Beecher and Henry W. Bellows began to preach against the vices of overwork and excessive devotion to business and moneymaking and to extol the virtues of regular vacations, outdoor sports, and recreation. Educators and public moralists like Charles Loring Brace steered vacationers away from fashionable resorts to the more mundane and healthy pleasures of mountain walks. At the same time, physicians began to warn of the physical and mental dangers of work unrelieved by recreation: neurasthenia, even addiction might result. As the physician and novelist Silas Weir Mitchell dramatically put it, the drug of work created "dollar fiends."[36] The logic was especially compelling to a people intensely devoted to commerce and getting rich.

This change in the moral conception of leisure also had religious roots. The early nineteenth century was a period of zealous religious enthusiasm and revival in the United States (and England). But by the 1860s puritanical fervor was regarded as excessive and out of date, and as new forms of sociability became popular, like clubs and athletics, religious moralists had to trim their rhetoric to a more secular and pleasure-seeking culture.[37] Post–Civil War economic changes also tipped the moral balance, as business boomed (and busted) and real incomes rose for all classes. Making money became the great concern of the "Gilded Age," and to clerical moralists it seemed that love of money was a more serious threat to religion than sports and leisure pursuits.

Besides, outdoor sports and recreations were becoming very popular in the 1850s, not just in the laboring classes but increasingly also among those in white-collar occupations and professions. Clergymen would hardly care to alienate their middle-class congregations by denouncing outdoor sport and recreation as immoral.

Fig. 2–3. "Before going to the Adirondacks" and "After going to the Adirondacks." A satirical image of the middle-class belief in the debilitating effects of overwork and the restorative powers of outdoor vacationing. In "The raquette club," *Harpers New Monthly Magazine* 41 (1870): 321–38, pp. 334 and 335.

Better to domesticate and tame the new customs by giving them a new moral sanction. Why should the worldly, the "fast men," be the only ones to enjoy active outdoor recreation, and not moral and religious people, asked one sport advocate: "Do not let us grow old and dyspeptic, because we are growing more religious." He recommended mountain walks, nature, and scenery.[38] In the same spirit, an English cleric drew up lists of pleasures that were morally sound and those that were not. Fishing was permissible, for example, because it was solitary and encouraged meditation and communion with nature; hunting was not, because it was done in groups and was often followed by conviviality and tippling.[39] Stripped of their association with idleness and vice, outdoor activities became an attractive addition to the cultural repertoire of a self-fashioning managerial and white-collar class. Idle and godless leisure was thus transformed into purposeful and improving recreation.[40]

This moral rationale for outdoor recreation became a standard litany after the Civil War, when vacationing resumed and became a widespread custom. Vacations were essential for physical and mental health especially for the intense "high quality" labor of commercial and professional workers—that was the idea. Regular vacations were good business, making work efficient the rest of the year and preventing nervous prostration and forced rest. Vacations were rest cures, informal alternatives to sanitoria for the nerve-worn. In the famous finale to his American tour in 1883, Herbert Spencer preached the "gospel of recreation" to his New York audience, warning that overwork could be the undoing of commercial civilization: "[W]e have had too much of the gospel of work," he proclaimed. "It is time to preach the gospel of relaxation."[41]

The "gospel of relaxation": in this one catchy phrase Spencer concentrated the cultural transformation of leisure into recreation. All the fashioners of modern middle-class life seemed to preach this gospel—cultural ideologues like the editor and moralist Edwin Lawrence Godkin; hotel and railroad promoters of the new vacation industry; and writers of resort guidebooks like John B. Bachelder. Stock stories circulated of fagged-out businessmen restored to vigor through camping vacations in the country. The Reverend E. H. Stokes, president of the Ocean Grove Association (a Christian resort community), pointed dramatically, indeed hysterically, to the familiar symptoms: "the haggard brow, the tottering, the irritated nerves, the sudden paralysis, the multiplying victims of the insane asylum." These were the stigmata of "*Over-work*!"[42] Later promoters of middle-class vacationing added to these stock themes the benefits to urbanites of consorting with hard-working and pious country folk: a moral type deeply rooted in American republican ideology. It is no surprise that work was so central to the new concept of recreation; work, after all, was how the middling and professional classes defined themselves.

Not coincidentally, the earliest and most vocal advocates of the gospel of recreation belonged to the few occupational groups who already integrated work and outdoor recreation: namely, teachers and clergymen. Professors always had annual summer vacations and were accustomed to working holidays. And clergymen were at some point granted annual paid vacations, though no one seems to know exactly when. It makes sense: like teaching, pastoral work is seasonal, and rural clergymen were as famously given to leisure-time naturalizing as academics were to summer scholarship. (Natural theology gave moral weight to this innocent pleasure: it was reading the Book of Nature.) Clergy are mentioned as frequenting country resorts at least as early as the 1850s: for example, the Boston Universalist minister Thomas Starr King, whose book on the White Mountains opened up this hinterland for tourists. In the 1880s faculty of the Harvard Divinity School were a distinct subset of the summer crowd at Randolph, New Hampshire, climbing and clearing mountain trails.[43] Nineteenth-century artists were another occupation that habitually spent the summer months in some rural place or picturesque village, painting landscapes and imbibing the exotic and authentic customs of local peasantries.[44] Wordsmiths—journalists, novelists, travel writers—likewise used vacations and travels to gather grist for their creative mills. As recreation was already bread and butter for these mental workers, they were naturally early and eager advocates of combining work and leisure.[45] Their own social customs encouraged the belief that a spell of outdoor recreation was essential to high-quality performance for all "brain-workers."

These working vacationers served as a vanguard of tourist economies in seaside and lake districts. Impecunious of necessity, being members of esteemed but mostly ill-paid occupations, teachers, clergymen, artists, and writers were often the ones to push out the fringes of fashionable tourist zones in search of cheap and interesting recreation: "middle-aged maiden ladies from university towns . . . college professors . . . literary men or women . . . clergymen," as one observer put it. As places got "discovered," thanks in part to the free publicity they themselves produced, these pioneers of middle-class vacationing were driven out by rising prices and cottagers, and forced to move farther up coastlines or deeper into rural and mountain areas, where the process would be repeated.[46] At least that was the stereotypical story, and there was doubtless some truth in it.

Tastes in vacationing also changed as the custom spread. Middle-class vacationers disdained the posh resort hotels frequented by merchant and planter gentries, flocking instead to modest family hotels, cottages, and boarding farms. This trend began in the

1850s, and by the 1870s the grand society hotels, with their competitive formal dressing and cliquish social rituals, were quite out of fashion with most vacationers, and objects of satire for vacation boosters. Vacationing middle-class families preferred informality, freedom from society and fashion, and the chance to experience country life and nature. New hotels like the Mohonk in the Catskills catered to middle-class families by offering lectures on topical social issues. Grand hotels constructed modest family-style cottages on their grounds; and resort towns sprang up that consisted almost entirely of small hotels and boarding houses. Modestly priced for young white-collar families, the new infrastructure of vacationing embodied their social values.[47] Where once a stay at a society hotel was essential to maintaining one's position in gentry culture, so in the 1870s and 1880s a sojourn in a humble cottage displayed one's commitment to the ideal of improving work and play.

Farm boarding was another way for middle-class families to escape the vexations of society hotels and have an improving vacation close to nature. It was cheap, fed urbanites' nostalgia for lost rural innocence, and associated them with a social type famous for their unrelenting work. It is unlikely that many farm boarders shared the long hours and hard labor of plowing, haying, milking, picking, and cooking. For most it was the symbolic association with honest labor that appealed. Farm boarding afforded virtue by association.

The custom of farm boarding was especially strong in the hill country of the Northeast, a region strong in Protestant ethic, and where tourism was the readiest replacement for a declining agricultural economy. Edward Hungerford, who systematically toured the back roads of deepest New England, reported that farm boarding was extremely widespread there by 1890, even in remote hamlets. In the White Mountains, boarding farms were a mainstay of the rural economy. In the Catskills a week in a boarding farm was attractively priced at six to twelve dollars, about the weekly wage of city clerks or young professionals.[48] Later, in the mountain West, ranchers in marginal range areas found dudes a more profitable and steady livelihood than steer. The census of 1930 reported 6,200 boarding farms: most of them in New England, but 366 western dude ranches.[49]

Fig. 2–4. "Waiting for an artist." This romanticized image captures the middle-class desire to buy an "abandoned farm" and return to nature and traditional rural virtues. In William H. Bishop, "Hunting an abandoned farm in upper New England," *Century* 48 (1894): 30–43, p. 36.

A related phenomenon was the new fashion for second, vacation homes. This too was a custom once limited to the wealthy, who congregated in private communities like Newport, Rhode Island, or in private hunting preserves and luxurious "cottages" in the Adirondacks. But as middle-class families became more prosperous and accustomed to annual vacations, they turned from farm boarding to buying marginal hill farms and converting them into modest summer homes. Boarders thus became cottagers.[50] This custom grew slowly and quietly in the 1870s and 1880s, especially in lake and seaside areas, then about 1890 erupted suddenly into public view as a rage for buying "abandoned farms." What unleashed this real-estate bubble was the publication by several New England state governments of lists of "abandoned" farm properties for sale at bargain prices. The idea was to revive depressed rural economies

Fig. 2–5. The unromantic realities of rural real estate during the fad for "abandoned farms." In E. P. Powell, "Simple vacation," *Independent* 55 (1903): 1320–27, p. 1321.

by attracting urban settlers (and taxpayers); it followed the failure of earlier schemes to entice Scandinavian immigrant farmers to upland areas (perhaps in the belief that they would be used to thin soils and a short growing season). Middle-class vacationers proved more susceptible to the propaganda. The idea of owning an "abandoned farm" captivated their imagination, and a mini-landrush ensued. The reality, of course, was not quite what country real estate agents made out: most farms were not abandoned at all but just underused, and few bargains remained after alert locals had skimmed the best.[51]

Nonetheless, images of abandoned farms, artistically rendered in popular magazines, offered consumers a nostalgic time-trip and the prospect of participating in the new fashion of vacationing. Boarding farms, cottages, and restored farmhouses were a visual embodiment of the ideals of the improving country vacation. It was in such places that leisure became recreation. (My Brooklyn-born grandparents were late participants in this cultural movement, buying a hill farm in the mid-1930s in the central Green Mountains. Unlike most such "farms," this one remains to this day a working dairy farm, making prizewinning artisanal cheeses.)[52]

Thus physically embodied and morally sanctioned, outdoor vacationing became, along with work, a defining activity of middle-class life. Stripped of its stigma of idleness, recreation became an essential complement to the work of shop and office, a kind of work. The vacation, with its rituals of transhumance, bestowed on middle-class townsfolk the moral virtues that were believed to reside in nature and a life close to nature. It was not that a preexisting middle class added vacationing to their social repertoire; rather, the middle class was reconstituted in part through the invention of the vacation, as it was also by new forms of business organization and managerial work.[53] Class and custom coevolved.

Recreation and Natural Science

But how exactly was vacationing connected to a naturalistic or scientific interest in nature? Through the concept of improving recreation, I suggest. The customs and infrastructure of outdoor vacationing channeled middle-class Americans into regular contact with nature, while the ideal of "working at play"—Cindy Aron's apt phrase—invited them to engage more actively and even scientifically with nature.[54] Natural history was an outdoor activity that particularly embodied the idea of active, improving recreation. It was a kind of work (and play) that was easily assimilated, morally and logistically, to the practices of middle-class vacationing. My thesis.

Physical work was initially favored by promoters of the vacation habit. " 'Camping out,' " John Bachelder wrote, "means living in a tent; sleeping on boughs or leaves; cooking your own meals; washing your own dishes, and clothes perhaps; getting up your own fuel; making your own fire; and foraging for your own provender. It means activity, variety, novelty, and fun alive."[55] There were provisions to be fetched from nearby farms, berries to be picked, fish to be caught and cleaned. As Cindy Aron writes, "camping afforded one of the surest ways to guard against the potential hazards of leisure: it furnished the opportunity, indeed sometimes even the necessity, of combining vacations and work."[56] Unlike sport, the physical work of camping or farm boarding was a moral enactment of a virtuous and democratic way of life for townsfolk who normally paid others to work for them.

Naturalizing was even better suited than brute physical work to the task of dignifying leisure: less onerous than hewing and hauling, and morally just as potent. Natural history, as the historian Mark Barrow has observed, was itself strenuous and exacting work: collecting specimens of birds and eggs, for example, "required both a great deal of physical effort and an intimate knowledge of the habits and haunts of the intended quarry. It therefore represented the epitome of the strenuous and purposeful activity that the anxious middle-class sought."[57] Actively observing nature also had powerful symbolic and moral implications. It was a pleasurable but not an idle pursuit, and a good way of distinguishing one's own virtuous recreations from the entertainments of society hotels. Nature was the cure for a surfeit of culture. For example, one John Todd, who kept a diary of his vacationing in the late 1870s, urged vacationers to follow his family's example and avoid the "vexatious comforts" of fashionable resorts and give themselves "wholly in trust to nature and the elements." A decade later the editor of the *Nation* pointed to the connection between "a growing national love of Nature . . . and . . . a certain shrinking away from the exactions of society and of fashion."[58] Pursuing the truths of nature was the very antithesis of social artifice: a symbolic guarantee of moral virtue. Actively observing nature was educational and self-improving, pleasant but not mere pleasure, and an ideal activity for a recreative vacation. No wonder that devotees of the "gospel of recreation" were so ready to see the virtues of amateur naturalizing.

We will never know how many turn-of-the-century vacationers actually engaged in natural history or collected specimens. But there is indirect evidence that many did. For example, the large sales of field guides to birds and local floras in this period suggests that amateur naturalizing was a widespread custom. As do also the numerous collections of specimens, many small but some quite substantial, that were donated or sold to natural history museums in the 1880s and 1890s.[59] Doubtless most campers and cottagers went to the country simply to unwind and enjoy its pleasures. But the moral imperative to make leisure self-improving was an incentive to take an intellectual interest in nature. It did not necessarily turn vacationers into naturalists; but it could amplify a casual interest into more active participation.

The connections between recreation and science were most explicitly drawn in the case of recreational walking, no doubt because so many walkers were collegians. Early walking clubs combined the ideal of physical culture and natural history: for example, the one at Williams College; or the Appalachian Mountain Club, whose founding members included leading Boston naturalists and veterans of geological surveys, and whose stated aims were scientific as well as recreational.[60] Walking and climbing were not scientific activities, but their aim of direct experience of nature—in contrast to the aesthetic and literary aims of earlier mountain tourists—was congruent with a scientific interest.

Natural science played a similar role in the cultural transformation of hunting and fishing into respectable middle-class recreations. Hunting, because of its association with aristocratic display and commercial foraging, had to be morally reconfigured to make it suitable for middle-class participation. Early advocates of sport hunting like "Frank Forester" (the pseudonym of Henry W. Herbert) took the English gentleman-sportsman as a model, exhorting Americans to adopt the exacting customs of sportsmanship, in order to shed any association with pothunting (that is, hunting for the biggest bag without regard to the rules of sport).[61] For many Americans, however, that meant an equally troubling association with aristocratic idleness. For hunting to become a truly respectable middle-class recreation it had to become a kind of improving work; and how better to accomplish that than to connect it with natural history. In the 1870s and 1880s middle-class advocates of sport hunting did precisely that.

Hunting and natural-history collecting were in practice quite similar. Naturalists had to be skilled shooters and trappers, as hunters did, and they had to have the same intimate and exacting knowledge of animal ecology and behavior if they were to get their quarry. The hunting promoter and writer Charles Bachelder saw "[not] much of a dividing line between ornithologists and sportsmen." The naturalist John Casin likewise regarded collecting as the "ultimate refinement—the *ne plus ultra* of all the sports of the field. It is attended with all the excitement, and requires all the skill of other shooting, with a much higher degree of theoretical information and consequent gratification in its exercises."[62] Hunters also had to know their quarry's haunts and habits, and such knowledge seemed a natural basis for a scientific interest. Connecting hunting and angling with natural history thus reinforced values of skill and knowledge that were central to middle-class work and identity. It helped dispel the image of hunting as a brutal pursuit of aristocratic and plebeian shooters.[63] As the historian John MacKenzie observes, it was not just manly courage and endurance that gave hunting its prestige, but also knowledge of natural history: "It was, indeed, that scientific dimension, the acquisition of zoological, botanical, meteorological and ballistical knowledge, and the associated ordering and classifying of natural phenomena, which helped to give hunting its supreme acceptability among late Victorians."[64]

This connection of hunting and naturalizing was embodied in the social type of the hunter-naturalist. Exemplified by such historic personages as John James Audubon, this role was an available cultural resource for those who, a generation later, set about transforming hunting and gathering into improving recreation. The ideal of the "hunter-naturalist" became widely popular in the 1880s and 1890s, exemplified by such well-known figures as George Bird Grinnell, William T. Hornaday, and of course Theodore Roosevelt, the most famous exemplar of the type.[65] Frank Chapman briefly adopted this persona on his way toward becoming a distinguished ornithologist, and it was a natural evolution in a naturalist's career.[66] Thus were hunting and fishing assimilated into the middle class's cultural repertoire of improving recreations.

Evidence of the connection between hunting and natural history is abundant, especially in the sporting magazines that sprang up in the 1870s. These served as both literary entertainment and a vehicle for reconstituting outdoor sport as a middle-class recreation, rather as the *Nation* did for the vacation. Sporting magazines like *Chicago Field* (later *Field and Stream*) and *Forest and Stream* are filled with natural history pieces, some of which are hardly distinguishable from the hunting stories that were their stock-in-trade. Distinguished American naturalists published regularly in these sporting journals, especially ornithologists, ichthyologists, and mammalogists, who dealt with the animals that most interested hunters and anglers. Frank Chapman recalled that *Forest and Stream* was "not only the leading journal for sportsmen but its high standing made it a recognized means of communication between naturalists." He

Fig. 2–6. The ornithologist Frank M. Chapman "still in the hunter stage" encamped on the Nueces River, Texas, in 1891. The pose and props embody the ideal hunter-naturalist type. In Frank M. Chapman, *Autobiography of a Bird-Lover* (New York: Appleton-Century, 1933), p. 99.

read every issue from cover to cover.[67] The leading ornithologists Elliott Coues and Robert Ridgway were regular contributors, as was David Starr Jordan, American's premier ichthyologist. Zoologists used these sporting magazines as outlets for getting their professional work to a wider audience. Coues, for example, mined his monograph *North American Mammals* for short pieces on subjects that would interest sportsmen: for example, habitats and ecology, feeding and defensive behaviors, and scientific names.

Turn-of-the-century hunters and anglers took a particular interest in taxonomic nomenclature, because the authority of scientific names and rules of identification could resolve disputes and avoid miscounts in competitive shooting and fishing. (However, game writers were less inclined to credit scientific authority if it contradicted long-standing custom.)[68] For these reasons sportsmen were keenly interested in the personalities and careers of leading naturalists.[69] Professional naturalists were not just anonymous experts who possessed useful information; the relation was based on participation in a shared way of life. In their practice and cultural meaning, natural science and field sport were sometimes hard to tell apart.

Women were especially associated with the intellectual and educational side of hunting, and sport writers appealed to them to take up hunting as an alternative to idle dissipation at fashionable resorts. It was, they argued, an opportunity to acquire improving knowledge and skill. "Many a crack shot . . . might well envy her accomplishments with the gun," wrote one journalist, "while the knowledge she has gathered in the great school house of nature would put to blush the professors of natural sciences in one-half the colleges and seminaries of the country."[70] Another advocate of female participation in sport hunting pointed to its cultural benefits: "Field sports . . . offer the largest opportunities for healthy culture, for deep study, and for the exercise of such accomplishments as drawing and painting."[71] Making hunting an activity for the whole family, rather than just for males, helped make it a middle-class recreation, and as the keepers of the educational side of family life, women were especially suited to connect sport to science.

Such propaganda may of course tell us more about reformers' ideals than the actual practices of ordinary hunters and anglers, most of whom doubtless went afield for pure pleasure or profit and cared little for purported moral or scientific benefits. Sporting magazines at the time ran numerous reports of "slob-hunters" who shot merely to kill as many animals as possible in the shortest time.[72] However, there is also evidence to suggest that some sportsmen did in fact practice what sport reformers preached. The "Cuvier Club" of Cincinnati (organized by 1871) had no trap shooting, and no bar, restaurant, or gambling in its club house, and it was well known for its fine natural history collection. It had a resident taxidermist and, from about 1880, a museum. Minnesota sportsmen organized a similar club, "The Lake Pepin Club," to encourage natural history and display its members' collections of scientific specimens. The Union Club of South Bend, Indiana, was organized in 1875 to a similar end, and natural history was a goal of the Michigan Sportsmen's Association (founded 1875), second only to game-law reform.[73] The ecologist Charles C. Adams was surprised and delighted at the response of a group of seventy-five Cincinatti men and boys to a lecture on the historical geography of their region: there were lots of questions, unfeigned interest in natural history, and an invitation to join the group for some collecting come the summer season.[74] Urban middle-class hunters and outdoorsmen wanted their sport to be respectable, and natural history was an agreeable and practical way of making it so.

Some of the most explicit evidence of the connection between vacationing and science comes from the side of science. Marine biological stations were usually situated in small resort communities like Woods Hole, on Cape Cod, so that scientific work could be mixed with the outdoor pleasures of seaside resort life. Biologists avoided large and fashionable resort areas, however: these were too crowded and expensive, and offered too many social distractions. The customs of inland field stations were likewise those of a working vacation.[75] The University of Michigan's field station was deliberately sited on a pleasant lake in the northern cutover, to provide summer students with the experience of a summer camp. Recreation made for better work, and work gave play a value that it did not have alone. As the ecologist Henry Gleason observed, "The student . . . not only learns his science thoroughly, but . . . has a vacation, and returns to his regular work in the fall completely recreated." Camping trips with guides were organized, and students were encouraged to become skillful in the practical side of camping and fieldwork: setting up tents, cooking, and so on.[76] The biologist Morton Elrod made a similar pitch for the summer courses of the University of Montana's station at Flathead Lake, in the beautiful Mission Mountains:

> Recreation and work will be combined, but study will always be foremost. . . . The Station is easy of access, the ranches nearby have an abundance of fruit, trails have been cut to several places of interest, and an automobile road is being constructed along the lake shore. Camp life, good food, [and] outdoor study . . . will give a tonic which will stimulate those attending for a long time.[77]

Collegians thus enacted in their daily routines the middle-class ideal of improving outdoor recreation. In such places it was hard to say where hiking, climbing, fishing, and camping ended and fieldwork began.

Nature Essay and Diorama

We have so far considered recreational activities that brought vacationers and sportsmen into actual contact with nature. Equally important, however, are the activities that brought them into imaginative or virtual contact: like reading nature essays or viewing museum exhibits and dioramas. If many families directly experienced nature as hunters, walkers, or campers, doubtless even more experienced it indirectly as cultural consumers. And because these virtual experiences of nature were calculated artifice, they are especially revealing of the affective and moral structure of nature-going.

Nature essays are the most familiar and well studied of these artistic products.[78] The classic nature essay was a short account of an author's experience of the natural world: typically a walk in rural woods or fields, or in a twilight zone between town and country. It was a naturalistic literary form, though not a scientific one, affording readers accurate knowledge of animals in their natural habitats, but in a poetic style that expressed the author's thoughts and feelings about his or her experience. (Many nature essayists were women.) The genre was a marked departure from the sentimental animal fables of the early nineteenth century, which made no pretense to naturalistic truth but delivered moral lessons using animals as Aesopian stand-ins for human types. The nature essay, in contrast, was designed to show readers that accurate knowledge of nature also gave emotional satisfaction, and that aesthetic experiences of nature were deepened by scientific knowledge. Natural knowledge was not something separate from everyday life but an integral part of it: that was the message of this medium.

The nature essay was famously the invention of John Burroughs, who began to write and publish in this mode in the mid-1860s. His books enjoyed a modest but growing popularity until the mid-1890s, when the genre erupted as a literary rage. One observer of the literary scene in 1902 marveled at American consumers' sudden appetite for nature books and at the dozens of authors who had joined John Burroughs in feeding the appetite: Ernest Ingersoll, Neltje Blanchan, Frank M. Chapman, Francis H. Herrick, Frances Theodora Parsons, Mabel Osgood Wright, F. Schuyler Mathews, Alice Morse Earle, Ernest Seton, James Lane Allen, Charles D. Roberts, John Henry Comstock, Anna Botsford Comstock, Mary Rogers Miller, Martha McCulloch Williams, A. R. Dugmore, John Muir, and J. P. Mowbray were just some of the better known.[79] A collection of Burroughs's essays for grade-schoolers cashed in on the nature-study movement and sold 300,000 copies between 1889 and 1906. New York and Boston publishers sold 70,000 birding books between 1893 and 1899.[80] Natural history periodicals appeared in the 1880s like mushrooms after a spring rain, multiplying from about ten in 1883 to almost forty in 1885, then receding to between fifteen and twenty in the 1890s. Naturalists listed in succeeding editions of Samuel Cassino's *Naturalists' Directory* went from 1,431 in 1877 to just over 6,000 in the early 1890s, and over one hundred commercial firms dealing in natural history specimens and collecting paraphernalia advertised in its pages.[81]

"Wilderness novels" became even bigger best sellers in the 1900s—novels of the inner frontiers, I would call them. The queen of the wilderness novelists was a dentist's wife and amateur nature photographer, Gene Stratton Porter. She wrote four of the five novels that between 1900 and 1930 sold over 1.5 million copies. The chief of these was *A Girl of the Limberlost* (1909), whose impoverished but spunky heroine, Elnora Comstock, paid her way through high school and became a teacher of natural history by selling natural history specimens collected in the "Limberlost" swamp, one of the fast dwindling remnants of the once vast wetlands of northern Indiana. Other wilderness romantics who took their readers to the "silent places" of inner frontiers included Harold Bell Wright, Stewart Edward White (the son of a Michigan lumberman), Mary Waller, James Oliver Curwood, and—most famously—Jack London.[82] Stories for children likewise began to take their young heroes and heroines afield, abandoning the traditional urban settings and mercantile adventures of Oliver Optic and Horatio Alger for outdoor adventures in vacation lands and inner frontiers.[83] Middle-class Americans' appetite for informative and entertaining books about nature and "wilderness" seemed insatiable.

This appetite coincided exactly with the fashion for camping, walking, and cottaging. No accident: nature essays and outdoor adventure stories were, in effect, the Baedekers of country vacationers—how-to guides to having an experience of nature that was both pleasurable and uplifting. They were the imaginative software of the vacation complex. And the custom of outdoor vacationing created the audience of consumers for these literary genres. John Burroughs consciously strove to make his essays seem like play; writing them, he alleged, was the final act of a country holiday:

> I cannot bring myself to think of my books as "works," because so little "work" has gone to the making of them. It has all been play. I

> have gone a-fishing, or camping, or canoeing, and new literary material has been the result. My corn has grown while I loitered or slept. The writing of the book was only a second and finer enjoyment of my holiday in the fields or woods.[84]

It was a crafty authorial guise designed to narrow the gap between the writer and his readers by implying that both were on a rural holiday. He cultivated a rustic public persona and lifestyle at his home, Riverby, on the Hudson River, and his writing retreat—Slabsides—resembled a vacation cabin. There is little doubt whom Burroughs saw as his intended audience, and how he meant his works to be read: by walkers, campers, and farm boarders, as guides to improving recreation.

A perceptive contemporary judged that the vacationing habit was behind the huge popularity of nature essays, or at least that the two phenomena had a common cause in Americans' longing for direct and vivid experiences of nature.[85] The two were mutually reinforcing: nature essays epitomized the values of improving recreation, and country vacations were the ideal venue for cultural consumption.

No less popular than nature essays were the new types of naturalistic animal display that began to appear in natural history museums in the 1890s. The most elaborate of these exhibits was the much-loved habitat diorama, which used visual tricks to create the illusion of actual places in nature. Its central feature was a lifelike group of animals arranged in naturalistic poses, at home and in action, exhibiting habits of feeding, predation and defense, and family life. In the foreground "accessory" materials were arranged: rocks, earth, grasses, and plants—either the real stuff collected in the field, or imitations skillfully fabricated out of wax and wood. In the background, artfully arranged to blend continuously with the actual artifacts, a painted backdrop gave an illusion of deep space. This backdrop was what distinguished habitat dioramas from simpler animal groups and panoramas. As Frank Chapman explained, the earlier "panorama" group and the habitat group both displayed groups of animals, say, birds, with nests and contents, vegetation all realistically reproduced. But where the habitat group's painted backdrop depicted a specific place, the panorama's was only of a generic scene: "In short, the first is painted from nature, the second largely from imagination or memory or from some locality other than the right one!"[86]

Habitat dioramas were designed to give viewers a virtual experience of being in a particular place viewing the animals that lived there as in life. It was an experience that was at once precisely observed, scientifically accurate, and aesthetically powerful.[87] The habitat diorama was the visual or sculptural equivalent of the nature essay: at once an aesthetically and emotionally engaging lesson in natural history, and a guide to the appropriate aesthetic and intellectual responses to an actual experience of nature. It was their vividly illusionistic art that made dioramas such a hit. But it was their truth to nature that gave them authenticity and made museum visits acts of improving recreation, not just idle entertainment.[88]

Unlike the nature essay, which appeared fully formed in Burroughs's essays, the elements of the habitat diorama came together gradually over several decades. This development began in the late 1860s and 1870s, when taxidermists, inspired by improvements in the craft of preparing and mounting animal skins, began to create animal sculptures that were at once more lifelike and more flamboyantly artistic, borrowing visual conventions from Romantic painters like Géricault. Jules Verreaux's *Arab Courier Attacked by Lions* (1867) exemplified this new mixed genre, and it was followed by others such as William Hornaday's dramatic orangutan group, *Fight in the Tree-Tops* (1879), and Frederick Webster's *The Flamingo at Home* (1880), a scene of a flamingos nesting, which was created from skins collected by John J. Audubon himself. These groups aimed for visual effect rather than scientific accuracy.[89] Some had details that were biologically wrong, and some were quite fanciful. Their purpose was to display the artist's skill and to catch the public eye, which they did by giving a new realism to traditional artistic subjects like animals in fierce but improbable combat. They represented both artistic and scientific truths, but the art came first.

By the early 1890s a variety of different types of animal displays vied for museum-goers' attention. Some were simply animals or animal groups mounted in dramatic poses and placed on pedestals without accessory material. Others were groups with some accessory materials but no painted backdrops, and displayed in freestanding glass boxes. These ranged in size from small groups of familiar birds—popular parlor decorations—to huge displays of large predators or game animals. The first buffalo group was a sensation when it opened to the public at the National Museum in 1888, and other museums quickly dispatched expeditions to get specimens before the species dwindled to extinction. The American Museum had eighteen glassed-in animal groups by 1888, mostly of common backyard birds.[90]

These groups were hugely popular with museum-goers and were widely imitated. The Field Museum in Chicago hired the renowned English taxidermists Mrs. E. S. Mogridge and her brother Henry Minturn to make a set of small groups of local Illinois birds—robins with apple blossoms, prairie chickens with nest and eggs—that especially appealed to local townsfolk and hunters. A duck and wildcat group was in an older romantic vein, and a group of Florida birds appealed to the public's taste for the exotic.[91] Other animal groups were more dramatic, like the orangutan group created in 1897–1898 by the great animal artist Carl Akeley. By enclosing groups in glass boxes, exhibit builders could include more realistic (and fragile) settings than was possible with open mounts.

In the 1890s these boxed-in groups became more elaborate and naturalistic, and any vestiges of earlier romantic art disappeared. Foreground materials became more complete representations of actual habitats. And, most important, painted backdrops were added as the back wall of exhibit boxes, depicting the habitats where the animals typically lived. Early examples of this new style were Akeley's Muskrat Group at the Milwaukee Museum (1889), which featured a cutaway of a pond and lodge; Frank Chapman's *Bird Rock Group* at the American Museum (1898), depicting a famous island bird roost in the St. Lawrence estuary; and Chapman's *Cobb's Island Group* (1903). This last was a watershed in diorama design. Again displaying a particular place (Cobb's Island, on the Virginia shore), this group featured a realistic, painted backdrop carefully integrated with the mounted specimens and foreground setting, giving an illusion of being at the place depicted. A sensation at the time, it was the model for the American Museum's great Hall of North American Birds. Carl Akeley also created displays of this type at the Field Museum in Chicago at about the same time: for example, groups of red deer in the four seasons of the year (1902), and of wild turkeys and other birds, and big bears, in the early 1910s.[92]

Fig. 2–7. The Field Museum's orangutan group of 1898 (created by Carl Akeley and reinstalled by Leon L. Pray). An example of the diorama in the middle period of its development: note the freestanding glass box, the dynamic poses, and the realistic but generic and contextless props. Courtesy Field Museum of Natural History, negative no. CSZ6235.

The painted backdrops were crucial, imparting a drama that groups alone with foreground settings could never have. These backdrops became even more effective when groups were mounted not in freestanding glass boxes but in wall niches, where paintings were not visibly part of a box but were integrated with the animals and the foreground props. Group, foreground, and background merged into a single natural scene.

Fig. 2–8. The Field Museum's wild turkey group of 1910, a fine example of the fully developed habitat diorama. Note the elaborate foreground material, artfully painted backdrop, and illusion of continuous deep space. In Field Museum *Annual Report* (1910), p. 100. Courtesy Field Museum of Natural History, image no. CSZ31630.

A midwestern zoologist touring eastern museums in 1895 saw an early form of such exhibits at Harvard's Museum of Comparative Zoology and realized at once that they would soon completely replace the traditional displays of stuffed animals arranged according to taxonomic principles. A room of bird groups, each set in a niche in a false wall, appeared from a distance to be a room of beautiful paintings, like in an art museum. But it was art science: "Each group . . . was a work of art in itself, and best of all, *each taught an important lesson*, in a way that no one could *look* at it and fail to see the point illustrated." The principles of variation and distribution were strikingly illustrated, and labels underlined the lessons.[93]

With most of its elements in place by about 1900, the habitat diorama took its final form in the following decade. Stages were raked upward and painted backdrops were curved to enhance the illusion of deep space, and both merged seamlessly with animal mounts and foreground props. Ceilings were hidden by projecting ledges, and the scenes were illuminated from above by hidden lights, creating a powerful dramatic effect. It was as if viewers were out there on the spot, observing animals in life from within a naturalist's blind. This artful device removed animal specimens from the indoor space of artistic display, opening virtual windows upon distant and exotic places.[94]

Dioramas were not generic scenes illustrating abstract scientific principles, but recreations of actual places, and it was this particularity of place that guaranteed their scientific—and artistic—authenticity. William Hornaday made this point in 1891 when he wrote of the advantages of a painted backdrop over the usual glass box: "The objects to be gained," he wrote, "are distance, airiness, and, above all, a knowledge of the country inhabited by the bird or mammal." The idea, as one curator put it, was to bring "whole living sections of nature" indoors.[95]

It did not take long for leading museums to envision series of habitat dioramas that would fill entire halls and represent the major bird and mammal fauna of entire regions or continents. At the American Museum, Joel Asaph Allen and Chapman were planning matching halls of North American Birds and Mammals after the success of the *Cobb's Island* group. So popular were these displays with museum-goers and donors that the museum's president, Morris K. Jesup, created a special fund to build a series of thirty-four dioramas. Completed in 1909, the Hall of North American Birds was followed by even more spectacular halls of North American and African mammals, which drew crowds of admiring viewers (and still do).[96] The evolution of the diorama tracked the development of a naturalistic literary and artistic taste.

These appealing and naturalistic representations of nature both drew upon and fed the public's interest in nature, and helped transform natural history museums into institutions of mass cultural consumption where viewers could absorb a bit of science and have a vivid virtual experience of nature, all without leaving town. Americans may have acquired the habits of nature-going and nature study as much by virtual participation in reading and viewing as by direct experience. As nature was physically accessible in the inner frontiers of North America, it became culturally and emotionally accessible in part through the efforts of the curators, taxidermists, and artists who created the habitat diorama.

Compounded equally of art and science, nature essays and dioramas were essential elements of the vacationing complex, whetting appetites for camping or farm boarding, informing novices of the behaviors and feelings proper to nature-going, and reminding vacationers of their summer pleasures. In turn, vacationing created the consuming audience for literary and visual artifacts. If Burroughs could pretend that the work of writing was holiday play, his readers could fancy that reading his essays was uplifting work. Nature art and nature recreation developed symbiotically.

The Science of Art

In what sense were dioramas and nature essays scientific as well as artistic artifacts? Should we see reading nature essays and viewing dioramas as a kind of virtual participation in science? The thought is not as farfetched as it may seem. After all, they were meant to be as faithful to nature as scientific collectors and observers could make them, and to convey scientific truths by making them pleasing to consume.

Burroughs, for example, aimed to give his readers accurate knowledge of nature by making it relevant to their own lives. He strived to capture not just the truths of nature but also the pleasure that comes from personally experiencing nature and grasping its truths. He prided himself on his scientific accuracy but insisted equally that his essays were not works of science but works of art. As honey was "the nectar of the flowers with the bee added," the nature essay was nature with the artist added:

> The literary naturalist does not take liberties with facts; facts are the flora upon which he lives. The more and the fresher the facts the better. I can do nothing without them, but I must give them my own flavor. I must impart to them a quality which heightens and intensifies them. To interpret Nature is not to improve on her: it is to draw her out; it is to have an emotional intercourse with her, absorb her, and reproduce her tinged with the colors of the spirit. If I name every bird I see in my walk, describe its color and ways, etc., give a lot of facts or details about the bird, it is doubtful if my reader is interested. But if I relate

> the bird in some way to human life, to my own life—show what it is to me and what it is in the landscape and the season—then do I give my reader a live bird and not a labeled specimen.[97]

In short, art connected scientific facts to daily life and thereby gave them meaning for those who might never pick up a work of science.

It is hard now to imagine just how fresh this genre must have seemed when it first appeared, at a time when science and art generally sought authenticity by quite opposite literary means. Scientific works strived for artlessness: it was a sign that the facts alone were strong enough to persuade, without the aid of literary artifice. (Contrariwise, sentimental "summer books" gave authenticity to social and moral truths by subordinating scientific veracity to moral ends.) Burroughs operated in a cultural space that was both scientific and aesthetic, and wrote as neither a moralist nor a natural historian. "I care little for the merely scientific aspect of things," he insisted, "and nothing for the ethical. I will not preach one word."[98] He was a literary naturalist, and his essays were artful science, or scientific art.

Burroughs's essays are in fact strikingly naturalistic and are even now surprisingly fresh and readable—and informative, full of facts about animal behavior and natural history. They are field guides for readers who might wish to recapitulate the author's experience, telling readers the likely places to observe birds and animals, what kinds of habitats to look for, and how to read telltale signs. (You find thrushes in the leaf litter, and different kinds of warblers at different levels: from ground level to three feet, from three to eleven feet, in lower branches, and in canopy—as an amateur birdwatcher I took note.)[99]

But Burroughs's essays were also guidebooks to the art of nature-going: directions for the right way to engage aesthetically and to have the appropriate feelings. A leitmotif in Burroughs's many books is the art of seeing, which was as much a matter of feeling as of science.[100] The nature essay was a field guide to outdoor vacationers for whom the moral qualities of improving recreation were still crucial. Modern field guides offer less art and more technique to nature watchers who no longer require moral sanction for their outdoor pleasures.

Just how seriously Burroughs regarded factual accuracy is apparent in his vigorous attacks on the new genre of sentimental nature fable that appeared in the 1890s and became hugely popular in the early 1900s. Purportedly based on recent scientific work on animal behavior, these heavily anthropomorphic treatments were actually more like a revival of the earlier moral fable in which animals played human parts and displayed human virtues, failings, and feelings. Burroughs was an early and an ardent critic of the "nature fakers," as they were dubbed by his friend and ally Theodore Roosevelt.[101] Scientific accuracy was vital to Burrough because his essays were not merely items of artistic consumption: the science made the art authentic.

Burroughs used a number of rhetorical devices to give his readers the illusion of virtual participation. Many essays are framed as a day's tramp, starting in the morning and ending at dusk, and apparently narrated in real time.[102] (They were in fact not real-time stories but composites of many trips to a place, combined later from field notes.) Burroughs also invoked the particularity of place to give readers a sense of being there. Essays describe in detail some particular place: a pocket of swampy woods near his home in the Hudson Valley, stranded between two roads and largely forgotten except by boys and birders; a forest clearing in the Adirondacks once farmed by a hunter-farmer named Hewett; an abandoned ironworks in the Adirondacks where a single family remained "to see that things were . . . allowed to go to decay properly and decently."[103] These were not the generic places that biologists invoke to make a point about animal ecology. Most modern sciences (including field sciences) adopt a generic, universalistic mode of expression to give authenticity to their reports. But in nature essays, as in hunting and fishing stories and travelogues, it is particularity of things seen and emotions felt that establishes trust by creating a sense of virtual participation.[104]

Burroughs seldom provided species with Linnaean binomials lest these make readers feel ignorant and mystified. But he understood the power of scientific names. When an ichthyologist friend informed Burroughs that what he took to be an odd fish was in fact a crustacean, *Eubranchipus vernalis* or "fairy shrimp," Burroughs was doubly charmed by the scientific and the common name: the Latin assured him that his "novel fish had been recognized and worthily named" and "dignified" by science; the poetic name reminded him of the emotional pleasure he had taken in its beguiling form. On another occasion he recalled how a guide he

Fig. 2–9. John Burroughs by the summerhouse of his farm, "Riverby," on the west bank of the Hudson River (late 1890s?). In Clara Barrus, *The Life and Letters of John Burroughs* (Boston: Houghton Mifflin, 1925), at p. 292.

knew in the Maine woods would always ask him for the scientific names of plants:

> [W]hen I would recall the full Latin term, it seemed overwhelmingly convincing and satisfying to him. It was a relief [to him] to know that these obscure plants of his native heath had been found worthy of a learned name, and that the Maine woods were not so uncivil and outlandish as they might at first seem; it was a comfort to him to know that he did not live beyond the reach of botany.[105]

Science gave a dignity and worth to ordinary things, and everyday experience gave scientific knowledge a depth of feeling and meaning that by itself it would not have. That was the essence of the nature essay.

The creators of habitat dioramas likewise combined science and art, only by visual devices rather than word craft. This was no small feat. Traditionally art had no place in natural history museums: it belonged in art museums. Natural history exhibits were constructed on strictly scientific principles—for example synoptic displays revealing the principles of phylogeny—and curators were not just indifferent but hostile to art. Judging a competition of taxidermists' art in 1880, for example, three naturalists passed over Frederick Webster's spectacular showpiece, *The Flamingo at Home*, awarding a prize instead to a single duck on a stand. Some of Frank Chapman's colleagues at the American Museum reacted no less strongly to their first sight of his *Cobb's Island* diorama in 1903. The bird mounts they liked, but the painted backdrop "they regarded as a violation of the museum tradition which made an exhibit . . . as formal as the papers read before their confrères. 'It is too dramatic,' said one." No matter that it was also scientifically accurate: the art made it suspect. Appeal was then made to President Jesup, who took one long look and pronounced the exhibit "*very* beautiful"—and that was that.[106]

It was not unreasonable to think that artfulness and drama might subvert scientific credibility. Commercial taxidermists, like the authors of summer essays, had long catered to the contemporary taste for sentimental kitsch by putting animals in human drag (frogs skating or fishing), or making them into ornaments (peacock fire screens) or sentimental tableaux (herons pierced by gilded arrows)—all with perfect indifference to how animals actually looked and lived.[107] It was no wonder that museum scientists preferred stuffed ducks in a row, or used "artistic" as a pejorative term.

Rapprochement between science and art occurred when artists and curators grasped the benefits of mixing scientific and artistic conventions. On the one side, taxidermists realized that truth to nature brought them influential customers and higher professional standing. On the other, curators realized that more artistic displays attracted paying visitors to museums and thus gave public standing to their science. Taxidermists in the 1870s were eager to place their masterpieces in museums for the same reasons that fine artists were: museums were public places, affording excellent (and free) advertisement and official endorsement of their products. It was career and commercial motives as much as anything that first inspired animal sculptors to make their work more naturalistic. When the U.S. National Museum bought William Hornaday's *Fight in the Tree Tops* in 1882, it hired Hornaday as a resident taxidermist. When the American Museum hired Mrs. Mogridge and Henry Minturn to make bird groups in 1885 it organized an in-house department of taxidermy—and also hired Frank Chapman as its first curator of birds.[108] Bringing art into the museum benefited both scientists and artists: appointments in public institutions removed the onus of commercial interest from animal artists, and curators could better control the scientific quality of exhibits if they were made under their watchful eye by museum taxidermists. Science and art advanced together in natural history museums.

Though symbiotic, art and science remained distinct categories. Animal sculptors did not aspire to become scientists any more than John Burroughs did. They regarded themselves as artists whose subjects and materials happened to be also those of science. They just appreciated the commercial and aesthetic value of a naturalistic genre. Science heightened the visual power and prestige of animal sculpture, and raised its value to upscale customers. For them, science served art. For curators, it was the other way around: for them diorama art made science more appealing to the museum-going public and raised their status as museum professionals. Dioramas became scientific representations that used artistic techniques to deliver the message. They never relapsed into the sentimental, moralizing mode, as nature writing did in the art of the "nature fakers."

Scientific aims and values prevailed in natural history museums largely because curators suceeded in wresting control of diorama design and construction from taxidermists, who then abandoned scientific ambitions, remaining a skilled craft guild.[109] Celebrated taxidermists like Carl Akeley might challenge curators for control of exhibits; but curators had the advantage of scientific authority, and exhibits were too important for their own careers as field collectors to relinquish control (more on that in the next chapter). In contrast, the freewheeling market for nature literature was beyond the control of any one group and thus more subject to changing public taste.

The history of dioramas and nature essays reveals how science and recreation came together as vital elements of middle-class life. Just as nature essays gave readers a mimetic walk through some twilight zone, so dioramas took viewers virtually into the inner frontiers, to places where large game still lived and birds flocked to breed—places few museum-goers could hope to go themselves. As nature essays guided readers to the feelings appropriate to a day's nature walk, so dioramas afforded a surrogate experience of serious nature travel. The American Museum's halls of North American birds and mammals were a virtual continental eco-tour, with stops at the special places where naturalists observed and collected, or big-game hunters pursued their quarry.

Nature essays and dioramas were how-to guides to experiencing nature both aesthetically and scientifically. One did not need to know science to enjoy them. But because essays and dioramas used naturalistic representations to heighten artistic effect, they could entice consumers to take an interest in natural science. In their mix of science and art, they gave countless consumers instructions in understanding nature in a naturalistic or scientific way. They were the cultural software of the middle-class outdoor vacation, giving a simulated experience of nature-going and endowing it with intellectual purpose.

Conclusion

If the landscapes and transportation nets of the inner frontiers gave middle-class urbanites the opportunity to experience nature, it was customs and values of outdoor recreation that enabled them to seize these opportunities. Access to nature is social and cultural as well as physical. The social customs of the vacation complex—camping, hiking, sport hunting and fishing, bird-watching, collecting, farm boarding and cottaging—enabled Americans to experience nature intellectually and emotionally in ways that suited middle-class values of self-improvement. Likewise, new genres of literary and artistic representation, especially the nature essay and diorama, created an infrastructure of feeling and pleasure as important to nature-goers as the infrastructure of rail, roads, and tourist accommodation, or the channels of social custom. Museum exhibits and nature literature lowered the threshold to nature-going, much as did cheap and easy travel. Cultural artifacts helped middle-class vacationers to know in advance how nature should be seen and experienced, just as earlier generations of tourists knew from their Baedekers the responses appropriate to natural wonders like Niagara Falls or the cultural icons of European cities, even before setting foot in these places.[110] The values implicit in cultural artifacts and customs sanctioned nature-going as recreative and educational. To poach a phrase of Steven Shapin's, the ways to the inner frontiers were made of iron rails and social conventions.[111]

The physical geography of the inner frontiers and the cultural geography of the vacation complex were compounded of nature and artifice. In the inner frontiers, nature was separate and semi-wild but also accessible to town dwellers with town habits and perceptions. Camping, sport hunting, and rural vacationing gave townsfolk an experience of nature that was not rough and raw but agreeably domesticated. Likewise, nature essays and museum dioramas made nature familiar and accessible (virtually) indoors and at home. The customs and values of the vacation complex were adapted to a landscape of inner frontiers. They were responses to changing physical environments, and in turn guided the human shaping of these landscapes.

Landscapes do not call social customs into being, to be sure, but they may selectively amplify them. The basic values of the vacation complex—a taste for muscular and improving recreation, and a naturalistic view of things—were rooted in the social and economic changes that transformed American society from the 1830s to the 1860s. They were part of the redrawing of class lines that accompanied industrial expansion, and of the proliferation of white-collar commercial occupations that put a premium on skill and knowledge, redefining the meaning of work and leisure. The accessibility of the inner frontiers helped define the social and cultural identity of the industrial middle class, making activities once confined to the gentry into activities for many or all. Twilight and tourist zones enabled Americans to enact en masse the ideals of recreative nature-going. In less accessible, truly wild landscapes, outdoor vacationing could well have remained the habit of a few, or accreted different cultural meanings.

We are dealing here less with cause and effect than with a feedback loop: physical accessibility drew vacationers and sportsmen to the inner frontiers, where their experiences fostered conceptions of nature that reinforced their social habits. Mass vacationing physically transformed the environment of tourist zones to fit conceptions of "nature" and "wilderness," as surely as commercial agriculture turned savannas and prairies into corn and wheat belts. Grand resort hotels were once islets of urban social life in an alien countryside, but they were replaced by a sprawling system of camps, trails, family hotels, and summer cottages. The inner frontiers became a landscape for nature-goers and nature-going, a physical and moral landscape of improving outdoor recreation for fagged-out professsionals and office workers—and, as we will see next, for survey scientists and collectors.

CHAPTER THREE

Patrons

THE CUSTOMS of outdoor recreation made it easy for amateur naturalists to collect and observe, but what about those who wished to make their avocation a vocation as professional collectors or survey naturalists? Unlike casual collecting, systematic survey required organization and resources. Survey collecting meant expeditions, which were well beyond the means of all but the most devoted and well-heeled amateurs. Scientific access to the inner frontiers had a higher threshold than did simple collecting.

The demands of intensive survey collecting were also beyond the means of the academic institutions in which most biologists earned their livelihoods. Most collegiate natural history collections were small and were mainly resources for teaching. Besides these practical impediments to expeditionary science, there were also social and cultural ones. Neither colleges nor civic museums were accustomed to seeing active collecting as an accepted, much less a required activity of their employees. If college teachers and museum curators collected, it was in the time-honored way: on unpaid working vacations.

To make collecting an accepted and regular part of their work as well as of their leisure, naturalists had to reconstitute social roles and invent cultural sanctions. They had to create an institutional infrastructure that would get them routinely and easily into the field. They had to assemble networks of patrons and invent compelling reasons for well-to-do families to underwrite collecting expeditions and for museums to house large research collections. To transform biological survey from an improvised activity of devotees to the business of a scientific occupation, institutions had to be persuaded to undertake fieldwork, and science had to be sold to those who experienced nature-going as recreation, not science. There were various potential sources of patronage in late nineteenth-century America, and generally an inverse relation between the amount of money available and the ease of getting it. But in one way or another, most forms of patronage were related to the vacation complex.

The most widespread form of underwriting was the summer vacation, which all academics and most government and museum employees enjoyed. Vacations afforded not money but time. An extended period of free time, at the season of the year when plants and animals are active, was a virtual endowment for collecting and fieldwork (though cash was still required for travel). So too was the free labor of undergraduate and graduate students, for whom summer fieldwork was holiday work, unpaid except in the coin of scientific publication. But these were limited resources. Working vacations were fine for recreational collecting but hardly sufficed for sustained scientific work in distant outbacks, or for comparative studies of different regions. And students, though a renewable labor resource, had always to be trained and were available for only a year or two before disappearing into other walks of life, just when they knew what they were doing. Local amateurs were likewise eager, but variable in their skill and hard to bend to the rigors of systematic survey. Professional survey collecting required skilled and salaried participants and steady institutional backing.

The vacation complex also afforded modest funding through the local natural history societies and nature clubs that thrived in many cities and college towns. Hunter-naturalists, some of whom collected on a substantial scale and were experts in the taxonomy of a particular flora or fauna, could be tapped for a few hundred dollars to pay the expenses of a survey party. Professional and amateur naturalists, as co-participants in the vacation complex, enjoyed a mutual trust rooted in shared values and activities. Natural history surveys differed mainly in scale and intensity from individual excursions, so private collectors understood their worth. Many local surveys were supported by amateur networks, though generally on a modest scale, since most amateur collectors pursued their avocations on shoestring budgets.

Public institutions had a far greater potential as patrons, and some did become major supporters of natural history surveys, especially state and federal governments and the larger natural history museums. But bigger money meant more complex politics and new problems of trust, since large-scale collecting expeditions were initially not part of the mission of either civic museums or government agencies. As public institutions their mandate was to provide economic or public services. Collecting masses of scientific specimens was not self-evidently a legitimate use of public funds: it had to be made self-evident, and that took work.

If the association of collecting with recreation helped secure private patrons, it may have made institutional patronage more difficult at first. Scientific collecting was just too much like camping and sport hunting to be taken seriously by guardians of the public purse—too much like plain fun. This may be one reason why Congress did not allow the curators of the National Museum of Natural History to organize expeditions, and why government agencies that did engage in fieldwork excluded curators' families from expeditions (at least they did in the late 1940s).[1] Also why museum trustees thought it improper for curators to do fieldwork on the job: they were paid to be at their desks, cataloging and maintaining collections or serving museum visitors, not off somewhere on expeditions that were all too reminiscent of trustees' own outdoor vacationing.

Another and a most enticing source of patronage was the private scientific endowments that began to appear in the early 1900s.[2] But the lower standing of field science vis-à-vis laboratory science, and its association with outdoor recreation, prevented field naturalists from tapping these funds. Efforts to persuade the newly established Carnegie Institution to underwrite biological surveys failed for just these reasons. Leading naturalists proposed some six survey schemes to the Institution's executive committee in 1902: a faunal survey of palearctic Eurasia, and another of Central and South America; a survey of Pacific fish; exploration of the tropical Pacific; Antarctic exploration; and an ecological survey of Hudson Bay.[3] Although the Institution seemed initially open to expeditionary ventures (it supported several archaeological digs), it concluded that limited funds were better spent on the experimental sciences. More precisely, John Shaw Billings decided, as the all-powerful chairman of the executive committee:

> I much prefer at present to give aid to experimental work . . . rather than to the collection of specimens, descriptions of species, etc., or to the preparation of local flora or fauna lists. I have known something of the joys of the collector, and [know] that there are many more persons who wish to make collections and are competent to make them and to describe new species, than there are persons competent to do experimental work, but the latter are the persons who, at present, most

> need aid and encouragement. . . . Governments, museums, botanical gardens, travellers, etc., are doing much in the collecting way, but very little in the other way.[4]

In fact, museums and government agencies were by then sponsoring survey collecting, and on an increasingly impressive scale. C. Hart Merriam's U.S. Biological Survey was dispatching up to a dozen field parties a year, mainly to the American West. Natural history surveys were or would soon be operating in six midwestern states (though on shoestring budgets), and informally and sporadically in several others. Annual collecting expeditions were also becoming a regular and visible activity of large civic museums like the American Museum of Natural History in New York, the Field Museum in Chicago, and the Academy of Natural Sciences of Philadelphia. In their heyday each would dispatch a dozen or more collecting parties per year. Research museums at several universities also engaged in large-scale collecting. Harvard's Museum of Comparative Zoology was the first, though it is not clear that it kept up the pace.[5] And it was surpassed in the early 1900s by the Museum of Vertebrate Zoology in Berkeley, California, and the University of Michigan Museum at Ann Arbor. In the 1920s organizations of various kinds would get into the expedition game: newspapers and magazines, the National Geographic Society, commercial biological supply companies, and so on. For some thirty years, patrons almost fell over each other to sponsor expeditions.

How this system of patronage was created, and how academics and curators became active field collectors, is the subject of this chapter. And the story turns on the association of scientific, survey collecting with the customs of the vacation complex. But first, a survey of natural history surveys.

Natural History Surveys

The U.S. Biological Survey was the first public institution devoted to natural history survey on a continental scale. In its salad years under C. Hart Merriam from 1887 to about 1910, it was to taxonomy and biogeography what the U.S. Geological Survey was to field geology and physiography under John Wesley Powell: an exemplar of organized, large-scale field science, a cornucopia of new knowledge and exact field methods, and the nursery of a generation of talented practitioners.[6]

One might have expected Powell to have also played nurse to natural history survey. The four western surveys of the 1860s and 1870s that were consolidated in 1879 had all included some biological work in their portfolios. But on Powell's recommendation, botany and zoology had been excluded from the combined Geological Survey on the grounds that they did not benefit the public economically and were already well supported privately.[7] (More to the point, their inclusion would have given Congressional budget watchdogs easy points of attack on his whole program.) The Department of Agriculture proved more hospitable, and what would become the U.S. Biological Survey was organized within its Division of Entomology.

This seems an odd arrangement, but like most federal science agencies Merriam's survey was a creature of politics, sneaked into existence in mufti, to avoid Congress's reflexive opposition to all new schemes for spending taxpayers' money (at least those that did not buy them votes). The occasion was the American Ornithologists' Union's project of mobilizing a continental network of private observers to census bird migrations. Merriam, as head of this project, sought the aid of New York senator Warner Miller (a family friend) in securing government funding, and to make the project acceptable to Congress they piggybacked wildlife census on an economic study of bird predation on insects. (The idea was that amateur ornithologists and farmers would send in bird stomachs, and survey biologists would analyze their contents to show that birds were not harmful, as farmers claimed, but rather kept insect pests in check.) The subterfuge worked, and in 1885 Congress appropriated $5,000 for the work, and a further $10,000 in 1886—large sums for that time.[8]

Merriam soon found ways of escaping from entomology. In 1886 he was put in charge of a new Division of Economic Ornithology and Mammalogy, and in 1890 he finally persuaded Congress to authorize a comprehensive biological survey of the vertebrates of the United States (though he got no extra funds for it).[9] It was what he had wanted from the start. As a private collector of mammals, he was used to working on a large scale, collecting large series rather than just a few representative specimens. Also, he was devoted to introducing exact scientific methods into collecting and

Fig. 3–1. C. Hart Merriam in 1887, age thirty-two, at the time he began to organize the U.S. Biological Survey. Courtesy American Heritage Center, Laramie, Wyoming, photo file B-M55ch.

preparing specimens. As director of the U.S. Biological Survey he would become famous for his obsession for order and method. By the mid-1880s Merriam had about reached the limits of what a private collector could do without institutional backing. He had completed a survey of the mammals of the Adirondacks and was contemplating a faunal survey of New York State. His collection of mammals alone contained some seven thousand specimens, housed in a home museum, and he had some twenty paid collectors working for him around the country![10] A national survey was the next logical step, and with Congress's approval he prepared to collect on a continental scale with a permanent corps of trained field naturalists.

The Biological Survey's official rationale remained economic: bounty and scalp laws on animals like the pocket gopher and ground squirrel probably wasted taxpayers' money, Merriam argued, because no one had ascertained if these animals really ate crops, as farmers believed. Also, a precise map of the continent's biogeographical "life zones" would serve as a practical guide to farmers in selecting crops that would survive in the semiarid homesteading regions. Taxonomy and biogeography were not arcane sciences, Merriam argued, but the foundations of practical agriculture. Politically these were winning arguments, and what was in name a department of "Economic Ornithology and Mammalogy" became in practice a national faunal survey.[11]

The Biological Survey, elevated to full Bureau status in 1905, was one of a handful of federal agencies with the authority and funding to carry out large-scale collecting.[12] In its heyday only the largest civic museums could match the scale of its operations. And because its massive collections were deposited in the U.S. National Museum, survey naturalists were unencumbered by the load of routine curation and exhibiting that kept museum curators at home indoors instead of in the field.

Merriam in his prime was a masterful survey chief, with a commanding and charismatic personality, social standing, and friends in high places. Though he hailed from the upper middle class, he was sociable and indifferent to upper-crust conventions. His openness, warmth, and plain speaking—as well as his exacting administrative habits—served him well in the capitol's egalitarian scientific culture.[13] He knew how to keep Congress off his back by doing basic science under a utilitarian umbrella. He made it a priority to publish large, beautifully colored maps of North American life zones (1890) and crop zones (1896)—visible tokens that the taxpayers money was well spent—and never neglected the practical work on animal diets, even as he devoted the survey's resources to systematics and biogeography. Younger field biologists looked back enviously at Merriam's organizational finesse: "Think," Joseph Grinnell recalled wistfully, "of the sound work it [the survey] did in both economics and pure science under C. Hart Merriam!"[14] Pure-science survey was the tail that wagged the economic-biology dog.

Others tried unsuccessfully to emulate Merriam's methods of institution building. John Merle Coulter tried repeatedly in the 1890s to get Congress to fund a national botanical survey. (Botanists were piqued by Merriam's use of plant species to define life zones: that was a task for botanists and for them alone.)[15] In 1891 the entomologist Clarence Weed promoted a system of federal grants to state agricultural experiment stations for state biological surveys.[16] But in agricultural politics, it was cash crops that mattered, not wildlife. Also, few possessed Merriam's combination of intellectual, social, and political skills, and even his political touch eventually failed him. In 1907–1909 changes in Washington politics and some political missteps resulted in his forced retirement in 1910 and the restricting of the Biological Survey's work to aiding state surveys and practical wildlife management.[17] But for almost twenty years Merriam's faunal survey set world standards of practice in survey collecting and systematics.

At the state level, patronage of biological survey was haphazard. States differed so greatly in their economies and political cultures, as well as their natural endowments, that a uniform system of surveys was never in the cards. The botanist Frederick Newcombe recognized three rationales for a state survey: economic, scientific, and educational. The economic rationale was potentially the most powerful politically, but Newcombe thought that economic justifications for natural history survey were just too far-fetched to work politically. The scientific rationale appealed to biologists but was politically a nonstarter. Education was the most plausible rationale, Newcombe thought: the argument that knowledge of a state's natural history would enrich high school and college teaching and give teachers at all educational levels a chance to participate in science. Voters were keenly interested in public education, especially in the upper Middle West, where education lobbies were well

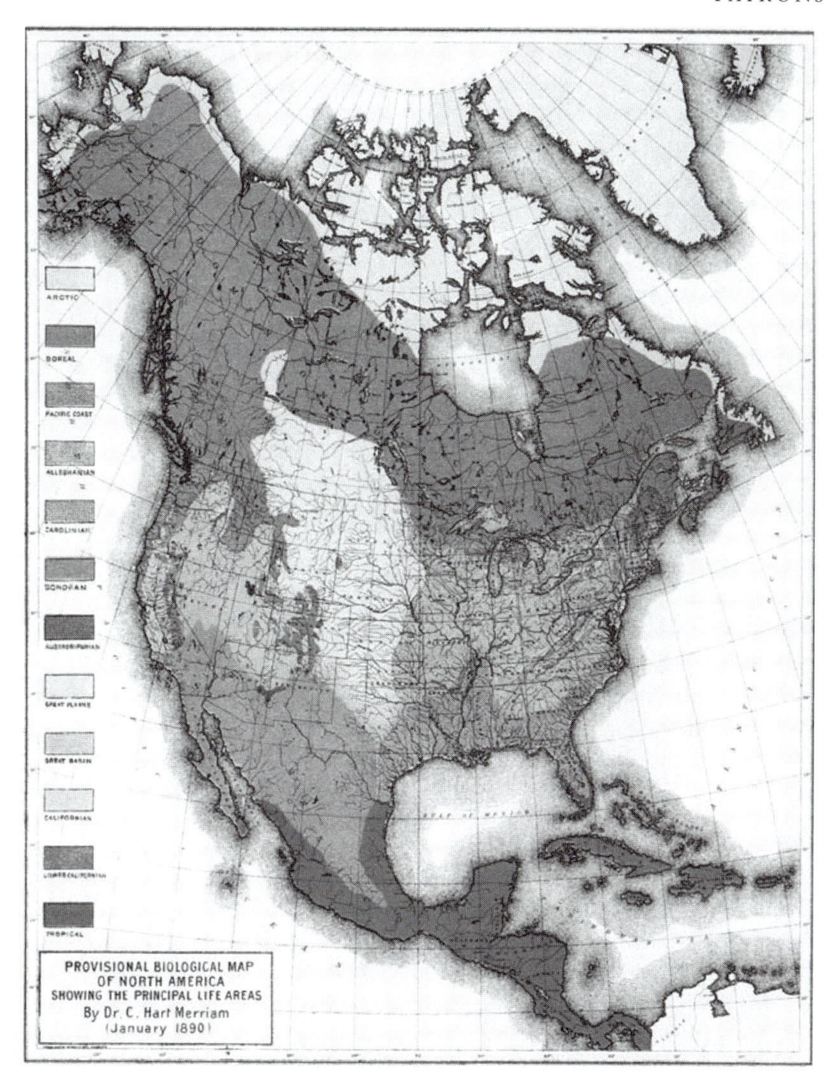

Fig. 3–2. C. Hart Merriam's life-zone map of North America, 1890. In C. Hart Merriam, *Results of a Biological Survey of the San Francisco Mountain Region and Desert of the Little Colorado in Arizona* (Washington: Government Printing Office, 1909), map 5.

organized and politically active. State teachers associations had been crucial in getting money for the Michigan Biological Survey in 1905, Newcombe recalled.[18] And the region's populist political culture strongly valued equal access to high culture. It is no accident that most state natural history surveys were in this region.

In practice, promoters of state surveys deployed all three rationales promiscuously to suit local circumstances, but especially the educational rationale. Stephen Forbes packaged the results of his survey of Illinois fishes "in a form suited to attract the interest of the intelligent citizen, to instruct the student, and to contribute to the economic welfare of the state."[19] He spent almost $1,000 a year to prepare natural history exhibits for use in schools. Zoologist Henry Nachtrieb considered school materials a better advertisement for a biological survey of Minnesota than research on the economics of fisheries.[20] Alexander Ruthven, in charge of natural history survey in Michigan, observed that the reports of a biological survey should be "of equal use to scientists, local naturalists, teachers, and private persons."[21] Conway MacMillan made sure that the report of his state botanical survey was handsomely illustrated, printed in a large edition, and distributed free of charge to interested Minnesotans.[22]

The seven more or less official state natural history surveys circa 1900—Minnesota, Illinois, Wisconsin, Indiana, Michigan, North Dakota, and Connecticut—were varied in organization and rationale, each the creature of local politics. In Minnesota, for example, ornithology and entomology had been included (after a fashion) in the state's geological survey in the 1870s.[23] However, the survey's aging director Newton H. Winchell stubbornly resisted Henry Nachtrieb and Conway MacMillan's efforts to get biology reinstated. In the mid-1890s they managed to get $2,500 each for fieldwork from the state legislature, but the money was restricted to economic work and student training and was far from the $22,500 they wanted for a full survey. When Winchell died in 1900 Macmillan managed to get hold of what was left of the survey's land-grant fund, but it was not much (from a few hundred to a few thousand dollars a year), and getting his hands on it required all-out lobbying of a recalcitrant legislature every two years. In 1906 the fund ran dry, and that was that.[24]

The history of the Michigan Biological Survey was no less frustrating. Biology was never part of the Michigan Geological Survey,

Fig. 3–3. Alexander G. Ruthven circa 1906–10. Ruthven Papers, box 64, folder "family and ceremonial." Courtesy Bentley Historical Library, Ann Arbor, Michigan.

but Alexander Ruthven, a young and ambitious zoologist at the University of Michigan, was eager to establish a biological survey and, frustrated by the university's lack of interest, turned to the legislature for support. In 1905 the legislature authorized the Geological Survey to undertake a biological survey and appropriated the modest sum of $1,000 a year for the purpose, renewable biennially. However, it declined to make this expanded brief official. (Ruthven referred to the "Geological and Biological Survey," but that was a fiction.) The grant was made indefinitely renewable in 1909, but still unofficial (no doubt to head off expected requests for increases). In 1913 Ruthven finally strong-armed the university into appropriating funds for biological survey through the University Museum. The scientific and educational rationales thus proved more effective than the utilitarian ones.[25]

The Illinois Natural History Survey was a hybrid of the museum and state agency models. Cobbled together by Stephen A. Forbes, it proved a notably stable and effective organization. Forbes began his career as the curator of the state's Natural History Museum (succeeding John Wesley Powell); and when the state museum was moved from Normal to Springfield in 1877, the old museum building was converted to a "Laboratory of Natural History," with Forbes in charge. In effect it was the research side of a museum without its collections and burden of routine maintenance. In 1884 the "Laboratory" was moved from Normal to Urbana, where it was associated with the university but was separate from any academic department. Forbes was thus protected from routine teaching and from interventions by professors who might imagine that their experimental or agricultural research was more worthy of state funding than were collecting and surveying. The next year, Forbes was authorized to undertake a survey of the state's fauna.[26] It was a nearly ideal arrangement for survey work. Though funding fluctuated, Forbes could count on having a few thousand dollars a year to send collectors into the field, and the survey's modest size helped it evade the notice of political budget cutters. In 1909 the survey consisted of Forbes, an entomologist, two zoological assistants, and support staff, with a budget of $8,000 for fieldwork plus $1,500 for publications and school exhibits. Few state biological surveys enjoyed reliable support on that scale.[27]

Most state surveys were volunteer operations, with the dignity of offical recognition but no state funding. Some were little more than an individual professor's summer research or class project. Land-grant universities might receive a few hundred dollars to enable students to take part in a "state" survey, but seldom more. Some of these projects produced important results with minimal means; most produced little. The first such survey, typical in its

Fig. 3–4. Stephen A. Forbes, director of the Illinois Natural History Survey (date unknown). Courtesy University of Illinois Archives, Urbana-Champaign, Illinois, Record Series 39/2/20, photo file 265B.

combination of grand aims and shoestring means, was conducted at Washburn University in Kansas between 1883 and 1891 by Francis W. Cragin, a newly fledged Harvard graduate who came to Washburn in 1882 as professor of natural history and director of a (somewhat grandiosely named) Laboratory of Natural History.

Ahead of its time in its institutional form and scientific aims (studies of plant communities, biogeography, variation), Cragin's survey depended entirely on student labor, petty cash from the college, and hopes of local volunteer participation around the state. Exactly what it achieved is not clear, but probably not much.[28]

The biological survey of Indiana exemplifies another, cooperative, mode of organization. Created by the Indiana Academy of Sciences in 1893, it depended entirely on the volunteer labor of amateur naturalists and high school teachers across the state. The survey had no state funding and no salaried staff and was directed by three professors at the University of Indiana—botanist Lucien M. Underwood, ichthyologist Carl Eigenmann, and geologist Vernon Marsters—whose task it was to collate and organize specimens sent in by their network of local collectors.[29] This mode of organization was popular in a state that greatly valued grassroots participation in high culture. In Illinois, Thomas Hankinson likewise hoped to use local teachers and amateurs for a fish survey, sending them directions, nets, tins, and preservatives and allowing them to keep duplicate specimens for their own collections.[30] Not surprisingly, the reality of carrying out systematic surveys with untrained, unpaid, and unsupervised volunteers seldom lived up to expectations.

But not all such efforts disappointed their founders' hopes. A notable success—it won fame as the nursery of American ecology—was the Botanical Survey of Nebraska, organized by the Harvard-trained botanist Charles E. Bessey. Like Francis Cragin's, Bessey's survey began as a classroom project. In 1886, just two years after arriving at Lincoln, Bessey began to take members of his undergraduate seminar on collecting excursions, and in 1892 these became a systematic state survey. The Botanical Survey was never a state agency and received no state funding beyond a small sum for publishing its report. Nor apparently did it ever rely on local amateurs for financial support. The only private aid was free railroad passes arranged by George W. Holdrege, the managing director of western lines for the Burlington Lines railroad. Free passes were as good as cash grants for travel, enabling survey parties to comb just about every corner of a large and botanically very diverse state.[31]

The Nebraska Botanical Survey was conducted not by Bessey himself, but by his brilliant and versatile student Roscoe Pound, who received his B.A. degree in 1888 (at the age of eighteen), a law degree in 1890 (from Harvard), and his Ph.D. in botany in 1897 for his work on the botanical survey. During the period of his leadership of the survey (1892–1903) he was a partner in a Lincoln law practice and, in his last two years, a judge on Nebraska's supreme court. His personal fieldwork was thus limited to weekends and vacations.[32]

As de facto head of the survey, Pound continued the practices of Bessey's "Bot Sem," though in a more intensive and organized form. Most of the actual field collecting was carried out by a succession of graduates of Bessey's seminar (about nine in all), who obtained their B.A. degrees between 1888 and 1897. But the survey was decidedly not an educational project: its purpose was to produce a professional, state-of-the-art inventory of Nebraska's flora. Distinct biogeographic regions were identified and systematically worked by Pound and his assistants, each of whom was assigned a piece of the territory. Botanically complex areas and boundary zones were visited repeatedly until they were thoroughly understood. An amateur naturalist in his career profile (the botany was a sideline), Pound operated more like a professional state botanist. He strove to make biogeographical survey exact and quantitative. (The invention of the quadrat method was an outcome of Pound's effort to create a quantitative scale of species abundance.) He never used untrained local collectors for fieldwork, but only trained and hand-picked—and supervised—apprentice botanists.[33] The Nebraska survey owed its success less to formal organization than to the qualities of participants like Pound, Herbert Webber, Per Rydberg, and Frederic Clements. Despite their very different career trajectories (the one away from science, the other toward it), Pound and C. Hart Merriam were kindred souls: talented and ambitious amateur naturalists who helped make survey collecting an exacting field science.

The Nebraska Botanical Survey had many imitators, as professors routinized student excursions and called them state biological surveys. Professors at Wesleyan University organized a biological survey of Connecticut in 1903 (though it is again not clear what actually was done). And Volney M. Spalding apparently oversaw a botanical survey of Michigan in the early 1900s. Charles W. Dodge, professor of biology at the University of Rochester, bruited a biological survey of New York in 1906. And Theodore Cockerell, on arriving at Colorado College in 1903, hatched an ambitious plan for a biological survey of Pike's Peak.[34] Other professors of botany who organized some kind of survey included Louis H. Pammel (Iowa), Albert S. Hitchcock (Kansas), Aven Nelson (Wyoming and Montana), William L. Bray (Texas), Francis Ramaley (Colorado), and a consortium of professors in Oklahoma. But it was difficult for individuals acting alone and without institutional backing to do the large-scale, intensive survey that Bessey, Pound, and their students achieved.[35]

Some more ambitious attempts to get funding for state-scale surveys never got off the drawing board. Take the case of New York, a progressive and wealthy state with an exemplary record in geological survey. Merriam lobbied New York's governor for a state ornithological survey in the late 1870s and, in the mid-1880s, tried to persuade James Hall and Charles D. Walcott to attach a biological survey to the state's Geological Survey.[36] Both efforts came to naught. In 1898 the New York State Museum invited the mammalogist Gerrit Miller to prepare a list of fauna as a first step toward a survey of the ecological effects of massive forest clearance. The survey never materialized. In 1906 biologists proposed a biological survey, especially of the changes in the environment since first settlement, and there was talk of such a survey again in 1908, but only talk.[37]

And yet, if there is one thing more striking than politicians' reluctance to spend money on natural history surveys, it is naturalists' persistent attempts to organize them and, failing political support, to mount surveys themselves on shoestring budgets. What we know of such efforts is probably just the tip of an iceberg of grassroots activity. In his appeal for a national botanical survey, John Merle Coulter pointed to environmental and cultural reasons for this activity. America was settled densely enough and botanists were sufficiently numerous that botanical exploration could be parceled out and pursued collectively in most locales: that was one incentive. Also, botanists believed that well-organized surveys would stop people from disparaging this science as mere collecting, and "train up a race of field-workers who shall follow their profession as distinctly and scientifically as the race of topographers."[38] If ready access to inner frontiers was the carrot to natural history survey, the stick was laboratory biologists' efforts to label fieldwork as less valuable than their own variety. In sum, natural history survey was well suited to a mosaic landscape, no portion of which was too remote from a state college or university, and to a boot-strapping academic culture with high ambitions, empty pockets, and an abundance of cheap, semi-trained labor.

Natural history surveys never became a regular feature of state government, unlike geological, topographical, agricultural, and land-use surveys.[39] The reasons, I think, are partly economic, partly unlucky timing. The knowledge produced by biological surveys enabled no significant social group to exploit their states' natural resources. Also, natural history surveys appeared on the scene toward the end of the period of extractive resource use, and there appears to have been little interest in using such surveys to help develop nascent recreational and tourist economies. I expected when I began my research to find evidence of such efforts, but the evidence was not there. Many state governments were investing in fish culture by 1910, and later in wildlife management. But few saw any political or economic advantage in faunal inventory and survey.

Museum Exhibition and Collecting

Museums, however, are a quite different story. No one, I think, would have predicted in 1880 that civic museums would soon be investing in natural history survey and collecting on the grand scale. It was not the custom for museums to sponsor fieldwork.[40] Their purpose was to keep and display collections acquired by gift, purchase, or duplicate exchange. Collecting was the work of professional (that is, freelance) collectors, amateur naturalists, and commercial supply houses like Ward's. Curators' work was indoor work: recording, cataloging, preserving, labeling, sorting, and writing checklists and descriptions. As Charles C. Adams quipped: "most of the museums make specimens of the curators and keep them out of the field—and this is the very worst enemy of a museum man."[41] Museum trustees, being men of business and not scholars, expected paid museum staff to be at their desks and visibly on the job, not outdoors having a good time. Curators did collect—but on their own time, on vacation; it was not something they were paid to do. Yet in the early 1900s larger museums were routinely dispatching collectors and curators on expeditions to far-flung corners of the globe.

What happened to turn curators into collectors and expeditioners, and to impel museums into the business of survey collecting?

In a word, it was changes in exhibit practices: first, the separation of study from exhibit collections, then the invention of the habitat diorama.

As we learned in chapter 2, it was the standard practice of nineteenth-century museums to keep as many of their specimens as possible mounted and on display. Many were not, of course, but the point is that the custom made no distinction between the aims of science and exhibit. And as museum collections grew larger, that practice became less viable, as a means both of exhibiting and of preserving scientific material. The public grew bored by rows of stuffed animals, many of them unprepossessing and mangy. And zoologists watched with dismay as type specimens, the unique and indispensable reference specimens of taxonomic science, gradually faded in the sunlight, lost their labels, and were eaten by moths and weevils. As Robert Ridgway put it, "a perfect chaos of birds on stands [was] neither a delight to the eye of the unprofessional visitor nor a satisfaction to the specialist."[42] The solution was to separate the aims of science and display.

Beginning in the late 1870s museum curators began to sort specimens into separate exhibit and "study" collections, dismounting most—starting with the irreplacable type specimens—and storing them as skins in insect-proof cabinets in private research rooms. Only the most attractive specimens were spruced up and kept on public display, usually those that had aesthetic appeal—colorful or elaborate plumage, striking form, large size—or that were exotic or rare, or that illustrated a biological principle. The segregation of display and study collections, discussed as early as the 1850s in European museums, became official policy at the Museum of Comparative Zoology in 1878, at the U.S. National Museum in 1881, at the Academy of Natural Science in 1889, and in the 1890s in museums everywhere.[43] By 1894 only about one-fifth of the bird specimens in the American Museum were on display, and only about 4,000 of the 135,059 specimens in the National Museum.[44]

Separating exhibit and study collections did not at first impel curators into the field. Quite the reverse: unstuffing, preserving, relabeling, and recataloging tens of thousands of specimens was a huge labor, and it kept curators indoors. Philadelphia's Academy of Natural Sciences began the process in the early 1890s with their large collection of birds, and their curators were still at it nearly a decade later.[45] However, the separation of exhibit and study collections opened the door to changes in the practices of both exhibiting and research that would in time draw museums and their curators into the expedition business. These changes were not intended or even foreseen, but were unanticipated consequences of this modest improvement in collection management. Separating the aims of exhibit and research liberated curators to build study collections to modern scientific standards. At the same time, it liberated curators and taxidermists to create ever more elaborate displays. Improved practices of exhibiting and taxonomic research both required specimens of a quality far better than what was usual in existing collections. And that gave curators and preparators a compelling argument for why they should be the ones collecting in the field, not amateurs or commercial collectors: only museum staffers knew the exacting requirements of the new taxonomy and the new art of display. Let us look more closely at how that happened.

As we have seen, taxidermists in the 1880s were already developing methods of making more lifelike and dramatic animal exhibits, and segregating study collections intensified this trend. Putting fewer but better specimens on display gave taxidermists the time and the professional incentive to refine their techniques. Instead of having to stuff entire collections, they could concentrate their efforts on a few elaborate and striking exhibits. The evolution of diorama art would probably have happened even if museums had not separated study and exhibit collections, though I doubt it would have occurred so fast or gone so far.

But it was the need for high-quality raw materials for dioramas that really made curators and taxidermists into field collectors. The animal skins and skeletons already in museum collections were just not suitable for the new naturalistic exhibits. Specimens purchased from local collectors or from commercial houses had been prepared by outmoded taxidermic methods and could not simply be remounted. As material for dioramas they were useless. Lifelike exhibits required fresh skins carefully prepared in the field according to the exacting standards of the best taxidermic practice. Animal groups and especially family groups had to be collected together, all from the same place at the same time, at the right season (usually the breeding season or in autumn, when pelage is lush and animal bodies in prime condition). Quality exhibiting material could only be acquired in the field, ideally by museum taxidermists and curators themselves, because they alone knew exactly what was needed.

It was not just the animal specimens that required in-house collecting, but also the accessory materials of diorama groups—the grasses, soil, rocks, and shrubbery that were crucial to the visual illusion and authenticity of diorama art. Such stuff was not something that sportsmen or commercial collectors would deign to gather. It was heavy and had no market value except for making dioramas; that is, it had value only when collected by the curators and taxidermists who knew how it would be used.

Also, since habitat dioramas depicted particular places, the material for making them had to be observed and collected on location, and that was best achieved by sending museum staff to observe, draw, photograph, and collect at particular sites. For example, when a curator from the Field Museum was diverted from his scientific collecting to gather accessory material for a capybara diorama, he was sent a preliminary painting of the backdrop to tell him exactly where to collect ("near some of the small lakes in the vicinity to Senor Tinendo's place which lies just across the lake from Maracaibo").[46] It was because the visual power of habitat dioramas was achieved by particularity and scientific accuracy that fieldwork by museum curators and taxidermists became an essential part of exhibit making. Collecting for dioramas required firsthand knowledge of the scientific and artistic principles of display; no such expertise was required for exhibits of stuffed animals in rows. Commercial collectors could acquire this knowledge, of course, by working closely with museum staff. But museum curators and taxidermists were there first and had compelling professional reasons to do the collecting themselves.

No one foresaw this turn of events, but curators were quick to make the most of it to expand their professional roles. "Specimens cannot be purchased," one curator informed his director, "for it is seldom [that] two are offered for sale that were killed at the same time, and the only way to procure them is to go after them, see them in their haunts, and while securing the animals, also obtain the accessories that will enable the groups to be restored within their natural surroundings."[47] "We have fully competent taxidermists," urged another, "but fieldwork is required in order to obtain samples of flora, photographs and skins. Fresh skins give far better results, and . . . we need the services of a first class accessory worker."[48] That is how curators became field scientists. And big time: Frank Chapman traveled some 65,000 miles in connection with the American Museum's hall of North American birds.[49]

Once afield, of course, it made sense for curators also to collect for study collections and for science. But it was dioramas, more than anything else, that transformed museums from passive recipients of objects into active sponsors of collecting, and curators from office workers to field collectors. If museum boards of trustees wanted state-of-the-art dioramas—and they all did, because donors and the public wanted to see them and would pay for the pleasure—they had to invest in expeditions manned by their own curators and preparators.

If diorama building was the main incentive for museums to engage in in-house collecting, it was not the only one. A scientific rationale for active collecting also became compelling, as taxonomic practices became more exacting and dependent on large series collections. Here again the separation of exhibit and study collections was the triggering event. So long as museums kept only a few representative examples of each species and exhibited and mounted every item they had, there was no reason for museums to collect in depth. (More rows of stuffed specimens were costly to mount and did not make exhibits better.) But once study collections were separate, the scientific reasons for survey collecting made more sense.

These developments further impelled museums into new large-scale collecting by devaluing existing collections. Older specimens purchased from local collectors or supply houses were not prepared to scientific standards (distorted skins, scrunched limbs), and they usually came without supporting field data. As taxonomic standards improved, whole collections became as worthless for research as they were for display. The only thing to do with them was to throw them away.[50] Even the best museums in the world found their collections thus devalued and were obliged to begin collecting again. Rebuilding collections did not necessarily require in-house expeditions, but it did make these a possible option—and, increasingly, the most attractive option.

Another unanticipated consequence of separating the functions of research and display was that older methods of building collections by exchange, donation, and purchase became less viable. Take exchange: when museums kept only representative specimens

Fig. 3–5. School class viewing a wapiti group exhibit. Courtesy Department of Library Services, American Museum of Natural History, New York, negative no. 33594.

of each species, extras were duplicates and available for trading. But in study collections these extras were not redundant but evidence of natural variation. "Duplicates" became too valuable scientifically to swap—they ceased to be duplicates. And as the meaning of collections changed, so too did the practices of collecting.

Donation by amateur collectors also became less viable, because separating exhibit and study collections caused the purposes of amateur and museum collecting to diverge. Amateur and freelance collectors continued to assemble synoptic collections of a few "typical" specimens of each species, whereas curators wanted large series of common animals that would reveal the extent of intraspecies variation. Amateur and freelance collectors could be taught the new methods of survey collecting, but in practice it was easier and less risky for museum curators to do the collecting themselves, since they knew exactly what was needed. As Alexander Ruthven observed: "[A] museum that depends upon donations must be content with casual acquisitions of specimens, often not in the best conditions, and rarely accompanied by detailed data, which is to say the least unsatisfactory." The advantages of putting curators in charge of in-house expeditions became more apparent, and the disadvantages less compelling.[51]

Finally, purchase became a less appealing means of large-scale collecting, because of changes in the market for natural history specimens. The market was probably anomalously favorable for buyers in the 1870s and 1880s, at least in the United States. Sudden access to game-rich inner frontiers made these decades a golden age of amateur and commercial collecting. Very large collections were formed, and in the the 1880s and early 1890s these flowed into museums as their owners ran out of space or gave up active collecting.[52] Of the 213 U.S. museums with zoological or botanical collections in 1903, 81 had more than 10,000 specimens in one of these categories, and about 31 had holdings in the range of 75,000 to 100,000 specimens or more.[53] The data are far from exact but a clear enough indicator of the extent of collecting at the time. However, the abundance of the 1870s and 1880s—like placer gold or old-growth forest—did not last long. As game declined and more museums entered the market for specimens, prices rose sharply, especially for rare and desirable species. Although at first in-house expeditions were more expensive than purchase, an inflationary commercial market changed the calculus.[54] In-house collecting became a cost effective as well as a qualitatively superior way of getting what museums needed for study collections and dioramas. For all these reasons, in-house expeditionary collecting became the method of recourse for large civic museums.

Curators, not surprisingly, were quicker to see the logic of the new situation than museum directors and trustees, who after all were the ones who would have to revise cherished views of curators' duties, and to come up with the money to pay for expeditions. What tipped the balance for museum directors was economics: it became more cost effective to collect by in-house expeditions than by traditional methods of exchange, donation, and purchase. For curators and taxidermists the technical superiority of in-house collecting was most compelling; for the businessmen who sat in front offices and on boards of directors, the economic argument was decisive.

Knowing that, curators often stressed the economic rationale in making their case. "The value of fieldwork has never been more demonstrated than during the past year," a curator at the Field Museum urged his boss. "I cannot too strongly emphasize the importance of fieldwork . . . as it is the cheapest way to obtain material, particularly of that kind and quality that cannot be purchased." He reckoned that a single full-time field collector could produce specimens worth three or four times his salary and impossible to buy on the open market.[55] No large collections remained in private hands, and as the museum's collections grew it would become increasingly difficult to fill in gaps by purchase, he predicted. Besides, only curators knew exactly what was needed and where to go to collect it. For all these reasons, he believed, "well directed and systematically ordered expeditions in the field will remain the almost sole method of development." Any other method, he thought, "will surely end in poor and unsatisfactory results, with both time and money wasted."[56]

Indeed, by the early 1900s it was clear to just about everyone that the money side of expeditions would offer no difficulties—quite the contrary. Trustees and well-to-do patrons were eager to underwrite expeditions, especially those sent out to collect specimens for dioramas, and most especially if they (the donors) got to go along and help collect. (Expeditions seeking big game in exotic places were especially popular.) Museum expeditions essentially paid for themselves by attracting crowds of paying visitors and wealthy patrons. Hardheaded museum trustees did not need to understand the scientific rationale for new, systematic, in-depth collecting: they understood well enough the logic of institutional growth and aggrandizement. As Charles C. Adams observed: "[N]o single factor does more to give a healthy tone to [museum] work than . . . exploration. . . . [F]ew subjects . . . arouse so much general interest in a manner to secure extensive cooperation as . . . an expedition."[57]

In-house expeditions led by curators and taxidermists thus became for everyone the preferred strategy of rebuilding museum collections. Museums did not disdain gifts and did make special purchases, some very substantial (like the American Museum's purchase of the immense Rothschild collection of birds). But expedition collecting remained the most economical and reliable method, for all the reasons discussed above: the scientific demands of the new taxonomy; the declining effectiveness of collecting by donation and exchange; the escalating prices of commercial specimens; and—most especially—the exacting demands of diorama building.

A few museum patrons were interested in the scientific side of museum work. For example, the president of the American Museum, Morris K. Jesup, wished to make the museum a place where research could be done, hired curators like Joel Asaph Allen and Frank Chapman, and supplied the money for them to collect in the field. In 1907 Jesup left the bulk of his substantial estate to an endowment specifically for museum expeditions.[58] In the early 1900s John L. Cadwalader, a New York lawyer, sportsman, and amateur ornithologist, financed Frank Chapman's fieldwork for the Hall of North American Birds—and later his numerous scientific expeditions to South America.[59] The Brooklyn Museum had a million-dollar endowment for fieldwork (the Woodward Fund) that provided curator Robert Cushman Murphy with a steady $4,000 to $5,000 a year for ornithological fieldwork. The museum's trustees were especially interested in dioramas, but were also conscious of the value of "the general expansion of the department as a result of the fieldwork."[60] That is, a flourishing scientific side was good for the institution generally. The Whitney family spent large sums of money over an extended period (1920–35) for comprehensive bird-collecting expeditions to the South Seas.[61] These were curators' best-loved patrons, but they were atypical.

A social historian has yet to study the people who poured money into collecting expeditions, but it is likely that they were professional or business families—urban gentry—with a taste for sport hunting and outdoor vacationing. Some were wealthy grand bourgeois and the yachting and safari set—the Roosevelts and Harrimans—and because of their social visibility it is easy to assume that all expedition sponsors were upper crust. But it is more likely that the majority of expedition sponsors were modestly well-to-do lawyers, judges, bankers, and merchants, whose cultural interests ran to improving outdoor recreation, rather than social display. In short, they may best be understood as participants in the middle-class vacation complex. The people who stepped up to pay for museum expeditions were probably the same ones who read nature essays and viewed dioramas (that for sure), camped and walked in lake and mountain districts, hunted and fished, and enjoyed recreational bird-watching or collecting.

The connection between expeditions and the vacation complex is most visible in expeditions in which patrons participated as hunters for diorama specimens. These expeditions strongly resembled traditional big-game safaris, with a mix of sport, science, and adventure. (Restrictions on big-game hunting were impeding private safaris, and it was easier to get a license if the hunt was for a public or scientific purpose.) Some were social and media events—museums welcomed and encouraged publicity—and were covered by the press in the same manner as it did travel and touring by the social elite. These were exceptional, but they remind us that patronage of the numerous more modest expeditions was also motivated by diorama building. The public's passion for dioramas and museums' newfound taste for expeditioning were a result of the same social and cultural imperatives that created the vacation complex. Scientific study collections and modern survey science were incidental beneficiaries.

Would civic museums have become major sponsors of expeditions if these had not been linked to building dioramas? Would scientific imperatives have sufficed? No doubt some in-house collecting would have been done, but probably on a far smaller scale. How many wealthy patrons would have turned out their pockets to subsidize curators on collecting jaunts, or to make large study collections even larger? A few did, but not many. Rooms of zinc boxes or shelves of "alcoholic" lizards and fish in jars were not things that trustees and patrons could put their names on and point to as visible evidence of their contribution to public education and civic society. But they lined up to pay for dioramas, and building these had to begin in the field, with expeditions led by the people who knew how to do the work—museum curators, preparators, and artists. That was what drew big money to collecting expeditions: sensational public exhibits and the prospect of scientific adventure. In research museums it was a different story, as we will see, but in the larger civic museums it was the diorama that made the expedition system work. Study collections made curators dream of going afield; dioramas made their dreams come true.

Museum Collecting: An Overview

Museum expeditions were variously ambitious, ranging from individual day trips to some urban fringe to set-piece expeditions in the grand imperial style. Some combined science with sport and adventure—quasi-scientific safaris—and were after diorama specimens. Some were dispatched to build collections of a region's plants and animals for both educational and scientific purposes. Local collecting by individual curators differed little from what amateur naturalists might do. Occasionally museums would send out large general-purpose expeditions for extended periods to gather omnivorously in some region of the world—like earlier voyages of exploration, except that curators were key players and not tag-alongs. Most expeditions, however, were organized by the curators of some particular department to address some particular taxonomic or biogeographical problem. For such purposes a small group was optimal, and shoe-string funding, to avoid the complications of patrons and the press. These were the expeditions curators liked best.

It would be hard to calculate how much money was spent, but it was doubtless in the millions of dollars for zoology alone, and certainly as much or more for anthropology, archaeology, and paleontology. The American Museum of Natural History was probably the premier investor in expeditions. In 1929 and 1930, at the peak of its activity, it had (respectively) thirty and twenty-seven collecting parties in the field, at a cost of $282,809.68 and $207,845.93 (museum accounting was precise).[62] A chronological file in the museum's archive lists 600 named and sponsored expeditions (in all fields) between 1887 and 1966 (409 through 1940), and the list is likely incomplete. The pace of expeditioning breaks into distinct periods: 1891–1901 (3.8 per year on average); 1906–25 (8.5 per year); and 1926–41 (13.9 per year, with peaks in the late 1920s and late 1930s). (The post–World War II average was 8.9 per year.) In vertebrate zoology alone the museum, by my count, fielded 206 parties between 1890 and 1940.[63] There was a small burst of activity just after 1900; then, after a lull, a much larger burst from 1908 to 1917, when a time-out was declared for World War I; then a period of sustained activity at a high level from 1921 to the outbreak of World War II. The Field Museum's expeditions followed roughly the same pattern, but at a considerably lower level—about 72 in vertebrate zoology between 1894 and 1928.[64]

The Academy of Natural Sciences of Philadelphia sent out about fifty-seven expeditions in vertebrate zoology between 1889 and 1930, but that number is a lower bound, because the Academy frequently used individual collectors, and it is difficult to distinguish between formal expeditions and semioffical private jaunts.[65] This habit of improvisation was probably an economic necessity. Philadelphia's old-money society was famously more cautious than New York's nouveau riches; but there was no shortage of local hunter-naturalists for scientific safaris. The National Museum in Washington, D.C., officially fielded no expeditions of its own, but its curators routinely joined expeditions sent out by other government agencies: at least eighty-eight in vertebrate zoology by my rough count between 1886 and 1927, not counting curators' vacation trips.[66] The data are imprecise but make the overall dimension of museum expeditioning clear enough: it was a very big business.

The transformation of civic museums from passive repositories to active collectors did not happen overnight. Museum directors and trustees had to comprehend the advantages that active collecting would bring to their institutions, and experience was required to see that fieldwork was legitimate work for museum curators, not paid vacation. Authority relations between curators and private patrons had to be worked out, and the logistics of expeditions mastered (more on all that in the next chapter). Museums began collecting in a small way in the 1890s, then expanded and systematized in the 1900s and 1910s. The system was in its heyday in the interwar decades, before being cut short by global war.

Museum annual reports from the 1890s reveal how at first one thing led to another. The first museum expeditions, dispatched by the American and National museums in 1887 and 1888 to get buffalo skins and skeletons, were an emergency response to a singular event: the anticipated extinction of an emblematic American species. But they unexpectedly proved to be "the event of the year," as Joel Asaph Allen put it, and the lesson was not lost on museum directors.[67] The American Museum organized similar expeditions for other big-game species in 1893 and 1894, and in 1895 President Jesup, remarking on the avid media interest in these events, noted presciently that "Such exploration could be greatly extended with advantage to the Museum, and it would be most gratifying if some of our wealthy citizens could be induced to contribute funds for this very essential part of the Museum's work." Potential donors were advised that sponsoring expeditions was a good way to get big, rare animals cheaply, and such exhortations became a regular feature of Jesup's reports.[68] Unluckily 1893–97 were depression years, and Frank Chapman was for a time the only curator who collected regularly—and he did so on his own time. Regular expeditions began at the American Museum in 1898, with a couple per year, doubling two years later and again in the early 1910s.

It was a similar story at the Academy of Natural Sciences, where curator Angelo Heilprin had his eyes opened to the potential of expeditions in 1888, when he took a group of college students to Bermuda to collect birds. The abundant material they brought back, with a modest outlay of money, Heilprin observed, "shows how much may be accomplished even with little effort . . . and encourages the hope that [such] researches . . . may hereafter be systematically conducted under the auspices of the Academy." Not everyone in the Academy was so foresighted, but Heilprin's hopes were finally realized in the late 1890s, when expeditions began to be dispatched on a regular basis.[69] The virtues of museum expeditions—their cost-effectiveness and media appeal—were at first less apparent to donors than to curators. But once expeditions caught the public's interest, the fever spread, and only war and depression could cool it.

It was also not clear at first that curators would be the ones to benefit from sponsored expeditions. Until the mid-1890s museums would hire professional collectors like Rollo Beck, perhaps because it was the larger game animals that were the first to become expensive to buy commercially and thus cheaper to acquire by in-house expeditions. And if museum staff were sent out to collect, it was more usually assistant curators, while senior curators remained at home to mind the institution's important business.[70] But that changed in the early 1900s, at least in larger civic museums. William Beutenmüller may have been the first curator from the American Museum to lead an expedition, in 1891–92. In 1894 it was reported that mammal and bird specimens were now mainly acquired by museum expeditions—though it seems mostly by attaching hired collectors to expeditions sent out by other departments, like archaeology or paleontology. Frank Chapman's annual field trips beginning in 1902 may indicate the turning point in the transformation of curators into field workers. By 1907 it was standard practice in the American Museum to send curators into the field.[71]

Other museums were slower to adopt new ways. At the Field Museum, for example, mammalogist Harry Swarth was confined to collecting in Chicago's exurban fringe for exhibits of common local fauna—an activity that he despised as mere weekend recreation. Serious scientific collecting he had to do on his vacation and on an unpaid leave. Both sides were wary of a closer patronage relation, as Swarth reported: "The Museum didn't care to have anything to do with [my] . . . expedition (I didn't ask them nor want them to) but hinted that if I was very successful they would pay my expenses for the material. That is their whole hearted way of doing business."[72] That was in 1907; but within a few years curators were regularly in the field.[73] Smaller museums took longer to follow suit. When he moved to the Los Angeles Museum in 1913 Swarth was again confined to routine labeling and installing, and the directors discouraged him from doing research even on his own time.[74] However, the idea that fieldwork was an expected part of the curatorial role ceased to be controversial. Once a rare adventure for curators, expeditions became routine. Alexander Ruthven recalled bumping into a colleague rushing out of the Michigan Museum: " 'Good-bye to you, too,' he said, and hurried on. 'Where are you off to this time?' he was asked. 'Just to Afghanistan,' he flung over his shoulder."[75]

The system of piggybacking scientific collecting on exhibit building became almost too successful. As diorama halls became more and more elaborate in the early 1920s, offers to sponsor expedi-

Fig. 3–6. Akeley Hall of African Mammals, American Museum of Natural History. Courtesy Department of Library Services, American Museum of Natural History, negative no. 328663.

tions—especially for big game in exotic places—came in so thick and fast that, as Wilfred Osgood put it, "every new one makes us gasp for breath."[76] Science suffered. "[A]lthough we are sending out more expeditions and getting more material than ever before, I have less and less time for study," Osgood complained to a British colleague.[77] When the crates of specimens arrived at the Field Museum from the Roosevelts' Himalayan expedition, Osgood had his "hands full in getting some of them [specimens] on exhibition in order to please the patrons who furnished the money."[78] At the American Museum, Frank Chapman had also become a full-time organizer of expeditions and exhibits: "You have my full and deep sympathy as an organizer of expeditions," he wrote Osgood, "for that has been my chief occupation for some time and under circumstances breeding many complications. . . . [A]ll these affairs arranged chiefly in the interests of others."[79]

Visiting the American Museum in 1916, Joseph Grinnell was impressed by its extravagant expenditure on expeditions and the pittance provided for actual research. Rollo Beck was in South America (again) with $1,000 a month just for expenses. But back home, hard-pressed curators were too busy managing expeditions to go afield themselves or even to study the material that was piling up: "often the work is saddled on someone who never went on the expedition, with the result that only a small part of the value of the collections is realized," Grinnell reported.[80] The "exhibitionist" school of museum work had accomplished much, he thought, but at the expense of museums' more important scientific purposes.

At a 1921 meeting of the American Association of Museums, Grinnell and Edward A. Goldman of the U.S. Biological Survey were the only two people there who wanted to talk about science. Everyone else could talk only of exhibits and expeditions: "Their speeches and conversation all show their minds to be full of 'big' things like getting more endowments by improved methods of impressing the public and legislatures," Grinnell acidly observed; "research detail they leave to their under-curators." It was much the same at the Field Museum, where Grinnell found Osgood swamped with administration, and the museum's director, Stanley Field, interested only in more exhibits, more buildings, and more expeditions, especially to exotic foreign places.[81] Between juggling a dozen expeditions and overseeing the construction of multiple exhibits, curators began to seem more the victims than the beneficiaries of museums' system of patronage.[82]

In the larger public museums, it appears, patronage and expeditions had become more ends in themselves than the means of furthering the science of natural history—or so it seemed to curators. It was a hazard inherent in a system of patronage that piggybacked scientific fieldwork on museum art and exhibition, and attracted patrons of expeditions by the appeal of sensational displays and adventure. The system worked brilliantly, but it was all too easy for the expeditionary tail to wag the scientific dog.

In the larger view, however, museums' patronage of collecting expeditions greatly benefited taxonomic science. Despite the demands it put on curators, museum patronage transformed their professional role from keepers of collections to active collectors and field scientists. Museum expeditions dwarfed even those of the U.S. Biological Survey, to say nothing of state surveys; and they were not restricted by state or national boundaries, but could operate on a world stage. Museum expeditions and patronage put more naturalists in the field, created more valuable natural history collections, and contributed more to our understanding of the world's biodiversity than any system of patronage since the grand imperial voyages of the age of exploration.

Research Museums and Their Patrons

Research museums were no less invested in expeditions, but lacking public exhibits they had to rely on patrons who were interested in science for its own sake. That meant a smaller pool of potential patrons and much smaller budgets than the large civic museums enjoyed. However, research museums tended to confine their activities to their own regions, and had no dioramas to eat up funds, so more modest budgets went much farther than in civic museums. They did very well on the support of networks of middle-class nature lovers with a serious scientific interest. Some of these part-time practitioners were respected experts in the taxonomy of their creatures of choice (shells, insects, plants, and birds were the easiest to preserve and store), and it was these practitioners who were the mainstay of research museums. We may see them as local, small-scale C. Hart Merriams, who could be counted on for the hundreds of dollars that it took to keep a party of unpaid collectors in the field for a month or so.

Whereas patronage in large civic museums depended on publicity and public service, in small research museums it was more usually based on close personal relations between curators and a few individual patrons. Patronage relations depended on a mutual trust based on personal acquaintance and shared experience and scientific interest. This is not to say that personal relations played no role in large museum expeditions; of course they did. But the absence in research museums of the politics of display meant that cultivating individual relationships was more crucial than selling institutional agendas to the public.

The University of Michigan Museum and the Museum of Vertebrate Zoology at Berkeley, which became premier centers of systematic zoology, illustrate how patronage worked in research museums. As curator and then head of the museum at Ann Arbor, Alexander Ruthven made deft use of his connections with amateur naturalists in the Michigan Academy of Science. Ruthven's talent for organizing expeditions and securing patrons was precocious. As a graduate student in 1904 he was already tapping his amateur networks for $75 or $100 for collecting projects, and within a few years he was an accomplished small-scale fund-raiser.[83] His circle of naturalist friends included Bryant Walker, an attorney and an expert malacologist; William W. Newcomb, physician and lepidopterist; Bradshaw H. Swales, attorney and ornithologist; Arthur W. Andrews, master cabinetmaker and coleopterist; and Charles K. Dodge, a customs collector and botanist. These men took part in collecting trips and wrote checks for expenses, and served as honorary curators of the museum. Ruthven never pushed and made a point of touching only those who he knew were interested in science.[84] Ruthven's circle of patrons was a microcosm of the vacationing, naturalizing middle class, and their modest but reliable support was crucial, because the state's support for natural history survey was erratic, and one season without field projects would give legislators an excuse to cut it off for good.[85]

Other support for collecting expeditions was made in kind by local businessmen or sportsmen who hoped that surveys would foster local economic development. The Isle Royale survey was supported by local inhabitants, who arranged cheap rail or ferry transport for field parties and made private camps or grounds available

Fig. 3–7. Members of the Detroit Naturalists Club, 1910 or 1911, and patrons of the Michigan natural history survey. Museum of Zoology (University of Michigan) records, box 4, “Detroit Naturalists Club,” Bently Historical Library, University of Michigan.

for field headquarters.[86] Ruthven was skilled in working local networks. At Bryant Walker's suggestion, for example, he wrote to a William B. Mershon, an amateur naturalist in Saginaw, who promised to give a hundred dollars toward a "Mershon Expedition" to the Charity Islands if his income permitted—he worked for a living, he explained, and had no copper mines or pineries in his family—and if a local friend would match his gift.[87]

A few years later Ruthven made a similar plea to George Shiras, the famous wildlife photographer, for $200 for an expedition to the Northern Peninsula, where Shiras had a family compound. It was a region that was dear to Shiras's heart, Ruthven knew, and affection for a place was often the key to getting a gift for a survey of its native fauna. Shiras was too committed to his own (elaborate and expensive) expeditions to contribute just then, but a year later he volunteered $200 for the Charity Islands expedition.[88] Ruthven kept his famous patron informed of the museum's work, and in 1915 Shiras offered to give $250 a year for surveys of the Northern Peninsula "before it is too late, or before outsiders undertake to do it." Shiras also shrewdly suggested that a survey of Lake Superior's north shore would afford a fruitful biogeographical comparison.[89] As a celebrity naturalist, Shiras was an atypical patron. But affection for their home state's outbacks and native wildlife was typically the currency of Ruthven's patronage network.

Ruthven attached his circle of local naturalist friends to the museum by a continual exchange of little gifts and favors: exchange of specimens for expert identifications; honorific expeditions and honorary curatorships; and cash gifts. Though based on shared values and experience, it was an emotionally delicate relation, as is any relation that mixes friendship and money. In the moral economy of small-scale patronage, material exchange of money and specimens was maintained by both sides as strictly voluntary and based on a shared intellectual concern for science. Any sense of material obligation might have been fatal to the relation.

With William Newcomb, for example, Ruthven maintained a fiction that his friend's gifts were entirely spontaneous. Ruthven kept Newcomb informed of the museum's expeditions and financial needs, but took care that such information would not appear as a request for funds. When a check arrived unexpectedly in the mail for a collecting trip to Nevada, Ruthven worried aloud that he might have hinted too openly of his need for funds. He wanted

Fig. 3–8. William Newcomb, Bryant Walker, and Percy Taverner at Point Pelee, Ontario, May 1909. Skilled amateur naturalists and patrons of the Michigan natural history survey. Museum of Zoology (University of Michigan) records, box 4, "Detroit Naturalists Club," Bently Historical Library, University of Michigan.

only his friend's interest and advice, he protested, and did not tell him of the museum's needs in the expectation of getting money: "Of course when you feel able and willing to assist any proposition financially I will not object but I want you to understand that you do this on your own initiative. So please never think I am suggesting financial aid when I tell you that we need it, for I want to be frank with you so that you will always know just where the museum is at."[90] If Newcomb wished, the party would collect for him as well. Newcomb reassured his friend that the gift was not prompted by any hints from Ruthven, or by the expectation of specimens for his own collection, though he did allow that "if you find it desirable in the interests of your expedition to collect some [coleoptera] I should, of course, be very glad to see it done."[91] It was a delicate little dance of autonomy and obligation that in one form or another structures all patronage systems.[92]

A similar moral dynamic is evident in Ruthven's relationship with Bryant Walker. Walker's gifts began early and increased gradually over the years, from $50 or $75 in the early 1900s to $500 or even $1,000 in the mid-1910s.[93] An avid collector of shells, Walker was a regular participant in expeditions within the state and would occasionally drop hints of areas he would like to have collected by out-of-state expeditions. But he was genuinely more interested in the museum than in his own collection. He paid Ruthven's expenses for a trip to Mexico, for example, so that Ruthven could take a vacation and spend his time collecting reptiles. Walker was more concerned that Ruthven succeed in his own work than he was to enlarge his own collection. "I was agreeably surprised you may be sure," Ruthven wrote his mentor Charles Adams, "for I supposed he was only interested in shells. He is certainly a prince."[94]

Walker's role in the Michigan Museum developed far beyond that of giver of small gifts. He became Ruthven's trusted advisor and confidant as they attempted to put the museum on a solid foundation, picking their way through the morass of academic and state politics. Walker advised on political strategy, sent well-timed and highly effective letters to the university's president, and lobbied state legislators. He jollied Ruthven into staying at Ann Arbor when university officials denied a request for a few hundred dollars, and again when Ruthven was tempted by the offer of a curatorship at the Carnegie Museum in Pittsburgh. He was instrumental in getting the state's museum moved to Ann Arbor in 1910 as a department of the university, and in the construction of a new museum building in 1928.[95] Walker's role as patron transcended the personal to become professional and institutional. Walker was delighted to be named honorary curator of molluscs in 1910, and Ruthven was not making an idle compliment when he told his friend that the best return he could make for his gifts was to train a first-rate man for the museum's staff in Walker's own field of malacology.[96] In research museums generally it was not public display that secured patrons' interest, but advancing the science of natural history.

Joseph Grinnell, the founding director of the Museum of Vertebrate Zoology, was also well connected to amateur networks—his own roots were there before he turned professional—but unlike Ruthven he depended on a single patron, Annie Alexander. Alexander was the daughter of a well-to-do Hawaiian sugar planter: intelligent, independent, and keenly interested in fossils and natural history, and with a taste for hunting and outdoor life.[97] These tastes were typical of her class and time, but for Alexander they were more than recreation. She conceived the idea of a natural history museum for West Coast fauna in 1906, as she was planning a hunting expedition to Alaska to acquire a collection of big game animals. For that she required a scientific hunting permit from the U.S. Biological Survey, and that in turn required that she have a public scientific repository for her collection. Writing to C. Hart Merriam, Alexander suggested founding a museum attached to her alma mater, the University of California: "The work is something that especially interests me," she wrote, "and I would like to make it a life work. I need only a little cooperation."[98] Merriam of course warmly supported her project (and got her the hunting permits). Alexander and Grinnell met a few months later through Grinnell's friend Frank Stephens, an accomplished amateur mammalogist who had collected for Merriam and knew that Grinnell could advise Alexander on Alaska, where he had collected for two years. Grinnell's advice was so informed and practical that Alexander decided then and there that he was the man to direct her museum. Grinnell, who was stuck teaching elementary biology at Throop College in Pasadena and longed to return to the field, accepted at once. The museum officially began its operations the next year.[99]

Like Merriam's survey, the museum was exclusively a research institution, with study collections only: no public exhibits. Alexander was initially intrigued by dioramas and had three built, but they were disappointing. The really good diorama makers like Carl Akeley were beyond her modest means, and she disliked anything second rate.[100] In fact it was science that really stirred her interest, as the collector Edmund Heller observed: "She is certainly much interested in systematic zoology and I doubt if the exhibition phase appeals to her so strongly. Her frankness and straight forwardness of manner I should think would make any relations with her very pleasant."[101]

Grinnell and Alexander's plan was to create a comprehensive collection of the vertebrate fauna of the Pacific slope from Alaska to Baja California, and from the coastal bluffs to the deserts of the Great Basin.[102] Alexander's annual allowance of $7,000 provided for three salaried staff (Grinnell and two assistants), with an additional $900 for field expenses—generous sums at the time. The university provided a building for the collections. For his assistants Grinnell selected men skilled and experienced in field collecting, preferably trained by himself.[103] It was in effect a scaled-down, regional version of Merriam's national survey.

Grinnell had full authority over the museum's scientific work, and Alexander took care to protect him from interference by the university's regents and zoology professors, who she was sure would find other uses for her money than collecting and taxonomy. (They did; natural history did not rank high with them.) She insisted that she and Grinnell have complete control of museum policies. "I may be asking too much," Alexander confided to Grinnell, "but it is absolutely necessary in that nest of schemers." The regents balked at this condition, alleging legal scruples, but promptly backed down when Alexander withdrew her offer, as she knew they would. Grinnell was relieved to be free of academic politicking and to be answerable to a person whose sole interest was the museum and its work.[104] As her biographer has observed, Alexander's business acumen, candor, and prudence were remarkable.[105] Looking back on their first seven years, Grinnell credited their success on Alexander's foresight in defining and delimiting the museum's agenda and assuring its independence.[106]

Alexander's personal interest in scientific collecting was a critical ingredient in her partnership with Grinnell. She enjoyed hunting big game animals as much as the sportsmen did who collected for the dioramas of eastern museums, but she was more interested in scientific collecting than trophy specimens.[107] Big bears were the

Fig 3–9. Annie Alexander, founder and patron of the Museum of Vertebrate Zoology, and skilled hunter and collector, Oregon, 1902. Courtesy University of California Museum of Paleontology, Berkeley, California.

chief object of her 1907 expedition to Alaska, but she already appreciated that small, unglamorous rodents were more important scientifically and volunteered to collect those herself, if Grinnell would teach her the tricks and work up her collections. (The bears she ended up leaving to hired Indian hunters.)[108] Grinnell suggested that Alexander herself work up the bear collection, assuring her that she would do creditable scientific work, but she did not rise to the bait. Probably, as her biographer suggests, Grinnell saw Alexander's talent for science more clearly than she did herself.[109]

Grinnell continued to nudge Alexander to expand her scientific work. In 1907 he advised her not to collect in the Galapagos Islands (exotic but overworked) but to make Alaska her special area: there everything would be new. In 1910 he urged her to undertake extensive series collecting in Arizona and, when she demurred, told her that she greatly underestimated her talent for scientific collecting. He was delighted when Alexander and her companion, Louise Kellogg, decided to collect on Vancouver Island, the "least worked locality you could pick out in western North America," and urged them to bring back large series for the museum's collection.[110] In 1911 Alexander and Kellogg collected in the Trinity Mountains, an important transitional zone that Grinnell was very keen to have worked, and with Grinnell's encouragement Kellogg herself prepared the faunal report.[111] Alexander and Kellogg were exceptionally skilled collectors and preparators. "I believe that so far Miss Alexander and Miss Kellogg have put up over a thousand specimens!" Harry Swarth reported to Grinnell from Beaver Creek, Vancouver Island. "They are wonders. You never saw anything like the way they work; I'm not in their class at all."[112] Grinnell's written directives to Alexander never defer or condescend, but address a professional equal.[113]

A long experiment in commercial farming between 1911 and the early 1920s prevented Alexander and Kellogg from serious collecting, but Grinnell never gave up trying to tempt them back. When the two women took a camping vacation in the Providence Mountains in 1916, Grinnell sent them a light collecting kit and directions for doing a visual animal census—a task more appropriate than series collecting for a recreational trip.[114] In 1918 he told his patron straight out that he thought she was wasting her time with asparagus and prize cattle, and was neglecting her true calling:

> I confess I am not altogether disinterested in my hope that you will travel and collect. For it would mean a big influx of material to the Museum. You have the talent to collect, comparable to Dr. [William L.] Abbott's of the National Museum—and it's a rare talent, too. Anyway, however your bent lies, I believe you ought to yield to it just as soon as fairly feasible.[115]

When Alexander and Kellogg did resume regular collecting for the museum in 1926–27, Grinnell set them the task of collecting topotypes from all over the West.[116] It was a kind of collecting that suited Alexander's taste for outdoor life, requiring extensive travel in remote places but not mass collecting, with its relentless routine of supply and transport.[117] The two women continued to collect in this way until Grinnell's death in 1939.

Grinnell and Alexander's partnership was based on personal sympathy and devotion to survey collecting. Both were strong-minded and outspoken, but also diplomatic and able to see issues through others' eyes. Grinnell accepted Alexander's overall financial control, and Alexander accepted Grinnell's scientific authority. Grinnell kept his patron informed of what was going on, and Alexander did not interfere. Grinnell never pressured Alexander to increase her annual support, which Alexander did anyway from time to time. She had hoped that Grinnell would raise funds from other patrons for expanded programs but never complained that he did not.[118] When university officials pestered Alexander with solicitations, Grinnell urged her to remember all she was doing already: "[Y]ou have always kept in mind the larger . . . aims of scientific accomplishment. . . . [Y]ou do not need to yield to the pressure for more and more giving. Be scientifically adamantine!"[119] And when university zoologists pressed Grinnell to teach introductory biology, Alexander reminded him firmly of the reasons why he had chosen museum work over teaching, and why they had insisted on the museum's independence from the department of zoology: "I hate to see you put your hands out to be shackled the very first thing by selfish men, who work on your enthusiasm and [illegible] under cover of a grand crusade for the New Zoology." Why sacrifice his field research to give the university a teacher of elementary biology at the museum's expense?[120]

It was also their shared joy in collecting and outdoor life that made their partnership so fruitful and untroubled. Here is Grinnell, in the Mohave desert, writing to Alexander, collecting in Alaska:

> Your own pleasure in carrying on the fieldwork this summer can hardly exceed ours down here. Although the conditions are diverse, the objects of our work are much the same. We have our poor catches; and then again something turns up which keeps us happy thru the next dull spell.[121]

Shared pleasure in outdoor work was as powerful a binding force in patronage relations as the more social aims of scientific safaris and diorama building.

Conclusion

I introduced this chapter by contrasting the irregular naturalizing of hunters, campers, and outdoor vacationers with the organized and systematic activity of survey collecting. I would like to conclude by making the opposite point: that natural history survey and collecting remained in some ways quite close to the activities of the vacation complex. This is no paradox, but simply a reflection of the dual character of survey as a mode of field practice. Its practices developed within the institutions and customs of amateur collecting and vacationing, and assimilated elements of that cultural system even as they became more exacting and rigorous. The connection between diorama building and survey collecting clearly displays this dual aspect; as do also the partnerships between research museums and patrons like Bryant Walker and Annie Alexander. It is clear too in the similar practices of survey collecting and recreational camping. Had expeditions not been sport and adventure as well as scientific research, it is unlikely that patrons and civic institutions would have sponsored them on such a scale. The cultural hybridity of natural history survey is a defining characteristic, differentiating it from both earlier and later collecting practices.

Survey collecting was complex and expensive enough to need support by nonscientific patrons, but its costs were still relatively modest, compared to either the grand multiyear explorations of an earlier age or the infrastructure of laboratory science. (Expeditions are seasonal and can be picked up and put down as finances permit or not; labs eat money year in year out.) As we have seen, many biological surveys were carried out by unpaid workers on shoestring budgets. Joseph Grinnell's annual budget ranged from $7,000 to $9,000 a year; and $15,000 kept a dozen of Merriam's collecting parties in the field for a season. With these modest sums the vertebrate species of North America were discovered, inventoried, mapped, and classified. Expeditions launched by the larger civic museums were more expensive, especially those sent to distant corners of the world to fetch examples of large predators or game animals, but they were less costly than exhibition halls and dioramas. A state-of-the-art diorama hall in the 1920s could cost a quarter of a million dollars, while $1,000 sufficed to keeping a collecting party in the field for a month. And study collections required little more in the way of infrastructure than a few hundred steel cases in basement rooms and some feet of field notes on library shelves—that was about it.

Survey collecting, though large in the aggregate, was in practice small-scale science, performed by small groups who worked and reworked faunal areas until they knew them intimately—almost residentially. Whereas imperial explorations had been affairs of state or transoceanic commerce, survey collecting, even when state sponsored, was an elaboration of the practices of outdoor sport and recreation. Because natural history surveys were conducted by many small parties from many institutions, they could improvise and take advantage of the free labor of those who pursued nature as an avocation, not a career. For the expensive and highly organized expeditions of the age of exploration, that was more difficult. The patrons of natural history survey were middle-class people acting out the ideals and fantasies of the culture of travel and vacationing. Government-sponsored surveys like Merriam's or Forbes's might depend politically on economic rationales, as geological and geographical surveys did. But in practice they were hardly economic at all but cultural and scientific. A mix of aesthetic and intellectual imperatives shaped the patronage of survey collecting, as it did the culture of amateur collecting and nature-going.

Organizationally the institutions of survey collecting were less formal than the large-scale institutions of modern states and corporations. Patronage of collecting expeditions depended on individual personalities and on relations that grew out of a shared love of nature and outdoor life. In the physical and biomedical sciences, in contrast, patronage relations more typically reflected shared interest in providing educational or social services. Those who paid to have laboratories built never meant to work in them, but the benefactors of museum exhibits and study collections could take part in stocking them. The relationships between Joseph Grinnell and Annie Alexander, or between Alexander Ruthven and Bryant Walker, were equally personal and institutional. The effective functioning of field parties depended on personalities, as we will see. The cultural and emotional infrastructure of survey collecting, as well as its institutions and economics, was fashioned in part from the customs of the vacation complex.

But survey collecting was also, and above all, organized and systematic—to return to my original theme. It was an exacting science. Unlike individual collectors or explorers, who gathered and recorded whatever serendipity offered up, museum and state surveys collected according to a plan, and in depth. And totally: they aimed to create collections that were not just samples of a region's fauna or flora but were comprehensive and complete embodiments of its biodiversity. Surveys aimed to make natural history exact and rigorous, by applying uniform collecting practices and meticulous data management. Inevitably, institutional affiliation gradually replaced personal character as the guarantee of authentic facts.

This difference in the moral economy of different kinds of work is evident in the contrasting narrative structure of modern scientific papers, and stories of recreational hunters and fishers. In the latter it is narrators' personal virtue and direct experience that lend authenticity to knowledge claims—that is why they are full of circumstantial detail. In the case of scientific reports, individuals and circumstances disappear into a universalized passive voice: it is the institutions and communal practices that are speaking and that give claims their authenticity.

We catch a little glimpse of this change in the moral economy of collecting, when one of Alexander Ruthven's volunteers objected to the modern practice of labeling specimens with the name of the expedition (that is, its sponsor) rather than the person who collected them. He objected "[b]ecause the name of the collector guarantees or does not guarantee the honesty of the data according to the integrity of the collector. And, furthermore, the man who has struggled to obtain the specimens is entitled to the credit."[122] That was true of amateur collecting, and of heroic exploration: individual virtue and fortitude made these activities credible.[123] However, what made survey collecting credible was not these personal and existential qualities, but the institutions behind the work, and the expedition itself: that exacting instrument, with its rules, customs, and quality controls.

The point here is that natural history survey mixed the customs of exact science with those of recreational culture. As middle-class vacationers made their play worklike and turned leisure into recreation, survey collectors retained older elements of aesthetics and play in their pursuit of an exact science of the field.

CHAPTER FOUR

Expedition

THE HEART of natural history survey was the expedition itself: the small groups of individuals who went forth into the inner frontiers to collect and observe. How were these ventures organized and managed? What customs guided their makeup and activities? And how were they experienced? These are the topics of this chapter and the next.

It is useful to think of the survey expedition as a kind of scientific instrument. That may seem odd to those familiar only with the physical instruments of laboratory or fieldwork—balances, galvanometers, barometers, sextants. But historians of science have recently been expanding the category to include laboratory animals and even human communities. Richard Sorrenson has suggested that the survey ships used for navigational and shoreline mapping in the eighteenth century can be conceptualized as scientific instruments.[1] And Mary P. Winsor has likewise proposed that natural history collections and even museums are the instruments of taxonomic science, analogous to the physical instruments of experimental labs. "Every workman must have his tools," Winsor writes, and "the tools of a zoologist are collections of natural objects systematically arranged." And museums: "Their storage cases, record keeping, and network of exchanges would be just as truly an instrument for scientific investigation as the cytologist's microscopy. . . . [T]he museum is an eminently respectable piece of scientific equipment, a tool literally rather than metaphorically."[2]

This view certainly applies to research museums, though is perhaps a stretch for the large civic museums with their mixed uses of display, research, and education. But study collections are unambiguously instruments. As collections of *Drosophila* or *Escherichia coli* mutants are instruments for producing genetic maps, so are specimen collections instruments for producing species classifications and biogeographic maps. And so too, arguably, are the expeditions that created study collections. It is not the physical instrument per se—the fly, or the specimen—that is the subject here, but the social systems of which such objects are key elements. A capacious macroconcept of an instrument will suit us best.

Sorrenson's concept of the ship as instrument is a surprisingly apt model for expedition parties. Survey ships were not passive platforms for instruments and observers. They actively shaped how science was conducted on their decks, and their design was vital to the geographical practice. Robust, broad-beamed coasting colliers were better than the more prestigious frigates or men-of-war, which could not get in close to shore, tended to run aground, and could not be careened on beaches for emergency repairs—potentially fatal shortcomings in emergencies far from home. Coasting colliers could travel far, yet hold known positions precisely for accurate mapping of coastlines—a feat as essential to geographical measurement as precise measurement was to laboratory science.[3]

Collecting expeditions were very much like Sorrenson's ships. Not merely vehicles for transporting collectors to their places of work, they were the instruments of a mode of collecting that was extensive and mobile, and also intensive and sessile—ships on land. (Recall the westering settlers' "prairie schooners.") Expeditionary parties, like survey ships, were best kept small, practical, and flexible. The larger parties of grand-scale explorations—to continue the nautical analogy—were less able to get in close to what they wanted to observe, were more likely to encounter obstacles, and were harder to adjust when things went wrong, as they inevitably did. Moving about and close observing were no less essential aspects of taxonomic science than preparing specimens and working out taxonomic categories. Survey naturalists took instruments with them on expeditions—traps, preservatives, books, scientific equipment—but the expedition itself was the chief instrument of survey collecting.

The analogy of ship and field party also highlights the differences between geographical exploration and natural history survey. Whereas the ship was self-contained and carried much of what it needed with it, survey expeditions traveled light and depended on local inhabitants and infrastructures. Whereas voyages of exploration were frequently out of touch for long periods, collecting expeditions maintained weekly or even daily contact with museum headquarters, sending back crates of specimens and in return receiving revised directives and supplies. Voyages of exploration were extensive; survey expeditions were both extensive and intensive—instruments of an exacting field science.

Collecting parties had to be the right size and have the right mix of skills and personalities. Itineraries were carefully mapped out in advance, but not so rigidly that collectors could not respond to unforeseen opportunities or setbacks. To combine intensive and extensive collecting, field parties always had to be choosing between staying put and moving on; between working remote areas, which were new but difficult, and those that were accessible but already collected. And where should parties stay: in or near towns for convenience, or in nature, where the animals were? Survey expeditions were complicated social instruments requiring care in their design and varied skills to make them work. How did they work?

The Field Party

Collecting parties ranged from single individuals to quite large and diverse groups. Vernon Bailey was on his own for a year or more in his first trips for Merriam in the late 1880s. Merriam's expedition to Death Valley, in contrast, was an elaborate affair, consisting of Merriam and his wife; three survey staff (Albert K. Fisher, Edward W. Nelson, and Theodore S. Palmer); the botanist Frederick V. Coville of the U.S. Department of Agriculture and his "assistant" (actually a U.S. senator's son on a junket); a topographer from the U.S. Geological Survey; and four teamsters, cooks, and packers. The outfit consisted of a big wagon (3,000 pounds) drawn by seven horses (ten would have been better, Bailey reckoned), plus five other horses, Bailey's small wagon, two buckboards for local work, and assorted pack animals—as well as two photographers and a reporter from a San Francisco newspaper.[4] Another of Merriam's expeditions, to the San Francisco Peaks in Arizona, was equally large and swelled by eight or ten local gentry who came up from Flagstaff on weekends.[5]

However, these examples are the extremes. Most survey work was done by small parties of between two and five or six people. Two was generally regarded as the absolute minimum for intensive collecting, and the standard party of the U.S. Biological Survey was three: one of Merriam's assistants, in charge; an experienced field

Fig. 4–1. U.S. Biological Survey party, Death Valley, California, June 1891. Left to right: Vernon Bailey, C. Hart Merriam, Thomas S. Palmer, and Albert K. Fisher. Albert K. Fisher Papers, box 50. Courtesy Manuscript Division, Library of Congress, Washington, D.C.

collector; and a combined cook, packer, teamster, and boatman. Small parties, Merriam found, could accomplish more in less time than could larger ones.[6] Fewer than two or three could just not manage the complex logistics of scouting, making and breaking camp, intensive collecting, getting shipments out, and staying in touch with home base.[7] Joseph Grinnell noted that two men could accomplish more than twice as much as one man working alone. Expenses per man were less, and two could work a locality more thoroughly; and the men, if congenial, were happier for having company. A third man (often Grinnell himself) remained at home to receive specimens and send out fresh supplies.[8] Some places, like the South American interior, were just too dangerous for a single man or even for parties of two.[9]

Field parties might also include green recruits learning the ropes from the more experienced hands. These could be museum volunteers, amateur naturalists on working vacations, or college students on a summer field course. The educational function of survey expeditions was crucial: good survey collectors were scarce, and university departments of biology had neither the means nor the inclination to raise them up. Their training was still an on-the-job, field apprenticeship. Merriam made sure that Bailey had a companion for a trip from Utah to California in 1888, partly for company, but also because the Biological Survey was desperately short of expert collectors. Merriam directed Bailey to make sure the new man could start off on his own immediately after the trip: "We want another *good man* in the field, but don't know just where to find him—so perhaps we had better *make* him." The apprentice was paid no salary, but Merriam thought the training worth more than money, because if he learned well he would likely be hired immediately by the Biological Survey. "What we want is a young man who has the right sort of stuff in him," Merriam enjoined Bailey, "who will profit by your experience and example, and be able to make trips independently."[10] For the same reason, the Field Museum sent a young man, Robert Becker, out to South America to assist Malcolm Anderson and learn the craft of series collecting. (Becker quickly proved himself an expert traveler and "rat peeler.")[11] Expeditions thus served the dual purpose of knowledge production and social reproduction.

Practical skills were required of a field party, and the men in charge of cooking, packing, and transport could make the difference between a pleasant and fruitful trip and one that was unproductive and miserable. Expert and reliable packers and muleteers were hard to find, expensive, and highly valued.[12] Bad ones could fail to show up after the customary predeparture binge, as happened to Wilfred Osgood in Peru, or could abscond without warning, making camp life a misery for everyone who remained. Joseph Grinnell ran into one of those, who deserted after a month leaving the group in the High Sierras to manage two burros and their own bad tempers.[13]

Some cooks and packers expanded their repertoire to making specimens and began to resemble professional collectors. Like the general handyman of an expedition party in Peru: "He is a working fool (like myself)," reported the biologist in charge, "and skins very fast now, almost as fast as I can stuff and do the necessary writing up in the Field Book each day. He is fine on the trail, a splendid packer and a regular slave driver. He gets more excited over new species than I do and is very good now at shooting.[14] A good example of the type is Segundo Garzon. This Ecuadoran learned to cook in a hotel, and as a jack-of-all-trades drifted to New York City, where he learned English and came to the attention of the collector George Cherrie, who knew a good man when he saw one and secured him for a forthcoming expedition to South America. It was a sagacious choice. Garzon was not only a first-rate cook but an excellent skinner, and he quickly learned to prepare fine bird and mammal specimens. He was expert at bargaining with local storekeepers; bossed porters and rivermen with elan; knew how to extract ticks; and played an excellent game of poker. With Colin Sanborn on a collecting trip, Garzon did much of the shooting (as he was not cooking for a large party).[15] Intelligent and practical men like Segundo Garzon were an indispensable ingredient in the mix of expedition parties.

Keeping order and control was another vital aspect of managing survey parties. Like ships at sea, expeditions were subject to the interpersonal tensions of any confined and isolated community. Like ships' companies, field parties were egalitarian but with a structure of authority that was clearly understood and respected. It was expected that everyone, including the party's leader, would share equally in the housekeeping routines of camp life. Those who shirked or groused were seldom invited back. Merriam was much respected for his willingness to share in the daily work of camping—as was Joseph Grinnell. At the same time, it was understood that the expedition leader was boss—an overland version of a ship's captain. This authority was usually tacit, but Charles C. Adams spelled it out for a survey party in the Porcupine Mountains of Michigan in 1904: The museum's taxidermist, Norman A. Woods, was responsible for keeping order, and anyone who refused to take his orders was liable to instant dismissal, Adams warned. Woods was also empowered to make field workers keep proper records, prepare specimens correctly, not exceed the quotas allowed by their game permits, and shoulder their share of camp work cheerfully and as directed. On the scientific side, Alexander Ruthven was in charge of work methods, and *all members* (Adams was emphatic) were subject to his orders.[16] Adams was more openly authoritarian here than was usual, perhaps because the party was all male and included several green students; but his concern for order was typical, and no doubt necessary.

Fig. 4–2. Native cook at work in a party of the American Museum of Natural History encamped near Verde Coche, Venezuela, 1922. Behind him, two naturalists prepare specimens. Photo by Frank Chapman. Courtesy Department of Library Services, American Museum of Natural History, negative no. 263994.

Multipurpose expeditions were especially prone to interpersonal tensions, because they were socially diverse. The conflicting demands of collecting for exhibits and study collections could pit taxidermists against curators. Getting specimens of rare or hard-to-get species for display took time that could be spent on scientific collecting and, because it had to be done in specific seasons, could disrupt a fine-tuned itinerary and create conflicts of authority.[17] As one museum director complained: "[M]useums are greatly in need of men who can be trusted on expeditions to secure both scientific collections and exhibition material. Most available men are either taxidermists who get only exhibition material or scientific collectors who get little besides material for study."[18] However, serious disputes could generally be avoided by tact and common sense.

Such tensions were less easily resolved when field parties included expedition sponsors. Most sponsors were content to leave decisions about itinerary and pacing to the scientists, but some regarded expeditions as exotic vacations with science as the pretext and felt that those who paid the piper should also call the tune. As one patron bluntly put it: "The naturalists [are] to be considered as our employees, and [are] expected to comply with our wishes."[19] In this particular case the two sponsors had a business interest in rubber and Brazil nuts and meant to do commercial exploration. Their business contacts in Amazonia could provide the field party with facilities and a launch, which the scientists needed. But Frank Chapman worried that the party would be too rushed to collect in depth, and that his two sponsors would want to overnight in towns, not camp in the forest. (In the event, heavy rains and flooding forced the party to stop and camp frequently, giving the curators plenty of time to collect.)[20] Other patrons regarded expeditions as eco-tours, and scientists as personal tour guides, as Frank Chapman feared would happen when a field party was augmented by an artist and a bird buff.[21] The more an expedition mixed science and tourism the more likely it was that personal relations could grow sour, as the zoologist Karl Schmidt discovered as a member of the Crane Pacific Expedition on the private yacht, *Illyria*. "[I]t seems to be a matter for a psychologist rather than a zoologist," a colleague quipped, "and perhaps it may be well to consider sending along one of that profession next time."[22]

Such complications were an inevitable consequence of a system of patronage that linked scientific collecting to diorama building and routinely assembled mixed parties of sportsmen, nature tourists, naturalists, taxidermists, and artists. So curators had to be able to judge quickly which prospective patrons would respect scientific aims and authority, and make agreeable field companions. Offers of sponsorship from patrons who were likely to be troublesome were politely parried. However, most were only too willing to follow curators' lead. And some, like Evelyn Field, the wife of Marshall Field, Jr., went out as tourists and returned as skilled collectors and specimen makers.[23] Museum archives reveal surprisingly few cases of actual disasters.

Forbearance and civility were highly valued qualities in expedition parties, as valuable as skill in collecting and observing. Alexander Ruthven spelled it out in his instructions to a working party in Texas:

> Each member is urged to bear in mind that in camp the close contact and tediousness of the work often induces strained personal relations, and that this can only be kept from interfering with the work by constant cheerfulness and unselfishness on the part of each person. Determine to overlook everything that arises to aggravate you and to preserve a constant show of good nature even in the most trying circumstances.

Desertions were not permitted except for illness, and accidents were individuals' own responsibility. (Ruthven recommended buying accident insurance.) Members of the group were enjoined to stick to their own special quarry and to assist colleagues in their special work: no poaching, no prima donnas.[24] Behavior in field parties was guided by such practical and largely tacit rules of moral economy, as it is in any small group of individuals working toward ends that were both personal and communal.[25] Assembling a field party required experience and the capacity to judge human strengths and frailties, and any mistakes would be paid for dearly in the field.

Field parties had also to be effectively deployed. Itineraries had to be decided: where to go, where to pause, and how to balance the opposing values of extensive coverage and in-depth local collecting. Explorers could meander if they felt like it, and resident naturalists could linger or leave a project for another and better day. But survey parties were on tight schedules and could do neither, so itineraries had to be meticulously (but not too rigidly) planned. Planners had to decide how far parties should venture beyond railroad nets and the infrastructure of towns and tourist resorts. Unsettled areas were the places where collectors were likely to find species "new to science," and such finds produced publications and rousing copy for annual reports. But such areas were also difficult of access and thus risky places for collectors. It was always a trade-off, as curator Daniel Elliott reported from the Olympic Mountains, where he and Carl Akeley were collecting for the Field Museum in 1898. It was the roughest and most difficult country he had ever seen—but potentially a gold mine: "A great portion of the Olympians [sic] is absolutely impassable, and we have reached a point beyond which nothing, unless provided with wings, can go. . . . Even if already known, specimens coming from such localities are of almost as much value as if undescribed."[26] The features of terrain and climate that had kept such places unexplored also made them worth collecting.

Collecting parties generally seized any chance to get into unworked territory, as Wilfred Osgood did in 1912 when he dispatched Malcolm Anderson to join a Peruvian government survey of the wild country at the headwaters of the Rio Negro. "The region is almost unexplored," Osgood advised his assistant. "It's exactly the country we want to get into and the chance of a lifetime to make a big collection."[27] When the Field Museum sent collectors to Rio de Janeiro in 1913, Osgood advised them to go where few had gone before: "Isolated desert regions, swamp areas or watered mountain ranges rising out of deserts are the best places for zoological novelties."[28] In the hinterland of Rio de Janiero, where travelers reported abundant animal life, Osgood anticipated that two-thirds of any species found would be undescribed, and he advised his assistant Robert Becker to go as far up the coastal rivers as he could by rail or steamer, then take pack trips of four or five days into the forests from river settlements. In the event, Becker found getting to these inaccessible places too expensive and time-consuming, and their inhabitants too rough and lawless for his liking.[29] Even places closer to home could be made inaccessible by their inhabitants' dislike of outsiders: for example, the forests of southern Appalachia. (The ecologist Lucy Braun worked this region, and Frederick Clements thought her very brave to do so.)[30]

But collecting parties did not have to head for inaccessible outbacks; there was important work to be done in areas that had already been collected. The great nineteenth-century collections were in European museums, and American museums were eager to have their own complete and state-of-the-art collections from important faunal areas, for convenience as well as national pride. The value of collecting in remote and dangerous places declined as such places became fewer and ever harder to reach. And the worth of modern study collections came to be measured less by the number of type specimens they contained than by their overall quality and scope. Topotypes (specimens collected from the same place as the original type specimens) served just as well for taxonomic research—or better, being usually prepared to higher taxonomic standards than the original types, many of which were old and decayed.

"I am not obsessed with the Thomasonian fever for types," Wilfred Osgood announced to his friend Oldfield Thomas, who as cu-

Fig. 4–3. Field party of the Illinois Natural History Survey at Thompson's Lake, on the Illinois River, a rich fishing and collecting area, 1894. Left to right: zoologists Frank Smith and Henry E. Summers, entomologist Charles A. Hart, and boatman Miles Newberry. Courtesy Illinois Natural History Survey, Champaign, Illinois, negative 882.

rator at the British Museum had described as many new species as any man alive. What Osgood wanted for the Field Museum was not rarities but a broadly useful working collection.[31] That meant sending field parties to rework known and relatively accessible places more intensively. For example, in 1913 Osgood advised his man in Rio not to skip any biogeographic areas as he proceeded down the coast, even though coastal faunas were well known. The museum would welcome anything he sent back.[32] And intensive collecting could and often did turn up things that explorers had overlooked. Grinnell's field parties were continually finding rare and new species in places thought to be worked out.[33]

Of course the appeal of new and exotic places never disappeared. Collecting close to home, however productive, could seem tame and too much like amateur collecting for curators' liking. Curators were not uncommonly required to collect locally, for displays of backyard animals or school exhibits, and their feelings about such work was mixed. On the one hand it was still possible to do serious collecting in urban twilight zones, and any fieldwork at all was better than being stuck indoors cataloging. Edmund Heller, an experienced collector, went out a few days a week into the country around Chicago and declared it well worth his while: "The state is practically virgin ground with perhaps two or three men in it who have any interest in mammalogy. These short trips are a good thing to take the edge off of a collector's imprisonment."[34] But if local work took the edge off curators' appetite for collecting, it seldom satisfied it. Harry Swarth spent two unhappy months in the country around Chicago collecting for an Illinois bird room: "I am rusticating in the country now," he reported to Joseph Grinnell, "supposed to be 'collecting,' but in a very genteel, ladylike way, which is not as productive of results as it might be. However we are obeying orders."[35]

Salvage collecting also drew collecting parties to the fringes of expanding towns or commercial agriculture. The central valley of California was such a place when Grinnell's field parties began to collect there in the early 1900s. They arrived early enough to see what the place was like before irrigation opened it to intensive settlement, but even as they worked it became clear that the entire region would soon be razed of its indigenous animals. "It's heart rending trying to find topotypes in a settled-up country!" Grinnell reported in 1911 from the field near Fresno. "The sandy part of the region is now the highly cultivated raisin belt, and subjected to frequent irrigation." Farmers told collectors that kangaroo rats, once plentiful, had not been seen for six years. Collecting was better farther west, near Tipton, where there was a large strip of uncultivated pasture, and even tiny islands of original prairie in the agricultural seas could yield valuable topotypes.[36]

Grinnell was always aware that he would be the last to collect such places, and the thought kept him at a less than pleasant task: "[T]here is nothing attractive about collecting in a settled-up, level country. But it ought to be done and the longer we wait, the fewer 'waste lots' there will be in which to trap for native animals." He made it a policy to collect first in areas that were being most rapidly overrun by humankind.[37] Even the Sierras were losing native fauna as overgrazing destroyed their habitats, but Grinnell consoled himself with the thought that his survey teams were gathering the data of a vast and fascinating experiment in evolution.[38]

System

We do not think of natural history as a precise science, but in its survey mode it had to be. Collecting large series of specimens from every corner of entire regions made it imperative to operate with uniform, even standardized procedures, if for no other reason than to be efficient and keep order in masses of complex data. And for modern systematics, system was an absolute imperative, especially the taxonomy of subspecies. Diagnosing species and subspecies often required exact measurements of large numbers of specimens, each of which had to be measured "in life"—that is, dead but before skinning and preserving—and had to be turned into specimens without distortion and in exactly the same way. Accurate observing was also required—locating precisely where specimens were taken and in what habitats—and records were kept on standard forms. And, since diagnoses involved comparisons of the fauna of neighboring regions, procedures had to be uniform among all who engaged in natural history survey.

Taxonomists collected large series to reveal the patterns and limits of variation within species. And rigorously uniform practices of collecting and preparing specimens assured taxonomists that observed variations were nature's, and not merely the variability of human skill and care. Variability is a problem in all empirical sciences, but different sciences solve the problem differently. Gene mappers solved it by constructing "standard" laboratory animals that had their natural variability bred out.[39] But whereas variability was a nuisance to geneticists, for taxonomists it was the object of study. It was to ensure that collections preserved the variability of nature that collecting practices were uniform and religiously observed. Of course, practices retained a degree of individuality; it was understood that specimens bore individual collectors' distinctive "make."[40] Individual skill was valued, but the ideal was a standard practice.

Many of the standard practices of collecting and preparing specimens were developed by C. Hart Merriam and his colleagues in the

U.S. Biological Survey. Merriam's letters to Vernon Bailey in the field are full of detailed instructions for improving the quality of Bailey's specimens (which were always rather good) and ensuring that they arrived in Washington in perfect shape. It was a complex series of steps, each of which had to be done just right: measuring "in life," skinning, degreasing (by rubbing with gasoline or cornmeal), preserving (with arsenic powder), drying, stuffing (not too loose, not too tight), smoothing pelage or feathers, arranging limbs or wings without bending or breaking, attaching skulls to skins, numbering, recording, labeling, packing (again, not too loose and not too tight), and shipping. Turning messy little corpses into scientific specimens was an intricate and exacting craft, and there were many ways of doing it wrong. For example, Harry Swarth had hard words to say about some specimens prepared by one of Grinnell's young apprentices, Walter Taylor: "They are decidedly punk in appearance," he expostulated. "The birds are ruffled and messy looking, whereas yours come in without a feather awry. His mammals are overstuffed, and . . . they all look as though they had run against a brick wall."[41]

It was survey leaders like Merriam and Grinnell who made the rules of specimen making exact and uniform. Their obsession with precision and rule should not be mistaken for the mania of authoritarian personalities: it was a trait widely shared by scientists who engaged in precision work, whether in the lab or in the field. Uniformity and precision were as essential to making taxonomy and biogeography rigorous field sciences, as precision measurement was to modern chemistry and physics. Taxonomists regarded Merriam's methods as perhaps his most important and lasting achievement, and they spread rapidly through the Biological Survey network—Merriam training Bailey, Bailey training Edward Nelson, and so on—and into museums everywhere.[42] When Wilfred Osgood left the survey to become a curator at the Field Museum, he introduced Merriam's methods there. The Museum of Vertebrate Zoology and the University of Michigan Museum also became centers of precise practice that spread with their alumni far and wide.[43]

Joseph Grinnell sometimes operated in "an intensive, specimen-production plan" where efficient mass collecting was required. "We are insatiable," he urged a volunteer collector. "[A]nimals are desirable in *series*, so that the full extent of individual, sexual, age, and seasonal variation can be ascertained."[44] For a transect of the

Sierras, for example, he required large series of specimens from many diverse environments and life zones. In Death Valley he had a long list of topotypes to get from places that were far apart and in a dangerously hostile environment. In such circumstances speed and efficiency were vital. In the "specimen-production" mode parties had to hustle, and Grinnell pushed them relentlessly. They kept a running schedule of species caught, from which they could see at a glance when it was time to let up on some and concentrate on others. In this intensive mode, Grinnell would use little tricks to keep his workers up to the mark. In the Sierras in 1911 he kept a running average of specimens caught to encourage competition among his collectors. It worked and, Grinnell felt, did no harm. He had feared it would make workers stop trying for difficult species for the sake of getting easy ones in larger numbers, but that never happened.[45]

Grinnell's specimen production was hardly factory discipline, but it was businesslike and does epitomize the intensive character of survey collecting. Survey teams had to worry about productivity: itineraries were long and seasons short, and only deep and comprehensive collections had scientific value.

Uniformity and precision were no less important in recording field data.[46] Here too Merriam and his men pioneered new standards of order. Merriam introduced standard field notebooks and printed forms, and insisted that field workers send their filled notebooks to headquarters at regular intervals. He also required monthly reports on where parties had been and what they collected, and detailed descriptions of terrain, habitats, and any other circumstantial data that might help in sorting and mapping fauna. Preparing these reports was the field worker's final task before leaving one collecting station and moving on to the next, and woe betide the man who failed to send his in. Merriam's letters to Vernon Bailey, especially when Bailey was still learning the ropes of survey work, are full of injunctions to keep more complete and precise records of his observations; as are Ruthven's to his young assistants in the field.[47]

In this matter too, Joseph Grinnell adopted Merriam's methods and improved upon them. For example, he modified Merriam's design of bound field notebooks by making them loose-leaf and numbered consecutively, so that notes could be sent to the museum more frequently and used to process shipments of specimens as they ar-

Fig. 4–4. Joseph Grinnell preparing specimens in the field, probably the 1920s. Courtesy Bancroft Library, Berkeley, California, Banc Pic M73.044-pic.

rived. Rapid feedback enabled field parties to take advantage of unexpected leads—another reason to insist on frequent reports. Grinnell was especially pleased with this little practical improvement.[48]

The Michigan natural history survey also pioneered exact field practices. Charles C. Adams enjoined his fieldworkers to never forget that field data "must be made *on the spot*"—not recollected in tranquillity (and imperfectly) back in camp or at home—and must include data on topography, vegetation, soils, fauna and flora, habits, and ecological relations. Heads of the Michigan survey's field parties were authorized to inspect notebooks regularly to make sure no one was cutting corners.[49] Taking photographs as records of environments was encouraged, and observing and collecting in *all* conditions—sun and rain, day and night.[50] Here are Alexander Ruthven's instructions to a party doing an ecological survey of the Texas coast in 1914:

1. Locate various habitats by the vegetation.
   a. Describe the habitats in notes and photograph them.
   b. Preserve specimens of plants.
   c. Determine physical conditions, e.g., soil, temperature, moisture, etc. in as much detail as possible.

2. Collect your groups in every habitat.
   a. Take series of each species in every habitat no matter how common the form.
   b. Find relations between the species and the conditions in each habitat. The local distribution is controlled by local conditions, and the manner can be determined by observation.
   c. Study habits, including food and breeding habits.
   d. Endeavor to determine the enemies of each species.
3. Secure series of specimens when possible.
4. Endeavor to preserve the specimens in the best way.[51]

Merriam also provided his fieldworkers with a standard checklist of things to observe and record, which included life-zone and crop-zone data, species abundances, altitudes and horizontal distances, slope exposures and distances, and stomach contents (for ascertaining diets).[52] Teaching apprentice survey naturalists the habit of note-taking was one of the most important elements of their training, and one of the most difficult, especially for enthusiasts who were used to regarding a large number of specimens as the sole indicator of a good day's work.[53]

Field notebooks were a tool of survey collecting no less essential than traps, guns, skinning knives, preservatives, and camping gear. "My notebook," Grinnell wrote Alexander from the coastal marshlands near Maryville, "will be, perhaps, the most important item of my equipment."[54] Vernon Bailey's 1932 handook for beginning survey collectors states that specimens were valued mainly as tangible supporting evidence for the written record of field notes.[55] Verbal description is as old as natural history collecting; but the field notebook as an exact instrument is perhaps only as old as late-nineteenth-century surveys.

The products of survey collecting were a complex package of physical, written, and visual evidence, all exactly recorded and preserved. Returning from an expedition to the Northwest coast in 1913, Grinnell proudly tallied the results: 2,177 specimens from a dozen stations; 717 pages of field notes; 100 photographs; topotypes of 7 species new to the museum, and 1 species new to science and several rare ones; 20 important modifications of life-zone boundaries; large series of birds, shrews, and chipmunks for studies of geographical variation; and a significant addition to the theory of life zones (on the combined effect of temperature and humidity on abundance).[56]

This is not to say that expedition camps were models of tidiness and antiseptic order; they were anything but that. Here is Grace Seton's (somewhat overdramatic) description of a party encamped near Therezopolis, in Brazil:

> It is night, after dinner. We are all seated around a folding table with a bright gasoline lamp in the center. [Evelyn] Field . . . is sewing a coat. [Karl] Schmidt is writing notes. [George] Cherrie is reading. Said he, "Do I smell formaline?" "You do" was the verdict. "The can you are using as a seat is leaking." Cherrie . . . transferred himself to the gasoline can just vacated by Sanborn and continued to read Nash's Brazil. . . . Cherrie has now [it was the next day] transferred his seat to the case of kerosine. . . . In turning it up some of it has spilled, so in the small room—if a dirt floor and pole walls and matting roof can be called a room—are several five and ten gallon cans of formaline, a large tin of gasoline, a case of kerosine, a dozen jars of alcohol. Also a lighted kerosine stove and the gasoline lantern, to say nothing of cans of carbide. And everybody smokes![57]

One is reminded of photographs of nineteenth-century machine shops: to outsiders, just chaotic jumbles of tools and metal; but to insiders, marvels of practical order. Out of the petrochemical mess of field camps came series of specimens neatly prepared, preserved, and available for exacting study in a tidy museum room.

Communication

Though we may think that success in collecting is serendipitous, in survey collecting it was less luck than cunning. Survey collecting was knowledge intensive. Collectors had to recognize plant and animal species, obviously—but also their variations, habits, geographical ranges, and ecological preferences. They had to know particular places as intimately as if they were themselves residents: the roads and trails; where to hire transport, board, and camp; the good places to collect. Such knowledge came from many sources: books and maps (the U.S. Geological Survey's topographical maps were indispensable); also prior experience, and local informants; and, most important—because zoological handbooks were too heavy to lug around—regular communication with survey headquarters.[58]

As thoroughly as collectors might prepare in advance in libraries and museum collections, they could never fully prepare for what they would find in the field. Places and animal distributions were always changing, and nothing was quite what was described in books. Survey collecting was thus an ongoing learning experience, and good communications were a crucial element of expedition practice. Survey parties in the field were enjoined and took pains never to be too far from a post office or railhead, where shipments of specimens and notebooks home could be exchanged for supplies and revised directives. With access to libraries, those at home were better able to identify specimens than were those in the field; but field parties had access to the new facts that were continually making books and maps out of date. Rapid circulation of knowledge was crucial.

For example, C. Hart Merriam knew the zoological literature well enough to tell Vernon Bailey what he could expect to find wherever he went.[59] But Bailey was always turning up new species and subspecies to correct Merriam's knowledge (thanks to his skill and their new Cyclone snap trap). The pace of change was especially quick in the heyday of survey collecting, when the empirical base of mammalian biogeography and taxonomy was being massively overhauled. Letters between Merriam and his field collectors circulated on a weekly and sometimes daily cycle. Expeditions were a flexible as well as a precise instrument, constantly being modified, adapted, and improved through use.

The dense correspondence between parties in the field and their home bases are a detailed and vivid record of this circulation of book and field knowledge. When Bailey first arrived in unfamiliar territory he depended heavily on Merriam's prior experience and encyclopedic learning. However, the balance between literary knowledge and immediate experience soon shifted to the field side. When Bailey's shipments and reports arrived at Washington, Merriam would immediately identify specimens and correct his species lists and maps. This knowledge would then be returned to Bailey in letters identifying and naming his specimens, with instructions on the species that Merriam expected Bailey to find at his next stop. Merriam once urged Bailey to reread his previous letters (like a reference book), and would occasionally return identified specimens for Bailey's education.[60] A similar system is evident in the correspondence between Joseph Grinnell and his field teams. For example, when Annie Alexander and Louise Kellogg were collecting in the Trinity Mountains, Grinnell would work up shipments when they arrived and recirculate the results to Alexander and Kellogg to guide their subsequent work.[61] It was by means of this continual circulation that knowledge was produced. Such exchanges were, in effect, taxonomic revisions on the installment plan.

Much effort went into keeping this communication system working smoothly. Itineraries were carefully designed so that field parties would stop regularly at places with postal service. When field parties were on their own and picking their own routes day by day, they were expected to telegraph their whereabouts and plans regularly to home base. Merriam would on occasion send duplicate directives to several places, enjoining Bailey to check for telegrams wherever he was near a Western Union office. Lapses in communication caused agitation on both ends of the long wire.[62] Though Bailey sometimes felt relieved to be beyond the reach of Merriam's interventions, he felt neglected when letters from Washington did not arrive. Collecting on his own in Mexico in 1889, Bailey decided that the way to get Merriam to write was to make him fret: "I haven't heard from Dr. M for a month," Bailey wrote his family. "Now I am going to punish him by not writing to him. When he doesn't hear from me for about a week I will begin to get letters every day. He seems to think I don't need any looking after." The flow of letters soon picked up.[63] Field camp and home base were the two ends of a scientific instrument that depended on the easy circulation of specimens, notebooks, letters, and directives.[64] It was good communications that made the machine work.

But the practice of collecting is always local, especially survey collecting, because plant and animal species are not evenly distributed throughout their ranges but are concentrated in favored local microhabitats. So in unfamiliar country, knowledge of books and maps was often a less useful guide to setting traps than local knowledge of the sort that comes from long residence in a particular place.

It may be useful here to distinguish generally between these two kinds of knowledge: cosmopolitan, and local or (more vividly) "residential." The one is acquired anywhere there are books and libraries, and is presumed to be true everywhere. The other is acquired by intimate experience of the sort that comes from living in a place, and may be true only for the present or only in that place. Cosmopolitan knowledge is more global and theoretical, residential more particular and immediate. Cosmopolitan knowledge, survey collectors carried with them. For residential knowledge, they would often turn to local residents.

Locating people who knew the country intimately and whose knowledge could be trusted was an essential skill of survey practice. Vernon Bailey bumped into such people with some regularity in the new homesteading areas of the northern plains. In Pierre, Dakota Territory, he was stopped from shooting in a brushy patch—which he was informed was the town park—by a man who turned out to be the mayor and a keen naturalist, and who invited Bailey home for dinner and scientific conversation. "He knows a lot," Bailey reported—his highest compliment.[65] In Fort Buford, he encountered a Scots farmer who was an avid naturalist and described a rare species so clearly (without knowing its scientific name) that Bailey knew exactly what it was, though he had failed to find it himself.[66] And in Kennedy, Nebraska, Bailey encountered an Episcopalian minister who was also a good botanist and passable birder. "He . . . has a very thorough education," Bailey reported, "and I got lots of information from him in the few days that we happened to be together." Another acquaintance was a postman who offered him a ride from Valentine, Nebraska, to his home in the sandhill country forty miles to the south:

> He proves to be a first rate fellow, well educated and interested in my work. His family are all pleasant people so I have a good place to stay for about a week. . . . Mr. King is an ex Professor of some academy or High school and introduces me to everyone as a United States Ornithologist, and I have had enough invitations to "come stay a week" to last all summer if I accepted them all.

Mrs. King was also well educated, Bailey noted: "[She] . . . speaks five languages and can quote from any author or write an eloquent essay, but I can beat her cooking." In fact she seemed unable to do any household work at all; one imagines she found her situation in that spare and lonely outback profoundly dispiriting.[67]

The best way to find knowledgeable residents was to tap the networks of amateur naturalists that extended even into the remotest corners of the inner frontiers. Merriam made particular use of the network of observers that he had set up in the 1880s for the American Ornithologists' Union's census of bird migrations. For example, when Vernon Bailey entered unfamiliar territory, Merriam would send him the names and addresses of the nearest AOU correspondents, who often proved valuable informants. In Cheyenne, Wyoming, Bailey was assisted by a Mr. Frank Bond, who knew all the local animals by their scientific names and was a pleasant hunting companion. In Tucson a Mr. Brown, the editor of a local newspaper, gave Bailey precise information about the best collecting places in the nearby mountains. In Kanab, Utah, Merriam heard from his contacts of an old man by the name of Adams who had been to the Grand Canyon and knew his way around the rough and forbidding canyon country. In Heron Lake, Minnesota, Merriam directed Bailey to look up Thomas Miller, a Scots farmer who had collected for him before and knew the area (but not scientific names, Merriam warned). Here is Bailey's account of their meeting:

> I went down to the lake yesterday and near the lake saw a man out in the garden weeding cabbages. He was a rustic looking fellow so I passed that way for a talk, and as we were talking of the weather, crops and cabbages he asked me where I was from. When I told him I was from Elk River he wanted to know if I knew a Vernon Bailey there. He was greatly surprised to see me and we were old friends at once. It was Thomas Miller, one of the corpse [i.e., Biological Survey corps]. I found him a very pleasant and intelligent scotchman and an ardent naturalist and geologist. He has promised me his assistance and I expect much help from him in my work, as he lives near the lake and is familiar with all the water birds.[68]

Joseph Grinnell made similar use of the southern California network of amateur naturalists of the Cooper Ornithological Club, of which he was a leading member. For example, Frank Stephens of Pasadena, a semiprofessional collector of birds and mammals, knew the deserts and mountains of southern California intimately from long years of hunting and collecting there. When Grinnell was planning a survey of the San Bernardino Mountains in 1906, Stephens sent him minutely detailed information about the place: the good wagon roads, campsites, and collecting grounds; what to look and listen for; resort areas for provisions; ranchers who would tolerate shooting if done respectfully; and the names of local residents who knew the country.[69] At Vacaville, California, Grinnell put Walter Taylor onto a Mr. George Sharpe, a well-to-do contractor by trade and a former member of the Cooper Ornithological Club: "He knows the country thoroughly and will direct you to a good camping place, and give you recommendations to ranch owners if need be."[70] A friend of one of Grinnell's circle of amateur mammalogists, a Mr. Ferguson, joined a collecting party in the Southern California mountains as hunter and introduced Swarth to local people. "[I]t was all that was needed to give us the freedom of the mountains!" Swarth rejoiced.[71] In the Carrizo plains of Kern County, Swarth bumped into one Ralph Miller, the general manager of a large commercial property and a member of Grinnell's naturalist network. Miller's letters of introduction to the company's local agents set Swarth's party up in a company cottage (free of charge) thirty miles from the railhead in an area that had no water or firewood and would otherwise have been impossible to work.[72]

Many people in rural or backwoods areas were devoted hunters, and their residential knowledge of local animals and plants made them potentially useful informants. Boarding with a family of orange farmers near Gainesville, Florida, Frank Chapman spotted one of the sons—an enthusiastic hunter—as a potential assistant. He was not disappointed: "Born and raised in the woods he is familiar with the ways of all the larger mammals . . . and also many of the smaller ones," Chapman reported. "[He] knows just how and where to place his traps and has the careful patient temperament so necessary for success in this kind of work. Above all he seems interested in my success, and . . . gives me abundant assistance and information."[73] How many of these practical naturalists there were in the inner frontiers we do not know, but certainly there were thousands.

People engaged in commercial harvesting of timber or wildlife were another sort who possessed intimate knowledge of their natural surroundings. Collectors for the Illinois fish survey used local fishermen as guides to the best collecting spots.[74] On Vancouver Island, Harry Swarth got useful information about the whereabouts and abundance of animals from the surveyors and timber cruisers who had crisscrossed the Island many times.[75] In Siskiyou County, California, Grinnell pumped a professional coyote hunter whom they chanced to meet for information about local animals:

"I've gotten lots of interesting information—some of it a bit shakey," he reported. "They all want to please." Having a member of the party who was a good talker and made friends easily was a great help in securing the trust and interest of local informants.[76]

Staying on good terms with residents was important for another reason. Because American land laws are so uncompromisingly biased toward the rights of private ownership, collecting parties were often illegal trespassers; so it was essential to keep landowners friendly and sympathetic.

Field parties would sometimes hire local hunters to collect larger game, though with mixed results. One man, highly recommended to Grinnell, got drunk on his first night in town and stayed drunk. Another trapped for three nights on a ridge and returned empty-handed, alleging that the animals he sought had moved to lower elevations, which Grinnell knew was a lie. The survey team had to do the work themselves—"as usual." Annie Alexander also used local trappers, though they had an unbreakable habit of whacking trapped animals on the head, crushing their skulls and ruining their value as scientific specimens (though not as pelts).[77] Collecting along the Mississippi River, Ulysses Cox persuaded commercial seiners to put fine collecting nets in their coarser ones and share the catch: the larger saleable fish went to market; the small ones were for science. Alfred Weed used a similar strategy to piggyback on commercial seining operations on the Texas coast.[78]

Fish and game wardens were another good source of residential knowledge, for obvious reasons. At Los Banos, California, Harry Swarth found the townspeople friendly and interested in their work, but unable to tell them what animals were about. So he paid a call on the resident game warden, whom he found shooting birds for another collector. He showed Swarth's team where the good birding places were, stopping on the way at the camp of a market hunter who had once collected for Rollo Beck and had specimens for sale. The warden also insisted that there were musk rats in the nearby river, but Swarth suspected they were plain Norway rats, and a deputy warden confirmed his suspicion.[79]

Survey teams could never take residential information for the gospel truth. Collecting fish in the Illinois River, Charles Kofoid found the residents of river towns friendly, but their reports of fish spawning and habitats unreliable.[80] Joseph Grinnell termed local information "hearsay"—meaning information possibly true and better than nothing, but provisional. Thus when Taylor failed to locate an expected faunal zone, Grinnell urged him to give up for the time being and "fill in the map . . . as best you can from hearsay."[81] Local knowledge was usable as a stopgap but lacked the authority of knowledge acquired by scientific collectors.[82]

This was not the snobbery of experts. Exact knowledge of species required a broad geographic perspective, which few local residents possessed. Annie Alexander marveled at how parochial local knowledge was: "People born and bred here are about as local in their habits as some animals and never venture beyond their home ranges." For example, a trapper in Weaverville knew nothing outside his locale and retailed her with "queer tales about mountain beavers" and other such wonders.[83] It was a distinctive feature of residential knowledge to be bounded geographically in a way that cosmopolitan knowledge was not. Experienced naturalists might know less about one place than another, but what they knew more or less, they knew in the same way: there was no geographical line where expertise gave way to folk knowledge.

However, survey biologists had to be tactful in contradicting local informants. For example, Merriam urged Bailey to tread carefully when dealing with the "Bald-Knobers" of the Arkansas Ozarks, recalling his own experience being shown the mummified remains of a remarkable prehistoric animal that locals had discovered in a cave deposit, and that had six pairs of legs, one growing from its jaws. Actually it was a flattened raccoon of recent vintage, but its fame had spread far and wide, and the family who owned it were exceedingly proud of their valuable scientific specimen. "I expressed my convictions a little too hastily," Merriam recalled, but "was fortunate enough to escape and am likely to repeat the tale to my grandchildren."[84]

The great advantage of dealing with members of amateur networks like the American Ornithologists' Union or Cooper Ornithological Club was that survey collectors knew something of individuals' reliability from prior experience in cooperative projects or specimen swaps. Participants in amateur networks were also more likely than most residents to have a cosmopolitan pespective, because they engaged in long-distance circulation of specimens and knowledge. Their collecting was both locally intensive and to a degree geographically extensive—not unlike professional survey collectors—though their translocal collecting tended to be secondhand, whereas survey collectors knew many places firsthand.

In time survey collectors could by repeated visits acquire a nearly residential knowledge of places.[85] But meanwhile, tapping the knowledge of locals was an efficient way to capture the advantages of residence without losing those of broad-ranging travel. This practice greatly eased the difficulties of conducting intensive survey collecting on a regional or even continental scale. We could say that surveys were an intensified form of collector networks, without the residential division of labor.

Infrastructure

Survey collectors also depended on local infrastructure for logistic support: railroads and wagon tracks, river and coastal steamers, logging and mining camps, and tourist resorts—the infrastructure of the inner frontiers. Earlier scientific travelers had been no less dependent on the infrastructure of empire and grand tours, as Alex Pang has shown in the case of eclipse expeditions, which resembled overseas package tours. Single heroic collectors had also depended on the social infrastructure of native villages and princely states, as Alfred Russel Wallace did in Malaya and Borneo.[86] However, the intensive character of survey collecting made it even more imperative that field parties stay within the networks of railroads, telegraphs, post offices, and rural towns: to receive directives and send back crates of specimens. Field parties had to know how to use this infrastructure to advantage.

A virtuoso knowledge of railroad maps and timetables was one essential skill. Railroads were usually glad to provide free or reduced-fare passes to expedition parties, believing that travelers who might attract publicity would also attract business. Field parties traveled with pockets stuffed with schedules and passes. For example, Charles C. Adams was able to get just about anywhere in the river systems of Southern Appalachia by railroad, and he found that a letter of introduction usually secured the help of station agents. To reach places not served by rail, Adams traveled by riverboat and horse-drawn wagon, but these conveyances were much less predictable than railroads, and he used them only as a last resort. Fixed rail schedules enabled him to plan in advance

Fig. 4–5. The professional collector Melbourne A. Carriker, Jr., and his son, Melbourne R. Carriker, starting down the Rio Bèni in a traditional "balsas" raft-boat, with native "balseros" in the stern, Bolivia, 1934. (The girl was saying goodbye.) In Melbourne A. Carriker, Jr., "Experiences of an Ornithologist along the Highways and Byways of Bolivia," photo no. 10, collection 900F. Courtesy Academy of Natural Sciences of Philadelphia Library and Special Collections.

exactly how much time it would take to work a given area; rivers and roads were less predictable.[87] Frederic and Edith Clements carried out an ecological survey of the high plains by working their way west on rail lines that zigzagged north and south, and going by horse and wagon where rails did not go.[88]

Good rail service was one of the attractions of working settled areas like the San Joaquin valley, Joseph Grinnell found. From a base in some local town he could dispatch assistants in different directions by rail or road and thus tap even the remotest parts of a large area in a reasonable time.[89] In Illinois, where a dense network of branch railroads covered the state like a seining net, field parties of the Natural History Survey would travel by rail to a rural town, set up headquarters in a hotel or boarding house, hire a horse and buggy, and make day collecting trips into the surrounding countryside.[90] Very efficient.

West of the Mississippi the three transcontinental railroads were in effect transects of some two thousand miles of diverse terrain and faunal zones—ideal for survey collecting. In Nebraska and the Dakotas in 1890, Vernon Bailey precisely located the biogeographical boundary between eastern and western faunas by working his way slowly westward, stopping at the small railroad towns that were interspersed at regular intervals along the route.[91] Created en masse by the railroads to attract settlers, these hamlets could as well have been designed by naturalists as collecting stations.[92] Working the Columbia River plain in 1897, Merriam was delighted to find that the region had a good network of branch lines and mail routes; he had expected to have to camp in a region where pack outfits were scarce and expensive.[93] The extensive but close-grained collecting that Merriam liked to do would have been impossible to carry out without this infrastructure.

The combination of transcontinental trunk lines and closely spaced branch lines serendipitously created just the right infrastructure for a kind of collecting that encompassed whole regions but was locally intensive. As laboratories were built to make delicate physical instruments function properly, so were survey expeditions an instrument that worked to best advantage in a semibuilt environment of transport webs. In the coastal forests of eastern South America, river steamers served collectors in the same way, though less predictably.[94]

In areas not served by railroads, survey parties used the vestiges of the stagecoach and wagon trails that preceded the railroads. In Kern County, California, an old stage road served Joseph Grinnell as a faunal transect across the Sierra Nevada. Once a major route between San Francisco and El Paso through Walker Pass, the road was abandoned but still quite usable, and pleasantly undisturbed. Grinnell's party saw just one other team in five days, and "Jack's Station," once a thriving way station, remained a small but welcome watering place, with no settled habitations for ten miles around.[95] These forgotten places of the inner frontiers were ideal collecting stations: abandoned to the local wildlife, but still accessible. Walker Pass also happened to be a major biogeographical spillway for species spreading north from Mexico into California, so Grinnell could easily trace the faunal dispersal. In this place the geography of human and faunal migrations coincided.

The remains of the infrastructure of military and commercial expansion also proved useful. Sojourning at Fort Sisseton, Dakota Territory, in 1887 Vernon Bailey was one of the last to use the network of military outposts that had been crucial for earlier naturalist explorers.[96] In Telegraph Creek, British Columbia, once a thriving way station for Yukon gold rushers, a collecting party made camp in one of the abandoned hotels (none were operating, and no cook was for hire anywhere), setting up their specimen-preparing shop in what was once a bar. In Glenora, a mining ghost town, the party made use of derelict dwellings that were, "[b]esides the said phantoms" and a horde of resident bushy-tailed rats, quite habitable.[97]

The infrastructure of mining, logging, shipping, and tourism could also be turned to a new use by collecting parties. Grinnell's faunal survey of Death Valley in 1917 made use of the Borax Company's mining outposts.[98] Railroad engineers were generally hospitable to visiting scientists in out-of-the-way places, providing shelter and transport.[99] In the dune area along Saginaw Bay, Michigan, survey parties set up their base camp in shacks belonging to a local fishing company.[100] Collecting in the Charity Islands of Lake Huron in 1910, Alexander Ruthven's field party made their headquarters in one of the many lighthouses that had protected the Great Lakes timber, wheat, and ore fleets in their nineteenth-century heyday and were then abandoned. A party surveying Isle Royale, in Lake Superior, established a base camp in an abandoned lighthouse and another on the grounds of a private sporting club—the old and the new economies of the Northern Peninsula.[101] In the cutover pineries of Wisconsin's northern lake distict, Edward Birge's lake survey teams made use of a railroad and travel infrastructure that was poised between a worked-out logging industry and an emerging tourist economy. Where there was no local hotel or transport for hire, survey teams just moved on to the next railroad stop. Their progress was more even in the southern lake area west of Milwaukee, with its resort hotels and convenient interurban trolley service to lake-port cities.[102]

For survey work in the mountains of southern California, Joseph Grinnell used the grounds of a fine resort hotel (Idyllwild Bungalow) as headquarters. "Swarth and I are now camped in luxury as compared with the six weeks' work in the desert east of here," he reported. "[B]irds are plentiful, mammals coming in steadily tho slowly, fresh provisions and P[ost] O[ffice] handy—nothing else we could wish for." Other survey parties used Yosemite Valley's tourist hotels for year-round collecting in the High Sierra.[103] Planning an expedition to the Southwest, Ruthven was advised that lodging in towns was easy and affordable, because many private homes catered to "lungers"; but harder in small settlements, where people had hardly enough for themselves. There it was better to carry provisions and camp out.[104] Collecting along the Gulf coast in Texas, a party of the Academy of Natural Sciences found local ranchers most hospitable, providing pleasant quarters in a ranch house with running water, electricity, and space to work—and all free of charge.[105] Collecting parties found South American planters similarly hospitable.[106] On the River Kāka in the deep Bolivian outback, the ornithologist and collector Melbourne A. Carriker, Jr., and his son, Melbourne R. Carriker, improvised a workshop in one of three open, thatch-roofed huts (the "chapel") belonging to an Aymara Indian man and wife who with their four children were discretely panning gold, fishing, and gardening in an out-of-the-way part of the forest. The friendly and industrious couple also served the Carrikers as guide and cook.[107]

Where residents catered to sportsmen, relations with survey parties were more commercial, even predatory. A collector for the Illinois fish survey was amazed by the prices in Rockford, a center of sport fishing: "[T]he people want the earth and if a stranger stays in the town very long they come very near to getting that stranger's share of it," he reported. (The only local man who knew the river demanded $2.50 to act as guide and refused to get wet.)[108] Harry Swarth was similarly impressed by the natives of Telegraph Creek, who had turned to tourist mining when the gold ran out. "The 'big game hunter' is one of this town's most highly prized assets," Swarth reported, "and the system of getting an income out of him is worked very thoroughly indeed." But scientific visitors may have been treated with more respect than wealthy sporting gents on a lark. The collector Edmund Heller, who had visited the area some years earlier, was remembered with affection as a celebrity.[109] Perhaps science gave status to the more practical, yet similar activities of hunting and tourism.

On the whole, though, local residents were helpful and interested in collectors' work. A party encamped in the Organ Mountains near Rio de Janiero was inundated by local Indians bringing in spiders for sale, but also presents of animals or flowers, because they were interested in what the visitors were doing and wanted

Fig. 4–6. Melbourne A. Carriker, Jr., and Melbourne R. Carriker encamped in quarters borrowed from an Indian family in the deep Bolivian forest, 1934. In Melbourne A. Carriker, Jr., "Experiences of an Ornithologist along the Highways and Byways of Bolivia," photo no. 4, collection 900F. Courtesy Academy of Natural Sciences of Philadelphia Library and Special Collections.

to help.[110] A photo of George Cherrie preparing specimens in Ecuador shows him surrounded by a small crowd of forest residents (figure 5–2). And in Baja California, Joseph Grinnell found the local inhabitants friendly and helpful, despite their evil reputation as thieves and bandits. It was their isolation that made them friendly, Grinnell surmised: "There is where you find hospitality undisturbed by the ordinary tourist. The people there have never been imposed upon, and travelers are so very few that they are welcomed."[111]

Large-scale survey would have been much harder without the infrastructure of travel and tourism. In real wilderness it was very expensive to keep an expedition in the field long enough to do intensive collecting, and parties had to move along and gather whatever they came upon along the way. Wilderness is a place for exploring, not survey. It was the accessibility of the inner frontiers, with their layers of vestigial and emerging infrastructure, that made survey collecting a practicable mode of fieldwork. In such landscapes collectors could linger and revisit—though never reside.

Knowing where to work along the gradient of humanized and natural environment was another essential field skill. Collectors had to balance the conveniences and comforts of towns against the direct experience of uninhabited nature. "Of course you will not find it profitable to put up at large hotels in large towns," Merriam enjoined Vernon Bailey. "Keep in the country all you can where you will be handy to your work." It was generally good advice. Lodging in Phoenix, Arizona, for example, Bailey found it an inconveniently long walk to his collecting sites.[112] On the other hand, camping required an outfit and a larger party (two or three plus cook and packers), and was often as expensive as staying in a hotel—and more labor intensive. Alexander Ruthven's expedition to the Southwest in 1906 made its headquarters at the Carnegie Institution's Desert Laboratory precisely to avoid camping. (Collecting reptiles required heavy tins and alcohol, which were heavy to lug around.)[113] Field parties of the Illinois Natural History Survey lived mostly on the move in the Survey's special wagon, or its river flatboat "lab," but would park or moor near towns to take advantage of hotel boarding.[114]

But most experienced collectors preferred camps to towns, especially where they were collecting series and working up specimens en masse—a business that would not appeal to town dwellers. "Your stand for camp life and economy suits me to a T," the collector Edmund Heller wrote Grinnell from the field in Louisiana. "If there is anything I detest . . . it is collecting from a hotel as a base. Give me freedom and lots of air where the odors of the catch won't annoy the olfactory organs of the uninitiated."[115] Ruthven believed that his willingness to camp in the bush in Colombia was why he got better collections than did workers from larger museums, who stayed in towns: "If they would send their men in the jungle instead of to the towns, and get them to work like they should for from fourteen to sixteen hours a day, they would get the stuff too."[116] For intensive systematic collecting, camping and total immersion were best.

Urban fringes often proved to be a good compromise between the convenience of town and intimacy with nature. Working Vancouver Island, Annie Alexander and Louise Kellogg found it saved time to camp in a clearing on the edge of town and board at a local hotel.[117] When Vernon Bailey fled Phoenix he boarded with a bachelor farmer living three miles out of town, where game was abundant and living was convenient, free-and-easy, and cheap. "It is just the place that suits me," he reported: "There is no one to complain if I make a dirt and no children to meddle with my specimens. I hunt where I please and come back when I get ready and can always find something to eat when hungry. I sleep on some straw by the wheat stack and it seems like camping out again. There is lots of game close by."[118]

It was the same at Fort Bridger, Wyoming, where Bailey boarded at a ranch five miles out of town: "The people are social, quiet, and neat, and only drink a moderate amount of whiskey. Anyway they don't get so 'galoriously' drunk as the old fellow was who brought me here. The board is good, and plain and healthy, and . . . there is a nice large creek of pure mountain water runs right by the house." Trout abounded, and the creek "attracts all the birds and animals of the region, making just the place for my work, and a pleasant place too."[119]

In the Black Hills, Bailey stayed in towns at first, but it was expensive, and the drinking and gambling was too much for his rather puritan sensibility, and he soon moved out. In the high country he boarded with an old couple at a ranch a mile from Custer, Dakota Territory. "I have . . . my office, workshop, and bedroom in the granary," he wrote his family, and enjoyed good fishing, clean water, fresh strawberries and vegetables, and abundant wildlife.[120]

Survey parties also depended on residents for transport and logistics, especially those whose occupation was moving people and equipment about in difficult terrain. Ranchers, who grazed cattle in mountain forests, packed in and out of the high country in an annual cycle, and could be hired. With a little luck: the good ones usually had their own business to attend to, and the ones that were for hire tended to be failed ranchers and unreliable. In Joseph Grinnell's experience, "the cow man who hasn't anything better to do than run a pack outfit is . . . not fit to do that properly."[121]

Dude ranchers were another group who packed people and supplies into the high country—a kind of recreational transhumance. But survey parties again had to find the good ones and win their favor, because demand was seasonal and competitive. In the Sierras one of Grinnell's field parties found themselves hanging on the decision of a Mrs. Kanawyer, who owned a dude ranch and monopolized all the local packers, who would not lift a finger for anyone without her say-so. However, she took Swarth's party under her wing, invited them to stay at her ranch, and arranged for packers to take them into the mountains, retrieve them in the fall, just ahead of early snows, and ferry them to collecting stations at lower elevations.[122]

Some farms and ranches catered to naturalists and became regular forward bases for survey parties. In California one chronicler could list half a dozen: Talley's and Warren's ranches in San Diego County, the Parish Ranch near San Bernardino, Duffield's in the Sierra foothills, and the Ricksecker Farm in Sonoma County—and these were just a sampling. Some of the proprietors of these establishments were themselves aspiring naturalists. Lucius E. Ricksecker, for example, was an amateur entomologist.[123] Similarly, a ranch operated by George B. West and his brothers in Dubois, Wyoming, catered to naturalists and writers. William Morton Wheeler and Carl Akeley were regulars there, and West hoped that Akeley would teach him taxidermy in exchange for his services as hunter and guide. West was also interested in natural history, and Wheeler sent him a copy of Darwin's *Origin of Species*, which West promised to read carefully "as I am interst [*sic*] in his works as you know."[124]

Another naturalists' resort was the Latham family homestead, Oak Lodge, situated in the palm and oak forest of a barrier island off Florida's Atlantic coast—prime collecting territory. The presiding spirit at Oak Lodge was "Ma" Latham, an intelligent and energetic woman who collected specimens for the National Museum (including a complete series of loggerhead turtle embryos she had acquired by lone daily treks to the seashore for two months in mosquito and bear season). As Frank Chapman reported on his first visit to Oak Lodge in 1889, she aspired "to make her house a resort for scientific people only and to offer them every facility for collecting." The whole family helped out. Mr. Latham helped run trap lines and piloted the sloop that both brought their guests from the mainland steamer stop and ferried them about to collecting sites along the sound (though he liked best to sit and smoke). The Latham son also hunted and trapped, and a nephew served as guide to visiting collectors.

But it was Mrs. Latham who ran the show. "What a woman!" Chapman recalled:

Fig. 4–7. Oak Lodge, in north Florida, which the Latham family turned into a "resort for naturalists." "Ma" Latham is third from left, and on her left, a visiting professor from Brown University, W. P. Jenks. Courtesy Department of Library Services, American Museum of Natural History, negative no. 119269.

> With half an opportunity she would have won a place among the foremost naturalists of her day. She had a keen, alert mind, broad interests, a consuming desire to know, fiery enthusiasm and dogged persistence, and fate had made her the wife of good-natured Charley Latham. . . . If Ma Latham was denied the privilege of self-development she did not spare herself in assisting others. No service within her power was refused to those she thought deserved it.

A combination of hosteler, companion, scientific collector, and guide, she took a lively personal interest in her guests' scientific discoveries: "To me she was a combination of mother and guide," Chapman recalled, "and when . . . my search for *Neofiber* [round-tailed muskrat] was rewarded I believe that her pleasure and excitement equaled my own. . . . I never lacked for a sharer of my joys."[125]

These glimpses of the uses to which collectors put the old and new infrastructure of the inner frontiers reminds us of how specific survey collecting was to that stage in the evolution of North American (and world) landscapes. It was only for a few brief decades, from the 1880s to the 1920s, that wild areas were both relatively unaltered by human settlement and accessible at modest cost. Native fauna could survive in landscapes not yet fragmented by sprawling suburbs and automobility. And by making use of the old infrastructure of boom-and-bust extraction, and the new one of tourist industries, survey parties could work the inner frontiers as widely and intensively as they wished. It was in such landscapes, and perhaps only there, that survey collecting could become a dominant mode of field science.

Mobility and Automobility

Survey expeditions were not only an exact but also a mobile instrument of field science. Whereas most scientific instruments must stay put to work properly, collecting parties had to keep moving. More precisely, they alternated between mobile and sessile modes, and judging when to stay and when to up-stakes and move on were crucial skills. How many stations to work and how much ground to cover in a limited season? How to balance the imperatives of geographical coverage and deep collecting? A survey party's success depended on how wisely it made these choices. Decisions to move on or stay put were not just practical but morally and emotionally fraught. Survey biologists could associate continual movement with tourists doing their Baedekers, or with amateur collectors rushing about for novelties. For example, the zoologist Henry Ward disparaged those who wrote monographs based on brief sojourns to many sites, comparing them to alpinists—*Hochtouristen*—who aimed to climb as many mountains as possible in the shortest possible time.[126] And an entomologist residing in the Philippines became vehement in his disapproval of a restless visitor:

> A German collector has just been here from Berlin—out for everything—and I showed him all our very best collecting grounds. After a few days he said that he had everything obtainable here and should have to move on! This is the way collecting has been done in these Malayan countries! After two years in this one locality I feel I have just begun to tap it. . . . I tell you it is actual residence that makes possible the real investigation of an insect fauna! I should like to spend the rest of my life out in these Malayan countries, putting in two or three years in a place.[127]

These were the extremes of collecting practice: long-term residence, and the scientific grand tour. Experienced collectors learned to balance mobility and residence.

Alexander Ruthven's practice with the Michigan Biological Survey is probably typical: field parties set up camp at a likely place and worked the surrounding region in several one- or two-day side trips. When that area seemed thoroughly combed, the party would move to a new station, often in another faunal zone, and repeat the process. Normally parties would spend one or two weeks in each place, but if a few days work showed little of interest the party would move on at once. This method of stop and start was far more productive than staying on the move and covering more territory, but more superficially.[128] The proper pacing of fieldwork was one of the first lessons that novice survey biologists received: "You have probably learned already," Wilfred Osgood wrote an apprentice on his first field trip, "that it pays to work two or three localities, perhaps only a few miles apart, in each region you are in."[129]

Though expedition itineraries were set in advance, it was understood that field parties were free to adapt schedules to what they found; but it was never easy to know just how much leeway to give, or claim. C. Hart Merriam was a meticulous planner of expeditions but knew from personal experience that the person on the spot knew best what places had to offer. So he carefully worded his official instructions to give field parties liberty to linger where collecting was good and pass over places that proved unexpectedly barren. His letters to Bailey in the field constantly enjoined him to use his own judgment. "We do not wish to hamper you by detailed instructions," he wrote. "You know the objects for which you are working, and are much better able to judge of the details of your movements than we are at this distance."[130] Field parties were expected to visit every station on their preset itineraries, but they decided how long to stay at each.

The number and spacing of collecting stations varied. For mapping life-zone boundaries many stations were best, widely dispersed but close enough for accurate mapping. Mapping species ranges required closely spaced stations, especially in zones of intergradation.[131] In complex terrain where different floras and faunas intermingle, collecting parties had to pause more often. In California's coastal ranges, with their intricate tangle of coastal and semiarid environments, Grinnell's field parties stopped to collect at five-mile intervals.[132]

For studies of some particular taxonomic problem, field parties might work just one or a few promising places very intensively rather than trying for wide coverage. Alexander Ruthven used this strategy to good effect on an expedition to South America. Making a virtue of necessity—he lacked the money for general collecting—Ruthven focused on a few kinds of animals in one area of special biogeographical interest. He was surprised and delighted on his return to find that his party's collections were judged more valuable scientifically than those made in the same area by parties from large eastern museums, which had collected more widely but less thoroughly. He had expected to have to seek help (as usual) from the curators of the National Museum, but found instead that they sought his help in working up their own collections, begging him to share his material, and practically admitting that their own fieldwork was not as good. The Michigan team came home "with such a swell opinion of ourselves that we can hardly wear our hats," Ruthven reported to Bryant Walker.[133]

For teams in the field, deciding to park or move could be a fretful business. On the one hand, they had to complete their itinerary before the first winter snows arrived, and that kept them on the move. But if they moved too fast they ran the risk of returning with incomplete collections from some places. In either case questions would be asked about their competence when they got home, and their on-site decisions could be challenged. It was safe to follow itineraries to the letter (thus shifting the blame from collectors to the planners), but that might mean fewer specimens, for which they could be blamed. And decisions to leave a place early or stay longer could to those at home look like lack of perseverance or laziness. Vernon Bailey, in the field near Ortonville, Minnesota, felt constantly torn between moving and staying. "I feel better satisfied with my work here," he wrote his family, "but have been altogether too long about it. Unless I work faster I will not get around before snow flies and I have not got into the more interesting part of the work yet."[134] It was a constant worry, and a characteristic one of survey work.

Generally, field parties tended to be too eager to move on to a new place and had to be reined in, especially volunteers who as private collectors were used to moving fast and skimming. Merriam was constantly urging Bailey not to rush: "Your letters . . .

from Kanab [Utah] gave me the bitterest disappointment I've had in many a day," he expostulated in 1889. "You must have forgotten your instructions . . . or you would not have hurried through Kanab country where we *particularly* wanted you to stay long enough to secure all of the evidence obtainable of the northward extension of the Southern fauna there." Bailey's slower progress across northern Arizona was more to Merriam's liking:

> Was much pleased with the last lot of specimens though they were not as *numerous* as expected. Now *drive slow*! You are in a mighty interesting region and I want you to appreciate the fact before you get out of it. . . . When you strike a good thing, as you did in those big eared hesperomys at Moccasin Spring, *stay there* long enough to get good series. Also look out for others, for where the local conditions are such as to produce variation in one animal they are likely to produce it in others. Keep your 'eyes peeled' for these things.[135]

It was the same a year later when Bailey was crossing Nevada toward a rendezvous at Death Valley: "I tried to impress upon you the importance of going slowly and making thorough collections from numerous points along your route," Merriam belabored his man in the field. "We have no material from this region."[136] Such injunctions are routine in letters to parties in the field. It was the alternation of moving and pausing that gave the work of natural history survey its distinctive rhythm and feel. Vernon Bailey described a day on the trail in 1890:

> Weekdays we get up before daylight and get breakfast and pack up and start on the road as early as possible, generally a few minutes after sunrise we are traveling. We eat a cold lunch and drive till we come to a good place to set traps and camp and then stop, sometimes at 2 or 3 o'clock and sometimes after dark. Henry [Vernon's younger brother] takes care of the horses and cooks up for supper and the next day and does lots of other small jobs about the camp. I generally take a bag of traps as soon as we stop and set traps till dark. Then if I catch a lot of things we have to stay half a day longer to get them put up. I put up a few specimens evenings, but it soon gets cold after sundown and we go to bed early. We get some chance to read as we go along the road, but there is always something new to see and we don't read much then. When we get down to the end of the journey and stop for a week or two in a place, there will be more time for reading and writing.[137]

This was not a typical day's work: the Baileys were moving right along to rendezvous with Merriam as scheduled and were not pausing to collect intensively. But Vernon's description gives a sense of the rhythm of expeditionary practice: days bracketed by setting and clearing traps and preparing specimens, and punctuated by visits to the nearest railroad station and post office to send out boxes of specimens and receive letters of praise or chastisement, plus instructions for the next leg of the trip.

When affordable automobiles came on the scene in the 1910s they were quickly adopted by survey collectors. In the 1910s survey parties in Wisconsin found they could hire a Ford as cheaply as a horse and buggy and do a lot more work with it.[138] (Fords moved faster than horses and did not have to be rested and fed at regular intervals.) In Death Valley in 1917, Joseph Grinnell found that a Ford saved much time "in spite of occasional shovel-work in sandy places."[139] Mammalogist Lee Dice bought a Ford in the early 1920s for his survey of North American animals and reported that it "is a great time saver in the field, and wonderfully expands one's range of action." The entomologist Alfred Kinsey reckoned that switching from railroads to autos doubled his rate of collecting—an important saving for a project that took him along 32,000 miles of back roads and across an entire continent.[140]

Automobiles gave collectors easy access to areas that were previously not practicable to reach by rail. In a three-week botanical survey of southern Florida in 1924, Roland Harper used trains and an auto (borrowed from a local friend) to reach areas that previous collectors could not reach in a day's walk from any town or railhead and thus remained prime collecting sites. With an auto Harper collected more new species in a single year than he had in the previous twenty and with much less effort than by rail and foot. In 1912 walking twenty-five miles cross-country between rail heads had taken him two days and a damp night sleeping rough in the woods.[141] The ecologist Victor Shelford estimated that autos and paved roads (it was 1930) gave field parties of the University of Illinois a reach of 100 miles from town on day trips, 160 on overnight trips, and 250 on weekend excursions. In 1937 another ecologist reported that few biologists used interurban railroads to get to field sites, and he suggested that future editions of ecological field guides drop rail for road maps.[142]

Fig. 4–8. Francis Sumner with his overland automobile, "Perodipus," collecting in the Panamint Mountains, California, April 1920. Francis B. Sumner Family Papers, folder 51. Courtesy Scripps Oceanographic Institution Library, La Jolla, California.

Autos were especially suited to a mode of collecting that was both extensive and intensive. For his botanical survey of the southwestern deserts, Daniel MacDougal found an auto ideal for work requiring "a bird's eye view of great areas of mountain slope and drainage basin, with the opportunity of making minute examination at widely separated localities." MacDougal surveyed southern California's complex faunal zones by making repeated east-west loops from the coast to the high mountain passes and back, thus acquiring "a comprehension of the vegetation not to be gained otherwise."[143] Alfred Kinsey collected gall wasps (which live in dense but scattered colonies), by driving slowly along rural roads and stopping every half mile to look for infested oaks.[144] The dense network of rural roads and the freedom from the fixed schedules and stops of rail travel made it easier to survey widely and minutely at the same time. And because they were not confined to rails, autos were good for overland travel in open and empty terrain—high plains, deserts, badlands—that was of scientific but not commercial interest, which railroads thus either went around or traversed nonstop. In the 1920s and 1930s the Clementses used autos to survey badlands and other empty places across the American West.

Not any auto would do for cross-country travel. Fords were a good choice for rough fieldwork—the only choice, Joseph Grinnell thought, because they were designed for farm use and do-it-yourself repair. For very rough country they were fitted out with heavy-duty springs and plenty of spare tires and other parts (including axles) and were lightly loaded.[145] (Joseph Grinnell's Ford, Perodipus, could carry one thousand pounds on paved roads and four hundred overland.) Collectors had to be good mechanics, and for trips to really isolated places like the Sonoran Desert or Death Valley, a backup auto was an insurance policy well worth the added expense. One could expect substantial repairs after such trips, but on the whole the autos of the day stood up remarkably well. Another of Joseph Grinnell's modified Fords, Dipodomys (named after the jumpy kangaroo rat) survived a field season with only a buckled steering rod suffered in an encounter with a nasty stretch of lava rock. Perodipus was less fortunate: it suffered a complete constitutional collapse and had to be sold off, like a broken-down horse, after a season in the Arizona desert.[146]

Autos were not just instruments of field survey but also cultural objects with associations of modernity, mobility, and adventure. Field scientists, like many others, became emotionally attached to their machines. C. Hart Merriam was passionately devoted to automobility, haunting repair shops and showrooms, attending auto shows, and taking his family on cross-continental auto trips, at a time when such trips were still newsworthy adventures. But not all were enamored. Frank Chapman congratulated Merriam on surviving his perilous continental crossing but felt no inclination to try it himself. (He had already broken his wrist cranking a car.)[147]

Frederick and Edith Clements bought their first auto in 1918—affectionately dubbing it Billy Buick—after several years of bad experiences of hiring rattletraps and terrifyingly awful drivers at local railroad stops. Frederic was all thumbs, but Edith was good at mechanical work and became (as she recalled) "an expert diagnostician of 'car-trouble,' of which there was much in the early days of motoring." She also did all the driving, because she liked it and did it well, and because Frederic was thereby free to study the ecological panorama passing by. For Edith, automobility was a sometimes scary but always thrilling experience, and a way of

Fig. 4–9. Edith Clements repairing the Clementses' field automobile, probably 1918. Frederic E. Clements Papers, box 121, folder "Edith." Courtesy American Heritage Center, Laramie, Wyoming.

getting male applause: as when she managed a mean curve and rocky hill with some fancy downshifting, and negotiated a "road" along a hogback with steep drops on both sides and impossible angles, ruts, and bumps. The "boys"—Frederick and his assistants—made picnic lunches, washed dishes, and pampered Edith, as a quid pro quo. They also let Edith deal with repairmen, as she recorded in her diary: "I have found that when I go to the garage and ask to have something done or start to do it myself, the men swarm around and do it for me. When the 'boys' go, it remains undone. So it seems a good move to have me take that over." Edith's diary of their years on the road—some 585,000 miles of road, she reckoned—is a nonstop adventure travelogue.[148]

However, when things went wrong, autos could be a particular misery, as Arthur Vestal discovered when he bought an auto (Frankie) to go collecting grasshoppers in the Southwest. It was a nightmare of rain and mud, bone-shaking roads, mechanical failures (magneto, cylinder, three windshield wipers), tows, and repairs that left him perilously short of funds. The experience produced Vestal's Rules of Auto Travel: (1) Bad roads are ten times worse than good ones. (2) Stay out of places like New Mexico in the rainy season, then get a horse or walk. (3) Start out with four times as much money as you think you'll need, not just two times. (4) Don't camp; you'll get there faster.[149]

In other ways as well, autos and roads proved versatile instruments of field research. Autos with speedometers were useful for measuring the flight speeds of birds. A closed car proved to be an excellent trap for fungus gnats (if that is what you wanted).[150] And Carl Hubbs discovered that a freshly tarred road was an excellent device for measuring the abundance of shy nocturnal desert animals: just count the corpses along a standard five-mile stretch of sticky road. It was an improvement on the usual method of driving slowly at night along a desert road and counting animals on the run or wing; though one had to arise very early in the morning to beat the ants and coyotes to the immovable feast.[151]

Conclusion

Survey collecting was an unusually complex variety of scientific work. Sandwiched historically between an age of exploration and an age of outdoor experiment, survey combined elements of both. Unlike exploration it was an exacting and semiquantitative science: intensive, organized, and methodical. It deployed residential knowledge of particular locales. But it was also mobile. Elaborate practices of communication and logistic support enabled survey parties to range widely without losing the advantages of intensive local work. Expeditions thus combined the global reach and adventure of exploration with the precision and control of laboratory science. Unlike garden or woodlot biology, which are fixed, survey science was science on the move—but not continually on the move, as explorers tended to be. Knowing when and how far to move were crucial expeditionary skills.

Survey work was special because it was both extensive and intensive (most sciences are one or the other) and conducted in places that were neither domesticated nor wild, but liminal—inner frontiers, twilight zones. Survey parties sought total inventories of faunas and floras but were obliged to work exactly and thoroughly in places more appropriate to outdoor recreation than to exact science. Survey work required physical qualities of fortitude and courage, as exploration did, but also cerebral qualities that we think of as typically scientific. Skills of close observation and analysis were as essential as those of hunting, camping, and hauling. Survcy collecting was laborious, messy, and smelly—but also intellectual and intuitive. Expedition parties had to plan to the smallest detail but had also to adapt and improvise. At a time when most sciences had become exclusively school subjects, natural history survey remained in part a craft learned through apprenticeship in the school of hard knocks.

Socially, too, collecting expeditions were an unusually complex kind of instrument. Because they required skills of very different kinds, they were also socially diverse, and participants with practical skills of packing, transport, and woodcraft were as critical as those who were scientifically trained. Survey parties interacted diversely and intimately with the varied inhabitants of inner frontiers, at a time when most scientists were insulated from the societies in which they operated, even scientists who did experiments outdoors, in experimental plots. Laboratories and experimental nature preserves also depend on social webs of support and supply, of course, but they have been institutionalized in a way that makes these connections relatively invisible—someone else's business. Expedition parties could be under no illusion of independence, because their dependence on local support and residential knowledge was palpably evident every day in the field. And whereas most other sciences divided labor, all members of a survey party might be called upon to do any kind of work. Much of modern science is done by homogeneous social groups in seeming isolation, but not survey science. Expeditions afforded participants a diverse occupational and social experience. That was one of its distinctive features, and one of its special charms.

CHAPTER FIVE

Work

We know nature through work, the environmental historian Richard White has observed.[1] And it follows that we know it as variously as we work: by foraging, farming, touring, timbering, exploring, mining, hunting, camping, perambulating, collecting, surveying, experimenting, naturalizing. The question for us is, what kind of work was natural history survey, and how was it different, say, from exploring or amateur collecting? What were its rules and customs, and how was it experienced?

The distinctive character of survey work derived from the basic fact that it was both science and recreation. As professional scientists, its practitioners were concerned with issues of skill, competence, and career, unlike casual collectors. But because fieldwork also resembled outdoor recreation, it was experienced not just as work but as a blend of work and play. It was methodical, intensive, and exacting, like laboratory science; but it also involved outdoor life and far-ranging travel—activities with quite different cultural connotations. The intellectual and the craft skills of survey collecting have quite different class associations, giving survey work a somewhat ambiguous social meaning. Career and recreation; science and craft; sessile and mobile; business and pleasure—the mixed character of survey work made it a special experience, with distinctive pleasures and pains.

I pay particular attention in this chapter to the affective side of survey work, an aspect of science just now beginning to get its due from historians.[2] Although we may regard feelings as individual, they are in fact no less socially structured by the customs of family, work, and life cycle than any aspect of our communal life. Expanding on Martin Rudwick's conception of "imaginative infrastructure," we might think of "affective infrastructures" that channel feelings much as transportation networks channel traffic—though the paths of feeling are less easily mapped.

Identity is another leitmotif.[3] Practitioners of most modern occupations acquire their social identity and standing from the character of their work, and survey collectors are no exception. However, their identity as scientists was complicated by the fact that they straddled the boundary between head and hand work, white and blue collar, craft and profession—a fundamental social boundary in a society that ranks some kinds of work more highly than others. Professional identity was also complicated by the social heterogeneity of the people who made up working field parties, and of the local residents on whom expeditions depended to sustain themselves in the field. Contrast that with the relative social homogeneity of laboratory settings (investigators and technicians, mainly, and they similarly trained). We experience identity as an individual attribute, but identity is constructed by participating in social relationships. And few sciences have a web of social relations as richly diverse as natural history survey.

Finally there is the issue of career. Work can be one of the great pleasures of life, but few work purely for the pleasure of it, or just for the livelihood. Especially in the middle-class occupations, people work to have careers, acquire greater independence, do more interesting work, and achieve upward mobility. We acquire skills as an investment in a future; and we exercise them in the expectation of acquiring a competence, a social identity, and an advancing life story. Survey naturalists are no exception to the rule, except that for them the social complexity of their work complicated the social logic of career. Early on, the practical craft of collecting, logistics, and preparing specimens was (or could be) an investment toward a scientific career. But as natural history survey became a more exacting science, academic training became the usual point of entry, not a field apprenticeship. Fieldwork per se became less an investment in a professional career than a skilled occupation: its social meaning changed. This transformation was never complete, but the trend is clear.

Practice, skill, affect, identity, career—these are the aspects of survey work that we need to address here.

Work and Skill

The work of natural history survey, like any complex occupation, was first and foremost an exercise of skills: and skills of both head and hand. As to the head, survey collectors had to know something of the complex rules of describing, naming, and classifying that distinguish one species from another, if only to know which ones were common and which were new or rare and worth collecting. This usually meant some knowledge of taxonomic handbooks, if not of the dense taxonomic literature that had been accumulating since the time of Linnaeus. Survey work also required practical craft skills that were not literary and were learned not indoors from books, but by active collecting in the field. Camping and woodcraft skills were imperative, for obvious reasons; also organizational and managerial skills, for keeping to schedules and managing the flow of information and material between field and home base. Survey collectors had to deal tactfully and firmly with different sorts of people and were expected to cope cheerfully with the discomforts and emergencies of expeditionary work.[4] Survey collectors also had to be skilled marksmen and trappers, and to be adept in turning dead animals into scientific specimens. Few occupations require such a range of skills from the arcane and bookish to the handy and practical, in a society that segregates the work of head and hand into distinct social categories.

The single most essential skill of survey collecting was knowing animals (and plants) and where to find them. That was what recommended a Mr. Farley to be a museum collector: "He is a good field man and good taxidermist. . . . He knows the birds and where to find them."[5] Finding the animals meant knowing their habitats, recognizing known and new species quickly and accurately, and—most especially—knowing how to catch them once you knew they were there. Partly these were museum and library skills: knowing what previous collectors had found, what kinds of habitats different species preferred, and their normal ranges and abundances—"cosmopolitian" knowledge, I have called it. But knowledge of "normal" range and prevalence was not always of much practical aid in finding the animals, because animals first inhabit the small particular places where food and cover are optimal, and such places change from year to year and may be hard to find. (Normal, after all, is a statistical reality, not a quality one observes.)

So finding the animals required not just a knowledge of geographical probabilities but also a local knowledge of where animals live here and now. It required visitors to have what I've termed "residential" knowledge—but without the aid of long-term residence. Local residents were often helpful informants, as we saw in the last chapter, but only up to a point. In the end, survey naturalists had themselves to have a knowledge of particular places sufficient to their task—and to acquire it on the move. Survey naturalists could not linger; nor could they just catch what was easily caught and move on. Because they aimed at a comprehenive stocktaking, they had to know, when they left a place, that they had found every species of animal living there. That required a universal knowledge of biogeography combined with a particular knowledge of this patch of woods and that farmer's hedgerow. Survey collectors could never reside, but with experience and a little help they could know places well enough to find the animals. It was science and know-how—a knack.

Vernon Bailey had the knack. As a young aspiring collector he astonished Merriam by sending him sixty specimens of a species of shrew that was seldom seen and believed to be uncommon in his area. Legend has it that Merriam had expected him to find a few, if any, and was startled when Bailey asked him how many specimens he would like to have. The trick, it turned out, was quite simple: Bailey first found out where shrews were active by setting out pieces of pork rind secured to pegs under logs, where shrews liked to run; where the bait was gnawed, there he set his traps. It was almost impossible to capture the creatures without first baiting them. Also, Bailey did not attempt to trap unless there was snow on the ground to betray where the secretive creatures lived and foraged. Wilfred Osgood used similar tricks and was famous for catching with a string of twenty-five traps what others did with a hundred.[6] He somehow knew where and how to set them.

Late in life Bailey revealed how he had acquired his skill of finding animals, as a farm boy on the prairie-forest edge in eastern Minnesota. One might expect that hunting was crucial, but it was not; hunting large game for the family larder was a serious job, one for his father and older brother. Rather, Vernon learned to find the animals by performing the humdrum chore of hunting the family cows, which grazed in the forest, swamps, and prairie openings in the area around the Baileys' frontier farm. Each day they had to be found and rounded up from the far corners of this varied terrain. It was a chore that Vernon enjoyed and always volunteered to do, and it was thus that he learned to know the animals and where to find them:

Fig. 5–1. Vernon Bailey returning to camp after a successful hunt, Squaw Creek, Idaho, 1898. Albert K. Fisher Papers, box 51. Courtesy Manuscript Division, Library of Congress, Washington, D.C.

> Hunting the cows made more than woodsmen of us, gave us some of the skill and understanding of the woods and trails and wind and sound of primative [*sic*] man and also took us into the heart of the wild life of the woods and swamps and prairies. We became familiar with every family of partridges and prairie chickens, with every duck pond and muskrat house for miles around, learned to read the tracks along the roads and trails, to know the plants and bushes and berry patches, the flowers and trees and where the hazlenuts grew biggest, and to locate the butternut trees that were going to yield a crop of rich nuts later on. . . . Much of my most intimate knowledge of the surrounding country and its plant and animal life came through hunting the cows, also some good woodcraft training where no [cow] bells could be heard[,] and only by following the freshest tracks or thinking why and where they could have disappeared, were they to be found.[7]

Bailey's skill as a scientific collector thus began with an intense practical interest in animals and an intimate residential knowledge of a particular place that gradually became more cosmopolitan and scientific.

The wildlife biologist Paul Errington acquired his skill in a similar way, as did Herbert Stoddard. Stoddard's self-confidence as a scientist derived, he felt, from his practical experience as a professional trapper and collector:

> I spent several years living outdoors and observing things[,] including turning the land in farming, trapping hundreds of small mammals for the Milwaukee Museum, and commercial trapping of fur bearers in the Prairie du Sac section, and even more intensive prying into our environment here [in a Georgia game preserve]. . . . Hence I had absolute faith in what I had said.[8]

Knowing the animals was an empathic ability to think or even act like the animals themselves. This was the point of an anecdote told by William Morton Wheeler, who was once asked by a herpetologist friend if entomologists studied the seasons of insects. He wanted to know when frogs were active and reasoned that it would be when insects were abundant. Wheeler replied that no scientist had studied the question, but that "collectors of insects . . . of course know very accurately the limits of the insect season since they have the same interest in the insects as have the frogs."[9] Fishing for trout in the alpine meadows of Mt. Lassen, Joseph Grinnell also illustrated this principle. Avoiding open streams, he trolled tiny, half-hidden rivulets with a simple hook and worm, which he let drift downstream around bends, so that the fish could not see him or his shadow. Trout had adapted to the hunting techniques of bears, racoons, and herons, Grinnell reasoned, and knowing that, he could elude their defenses. Thus reasoning, Grinnell did a great deal better than the gents with waders and fly rods fishing open streams looking to any fish like herons—and to Grinnell like illustrations from sporting magazines.[10] Hunters as herons; collectors as frogs.

As a more famous naturalist observed a century earlier: "There is a delicate empiricism which so intimately involves itself with the object that it becomes true theory."[11] In this lapidary remark Johann Wolfgang von Goethe captures the essence of all the sciences that depend on sustained, close observation in particular places.

Active collecting was indispensable to a survey naturalist's education, many believed—Joseph Grinnell, for one. "The process of hunting, and personal preparation of bird skins bring a knowledge of the characters of birds . . . which can be secured in no other way," he declared. "The making of natural-history collections is useful as a developmental factor, even if dropped after a few of the earlier years in a man's career." Because it combined intense physical and mental exertion devoted to an intensely desired end, collecting drew men to fieldwork in a way that book learning never could. Collecting was too valuable to science to be allowed to become a mere recreation "indulged in only in superficial way by amateurs or dilettantes."[12] C. Hart Merriam agreed and reacted vehemently when his friend Witmer Stone tried to argue that ornithology was better pursued by observing living birds than by collecting them and turning them into specimens:

> I do not believe that a naturalist was ever made without the actual field experience of collecting specimens. The idea that a man can become a naturalist by studying specimens in the museum and books in the library is to me altogether preposterous. Without the enthusiasm generated by actual field collecting and the exhilaration and thrills incident to the capture of a rare specimen, how can one even hope to become more than the mere vacant shadow of an amateur?[13]

In fact, Grinnell found that students who came to survey work from academic studies tended eventually to leave it for more secure and prestigious careers in laboratory biology. That is why he preferred to invest in "the seasoned, tried, experienced type of man."[14] Early and intimate experience with animals in nature was the best assurance that aspiring survey naturalists would persist in the work, in a world that regarded it as second-rate science. Hunting and collecting not only developed requisite skills, but created a structure of feeling that was as essential to good work and as powerful in shaping life choices as the intellectual structure of classroom learning. For those who came to science from the field, the pleasures and excitement of fieldwork was what made knowledge worth having and pursuing.

Knowing the animals and where to find them were necessary but not sufficient qualifications for survey collecting: collectors also had to work expeditiously. Mass collecting was not a leisurely pursuit but a constant race against time and the passing seasons. Collecting parties were under constant pressure to collect quickly and efficiently, complete itineraries, and fill quotas. Their daily routine was an unrelenting cycle of trapping, preparing specimens, packing, and shipping. Vernon Bailey liked to hunt and clear his traps in the morning, skin and prepare specimens in the afternoon, and then go out hunting again after supper.[15] Others spent the whole day shooting and evenings preparing specimens—and sometimes much of the night as well, if the day's work had been successful. Grace Thompson Seton described a party encamped in the coastal mountain forest of Brazil:

> [George] Cherrie collected ten new birds today and made up 11. Yesterday was even heavier. He made up 23 skins in a stiff afternoon's work, while [Colin] Sanborn toiled over his 30 odd mice and bat skins. He averaged 10 minutes to a skin. Thus it took him till ten at night before the last one was labeled and pinned into place on specially designed trays in a collecting trunk. The eaves of our wattled dining room . . . is [*sic*] beginning to be festooned with strings of skulls, shining white with borax, and with mammal skins prepared and waiting to be made up.[16]

Collecting at a leisurely pace for oneself was one thing—survey collecting, quite another.

If dexterity and speed in preparing specimens were essential to survey work, it was no less essential to work exactly and never sacrifice quality for speed. This was because only specimens prepared by exact and uniform methods had scientific value. Vernon Bailey described the standard training of recruits to the U.S. Biological Survey: "With beginners we always insist on careful work at first until the quality of work is satisfactory, and later on practice for rapid work in skinning and making up specimens."[17] It was speed and precision that distinguished professionals from ordinary collectors, and that gave the experience of survey work the feel of an intense, exacting science.

Good survey collectors also had to be able to extend their knowledge of familiar places to those in which they had never set foot. Finding the animals in a place one knew intimately was a relatively easy, residential skill; but survey expeditions were a ceaseless round of novelty, taking collectors to new places every few days or weeks

Fig. 5–2. George Cherrie preparing bird specimens near Pallatanga, Ecuador, with native onlookers. In Frank M. Chapman, "The Distribution of Bird-Life in Ecuador," *Bulletin of the American Museum of Natural History* 55 (1926), plate II at p. 6.

that had to be learned quickly to keep the flow of specimens coming in. Collectors could linger or return, but never reside.

For example, on his first trips for Merriam, Vernon Bailey was often distressed at his ignorance of unfamiliar animals and depended on Merriam to tell him what he could expect to find. "If I had known where to look for *Onychomys* I would have found them at Sisseton," he wrote Merriam after a rather unproductive stay, "but with the mammals that I have not been acquainted with I can not tell where to look for them, nor how to trap them, so that any information which you can give me in regard to habits will help me to find them." However, he proved a quick study, and having identified unfamiliar species from Merriam's descriptions and observed their habits, he began to catch them almost at once.[18] The experience of expedition collecting was like that: knowing how to find and catch one kind of creature, collectors learned to find and catch others. Knowing how to work places they knew well, they learned to work places they did not know at all.

Not all good field naturalists made good survey collectors. Some were just too slow or ham-handed in making specimens; others lacked the desire or discipline to keep up the relentless pace of survey work. That was the case with William Lloyd, an experienced naturalist and member of the American Ornithologists' Union, who lived with his Mexican wife and family in Texas and had collected for the British Museum in Mexico (where he had also spent a year in jail for killing a man). Directed by Merriam to recruit him to the survey, Bailey tried to teach him to trap and make good skins, but concluded that it was hopeless. "He is slow and not skilful [*sic*]," Bailey reported, "though a thorough naturalist."[19] Bailey tried hard to recruit his younger brother Henry to a career as collector, taking him on as cook and teaching him to prepare specimens and act as his assistant. Though Henry had often helped Vernon find the cows, he proved too slow and clumsy with his hands to be a successful survey worker. (He also preferred the settled life of a farmer to one constantly in the wild and on the move.)[20]

Nor did series collecting necessarily appeal to naturalists accustomed to collecting for themselves: it was rather too much like plain work, or even commercial hunting—so Frank Chapman decided when he began to collect on a large scale for the American Museum in the late 1880s. "[U]nless you have tried it," he complained to a colleague, "you cannot imagine what a difference there is between collecting for a museum or yourself. There is just exactly the difference between market shooting or shooting for pleasure. Notes on habits has [*sic*] no value at all in dollars and cents."[21] And to the same friend a few months later:

> This miserable collecting. It is the curse of all higher feeling, it lowers a true love of nature through a desire for gain. I don't mean a specimen here and there, but this shooting right and left, this boasting of how many skins have been made in a day or a season. We are becoming pot-hunters. We proclaim how little we know of the habits of birds and then kill them at sight. . . . The American Museum desires specimens, not notes.[22]

The case is perhaps atypical: the American Museum was unusually aggressive in pursuing large collections, and Chapman unusually interested in animal behavior. But his experience suggests that the transition from recreational to survey collecting was not for everyone an unalloyed gain. The aesthetics of recreational collecting could make survey collecting seem by comparison crassly business-like. Of course Chapman did not give up series collecting—at least not until much later in his career: modern taxonomy and biogeography could be pursued in no other way.

Pleasures

The pleasures and pains of different kinds of work are as distinctive as their skills, and indeed often derive from them. The special pleasures and pains of survey collecting reflected its dual character as outdoor recreation and exacting labor. The physical pleasures and pains of survey especially resembled those of camping and hunting—no surprise, since the activities are so similar. But its psychic pleasures and anxieties derived more from the successful or unsuccessful exercise of skills.

The pleasures first. The experience of survey collecting was generally pleasurable, and often afforded the intense pleasure that arises from vigorous physical activity directed toward some ardently desired goal. It is not a mysterious sensation; field sports are arranged to induce that psychosomatic state. For some, collecting was almost an addiction, recollected with pleasure and keenly anticipated as the coming of spring heralded the beginning of the field season. Here is Ulysses O. Cox, a young field worker, writing in May 1901 to the head of the Minnesota Natural History Survey: "This hot weather, fresh vegetation, migrating birds and spring fishing arouse a person's camping instincts, so I am wondering what plans are being developed for survey work." It was part of an annual cycle of feeling, as characteristic as the cycle of planning and preparing expeditions: a bit of affective infrastructure. "It was just a year ago today," Cox wrote again, "that we set sail on the Megalops [the survey's river flatboat] on that delightful and successful voyage down the Minnesota and Mississippi. I have been thinking of it frequently during the past week and have been wondering what you have planned for the coming summer. I am anxious to get out in the field again."[23] It is a feeling with a venerable history: come April showers, folk long for pilgrimages, especially collector folk.[24] "I just *long* for the mountains," Joseph Grinnell wrote an academic colleague in June 1907, "and just as the school routine ends, I will be there."[25]

Fig. 5–3. Field party of the Minnesota Natural History Survey on the Minnesota River at Sebastapol, Minnesota, 1902. Henry Nachtrieb and his wife are on the left, with the survey flatboat *Megalops* in the background. UM, photo files, folder "zoology department." Courtesy University of Minnesota Archives, Minneapolis, Minnesota.

The pleasures of fieldwork were often linked to physical and mental health, just as feelings of physical and mental exhaustion were linked to sedentary indoor work like teaching or curating. Collecting was commonly recommended as a cure for nervous exhaustion, as outdoor recreation was for the mental exhaustion of commercial and professional work. It was the familiar trope of vacation culture. Wilfred Osgood, a colleague recalled, was two different people in winter and summer: as the curator of mammals at the Field Museum, he was preoccupied and plagued by chronic stomach trouble; in the field he was healthy, happy, and carefree. When Walter P. Taylor, a student of Grinnell's, was forced by insomnia and mental exhaustion to drop out of school for a time, he looked for a cure to a season or two in the field with collecting expeditions. Hike for the woods and stay there for a while, his doctor prescribed. It was about the only work Taylor could tolerate in his condition; even routine museum curating was unbearable.[26]

Grinnell and Harry Swarth, who each took turns minding the shop back home when the other was afield, also took turns feeling neuresthenic and envying the other for having "nothing to do but collect specimens." It got worse for Swarth when he became the assistant director of the Los Angeles Museum, where endless curatorial routine and politics with no fieldwork made him physically ill. "Seriously," he wrote his mentor, "I do believe that my physical well being is somewhat affected by continuing sedentary work. . . . I have headaches a week at a time, so that evenings I am good for nothing." By the end of his stint he was a nervous wreck: "I want to get out in the desert and collect for awhile. It's what I need."[27] Grinnell's schedule of fieldwork also followed this cycle of malaise and restoration. If he felt "in trim" he would stay and do necessary office work: "But I might reach a stage of fatigue where at least a six weeks (or more) period of strenuous field work would be best, both for me and the Museum." The ideal was to alternate field and museum work: their different perspectives and demands kept him intellectually alert and active, he felt.[28] As a steady diet, either collecting or curating would grow stale, but in seasonal alternation they provided both intellectual stimulation and physical well-being. It was the characteristic emotional rhythm of expeditionary work.

The symptoms described were physiologically real, to be sure: no one who has spent long days at a desk will doubt that. But the formulaic quality of such testimonials suggests that these feelings of discomfort and relief were also cut to a tacit cultural template. It was not just survey naturalists who expected that professional work caused neuresthenic symptoms, and that outdoor recreation would bring relief. These were stock beliefs of the vacation complex, as we have seen; they were elements of the affective infrastructure of middle-class work and life. Symptoms of stress and overwork were manifest signs of virtue in the class that defined itself by its work, and restoring one's health and capacity for work by going fishing or walking in the country was a moral act as well. It would be strange if the feelings generated by collecting and fieldwork did not flow in the well-worn channels of middle-class culture. After all, most survey naturalists came from that social stratum. When middle-class vacationers engaged in recreational hunting or bird-watching they worked at play, as Cindy Aron put it. For survey naturalists these same pursuits were recreative work.

So many of the skills and routines of survey work—hunting, foraging, camping, observing, collecting—were physically the same as outdoor recreations that it would be surprising if they were not experienced as a kind of recreation. Reports from collecting parties in the field describe the work in terms that recreational campers would use: the pleasure of finding good campsites, supplies of clean and tasty water, abundant fish and game, and terrain that was pleasant and easily negotiated.[29] Francis Sumner recalled how his first field collecting with Henry Nachtrieb's river survey "combined the sport of the angler with the scientific interest of the zoologist." Less agreeably, seining spoiled him forever for sport fishing with hook and line, with its relatively—and deliberately—meager rewards.[30] Work thus became more pleasurable than play.

The deepest and most sustaining pleasures of fieldwork were those that attended the successful excercise of scientific skills. Survey naturalists' letters from the field exude the intense, euphoric pleasure of working in good country where traps are full every morning and crates of specimens are filling up. Joseph Grinnell thrilled to Annie Alexander's reports of fine weather and good hunting: "There is nothing more pleasurable than field-collecting under favorable conditions," he rhapsodized, "at least to those of us with a love for such a pursuit."[31] From Yuma, Arizona, Vernon Bailey sent his family a vivid vignette of collecting at its best:

> My drying box is full of skins, there are a lot of boxes full and a level place in the shade about 2 yards square is covered with skins drying. This is panning out the best of any place I have found for a long time. . . . [T]here is the satisfaction in seeing the result of my work lying all around. The neighbors don't trouble me at all. Haven't seen anyone for a week except a few Indians and the day I was in town. Here comes an Indian. I hear him singing.[32]

Perhaps the greatest pleasure of all was the thrill of finally finding an animal that one knew was around but that had eluded one's traps. As Vernon Bailey reported from his first trip to eastern Dakota Territory:

> While out to my traps tonight after supper, I caught a not full grown *Cricetodipus flavus* which is the first mammal really new to me that I have found. . . . Did you ever get a specimen that you had been hunting for for weeks and never had seen? I think I never felt more pleasure in

Fig. 5–4. Collector with skins drying in preparation for packing and shipping, Escuinapa, Mexico, January 1904. Courtesy Department of Library Services, American Museum of Natural History, negative no. 23493.

> making the acquaintance of a species. I have it laying [*sic*] on the paper as I write and spend more time with it than with my pen.[33]

The little creature had rashly dashed across the road in front of him, but Bailey was sure he would find more once he got to know their habits. The little corpse on his letter paper embodied all the intense anxieties and pleasures of exercising skills. Finding what Merriam expected him to find restored his self-confidence, and the next day he found two species new to him.

And here is Frank Chapman's exaltation when after weeks of empty traps, he finally trapped a shy water rat he knew must be there, from seeing its nests and trails:

> Hurrah, Ring the bells and sound the horn; today is a day of general rejoicing at Oak Lodge, with myself as chief rejoicer, for at last I have succeeded in securing the long sought for specimen of *Neofiber alleni* . . . the fourth known example. . . . [A]s usual I went the rounds with the same discouraging results, until I came to a platform trap set five

> days since and left undisturbed, as I approached I saw something black through the grass, I feared *Sigmodon* [wood rat] but hoped *Neofiber* and when I found my hopes realized there was not a millionare in the world who could have bought out my stock of joy.

Even greater was his joy on another occasion, when what seemed to be an odd specimen of wood rat turned out to be a species new to science: "Even *Neofiber* was eclipsed," Chapman exulted, "and my general elation causes them [the Lathams] to remark . . . that they will have to put a brick on my head to hold me down."[34]

Walter Taylor had the same experience when he finally found the exact boundary between two subspecies of ground squirrel, just where Grinnell had predicted it would be. His pleasure in the achievement fired his determination to work out its complete biogeography in the brief time that remained to him in that place. His success was especially welcome as it followed a depressingly meager spell in Owens Valley.[35] Periods of abundance and dearth tended to alternate in survey work, and good ones kept collectors happy through the dry spells. "I wish you could look into our camp," Grinnell wrote Annie Alexander from Mecca, California. "It is surely a lively place." Forays across the desert to check out rumors of bat caves, animals captured that they had not seen in life before, large series of specimens, and reports of desert sheep nearby—the pleasure of finding the animals was addictive.[36] As Francis Sumner observed, collecting mice in the California deserts: "This business has much of the excitement of gambling."[37]

Pains

The distinctive pains and disappointments of survey collecting likewise derived from its dual character as recreation and as exacting, skilled labor. There were the physical hardships of camping out in bad weather and poor conditions familiar to all outdoorsmen. But worse were the gnawing doubts about one's skill and competence when collecting was bad: these were peculiar to survey work, with its itineraries and quotas.

Aspiring survey collectors often assumed that the work was like recreational collecting, but in fact it was strenuous, unrelenting labor often in trying circumstances. As Francis Sumner observed, the daily grind of collecting and preparing was anything but fun.[38] In 1922 one hundred people applied for jobs with Field Museum expeditions, thinking it a "a very delightful occupation, giving opportunity for travel in a most charming and comfortable way." But only one of that hundred was up to its rigors.[39]

Grace Thompson Seton described one day's work by the herpetologist Karl Schmidt in the Honduran rain forest. He was up at four in the morning, climbed 4,500 vertical feet in four hours (a steep, hard climb short on water), made a breakfast of coffee and bacon at the top, chopped down trees for eight hours straight to get salamanders, then down again skipping and sliding down the grade for two hours, and back in camp fourteen hours after setting out. It was "a fair sample of what collecting means in the way of hard work." What person passing idly among museum dioramas could imagine the hard labor that went into making them, Seton reflected.[40] Ulysses Cox likewise warned his field assistants not to expect a "soft snap" on an expedition to Minnesota's North Woods: it would require portaging boats and heavy equipment, including a cook stove essential for drying tents and specimens, and eating what they could pick, dig, or shoot.[41] In the deserts of southern California Harry Swarth endured extreme heat, dust, scorpions, and swarms of flies. But he was philosophical: "[L]ike a singed cat," the place was "not as bad as it looks." On another trip, in the nearly impenetrable wet wilderness of Vancouver Island, he struggled for three miserable days on short rations just to get a few birds.[42]

Malcolm Anderson, in South America in flood season, was prevented from getting around in a boat by near-solid flotsam; he had to collect by swimming and crawling over floating islands of vegetation infested with biting ants, until his body was a mass of bites and scars.[43] In the Andes, where rainfall was frequent but unpredictable, Melbourne Carriker had to rush out between showers and collect furiously, then spend most of the night preparing specimens, in a cloud of biting insects.[44] Frank Chapman, collecting insects in Florida's coastal marshes, was less predator than prey: "[H]ere the bugs turn collectors and I have been collected by everything," he quipped. "Each day I gather a fresh crop of fleas, ticks, lice, redbugs, etc., which . . . map out tracks across my poor body, but I am getting used to them."[45]

Grace Seton described a party returning by steamer from the interior of Paraguay and stuck fast in flotsam: herself and Evelyn Field covered with bites of mosquitoes, fleas, spiders, and ants; Field with bronchitis; Walter Taylor with hives; Colin Sanborn with cut fingers; Karl Schmidt with hand boils. "Only Mr. Cherrie [George Cherrie, the chief collector] smiles serenely . . . as he stands in [the] bow and surveys the mass of 'Floating Island' in front of our stationary launch."[46] Fortitude was a virtue as valued by survey collectors as by hunters and outdoorsmen, and it was displayed—as exemplified by Cherrie—as serenity in trying circumstances.

Just the routine hazards of moving cross-country with gear and boxes of specimens could be trying. On the coasts of Vancouver Island a series of minor mishaps—boxes lost overboard, crews refusing to land the party—made Harry Swarth wonder superstitiously what worse calamities were waiting ahead.[47] Floating down the Minnesota River on the boat-lab *Megalops* could be an idyll, but spells of high or low water could make it a misery of dodging uprooted, floating trees or fighting free of sandbars.[48] Surveying the fish fauna of Illinois's streams and rivers with horse and wagon, everything went wrong for Thomas Large that could go wrong, when a period of heavy rain and floods set in, turning streams to torrents and dirt roads to mud. His horse got a bad foot, then a sore neck and a burned heel. He lost his camp stove on a bumpy road and had to retrace his steps, only to find it stolen. He cut his ankle on barbed wire; his helper got too sick to work; and Large was too broke from paying for room and board to hire a replacement helper (it was too wet to camp and cook for himself). The wagon suffered a sprung axle and a broken yoke, and the harness badly needed repair. Bridges were washed out, and streams were too high for collecting; and Large got down to his last thirty-five cents when a scheming postmaster neglected to forward a letter of credit in the hope of getting a paying customer for his hotel. Leaky boots, shortages of preservative, bad hotels and worse food, and cheating horse traders did not complete the list of Large's miseries.[49]

In some places expedition parties faced more than mere discomforts. A member of a Field Museum team was murdered in the Philippines, another died of fever, and several suffered lasting damage to their health.[50] One member of a University of Michigan field party returned on an improvised litter strapped to a mule, and

Ruthven himself came home from a trip to the West Indies with a dose of malaria.[51] The human inhabitants of the inner frontiers could also prove dangerous: like the men in a car who followed Frederic and Edith Clements and, when asked directions, led them away from their destination and off the road into the desert, at which point the Clementses made a quick U-turn and fled.[52] Or the hobos who took to roads and trails when kept off railroads, whining and begging when sober and threatening when drunk. Grinnell feared them and kept his rifle close when checking his trap lines.[53]

Then there was the man who accosted Frank Daggett while he was collecting along the Illinois Drainage Canal, flashed a special police badge, and told him he was shooting illegally in the village of Lyons and would have to appear in court and pay a fine. Back in Chicago, Daggett discovered that Lyons was a "whisky town" of saloons and dance halls controlled by a local gang that used the court to rake off fines, with the connivance of the man that Daggett had encountered. The man, named Fonter, ran an illegal saloon, or "blind pig"; drugged and robbed his customers; and generally terrorized the countryside. Officials had been trying to nail him for years, and with Daggett's testimony they finally succeeded, making Daggett a hero to the area's rough saloonkeepers and petty politicians. "We can get a free drink in any saloon along the Des Plains River Diversion," Daggett wrote Grinnell, "as 'the two fellows that put Fonter up for thirty days.' "[54] Fortunately, such adventures were an exceptional experience for survey parties.

Of course, a certain amount of hardship was not an unwelcome feature of the collecting experience—"just enough to emphasize the element of adventure," as Grinnell put it.[55] Difficulties endured and overcome made good stories and displayed collectors' fortitude and skill in coping with adversity. It was the same with survey collecting as with recreational camping and hunting—trials and adventure were essential to good trail stories. Of course, too much adventure could be read as a sign of bad planning and incompetence. Survey expeditions were science first and recreation second, and they were supposed to be professionally and efficiently managed.

Far worse than bad weather or black flies for survey collectors was the fear that one's skill was inadequate to the task. The pain that gnawed most sharply was not knowing whether empty traps were a sign of their own lack of skill and perseverence, or simply that the animals were not there. Vernon Bailey felt personally responsible when he failed to capture animals that local residents said were present. "Don't envy me," he wrote Merriam on one such occasion. "It has been anything but enjoyment for me here: like riding a dead horse, no fun and don't get there. If I was accomplishing anything I could enjoy it thoroughly, but hope to have better success somewhere else so I can think it was not all my fault that had none here." Leaving the place empty-handed made him "feel much like a deserter, but why stay?"[56] Or as Annie Alexander reflected at a collecting ebb: "The gun feels heavy when birds are scarce."[57]

At another stop Bailey apologized for not getting the pocket gophers that Merriam had directed him to find, but "they have been unusually hard to catch or I have lost my ability to catch them as usual. Night after night they sprung my traps without being caught, in spite of all I could do." Years later, when Bailey was the one telling field parties what they should be finding, one of his young assistants became so disheartened by his failure to catch lots of animals that he offered to resign from the survey. In fact, it was clear to the experienced men back in Washington that he had just struck a poor place for collecting and needed to be reassigned to a better one to regain his self-confidence.[58]

Field parties were bombarded by detailed directives from curators who knew precisely what species they wanted and expected to be found. Here is Grinnell spurring on Harry Swarth:

> Watch out for *Citullus mollis stephensi* . . . at about the 5500-foot level. It ought to be there. . . . Get *enormous* series of all the chipmunks (*Eutamias*) from the bottom to the top of the mountains. I figure that the following *may* occur. . . . Watch out for *Citullus beldingi* in Canadian [zone]; we failed to get it last year. It is *very important* to get *gophers* in series at near intervals all the way up. Also we *need Ochotona*, *Teonoma*, and *shrews*. Supply yourselves with *good grub*, camp comfortably, keep everyone in good spirits, ambitious to establish and maintain an "average," and *work hard*.[59]

Unfortunately for collectors, places changed, and animals that were abundant one year might well be scarce or absent the next. Established knowledge and on-the-spot experience thus became competing sources of authority, and field-workers were never sure just how far to depart from home-based directives.

It was the uncertainty of who was to blame when collecting was poor—the planners, the collectors, or just nature—that made psychological pain more acute than mere physical discomfort. Managing failure is a problem with any kind of skilled work, but the moral economy of fault was especially complex in survey work, because survey was methodical and exacting work performed in unpredictable and uncontrollable conditions. Expectations were high, and so were the chances that something would go wrong. (Risk and expectation were more congruent in the controlled environment of labs or experimental gardens: one reason, perhaps, for the higher standing of experimental science.)

Recreational hunters and collectors also faced this conflict, but it was worse for survey naturalists because failure to find the animals reflected on their professional competence and self-worth. Sport hunters and amateur collectors, when they failed, could appeal to bad luck, which falls on the skilled and unskilled alike and cannot be remedied, and so is blameless. Survey naturalists could not appeal to luck so easily, because they were expected to be reliably productive whatever the circumstances: that was what made them professionals. In the moral economy of survey work, luck afforded cold comfort.

Topotype collecting was often the most anxious work of all, because topotypes must be taken from the exact area where the original type specimen had been collected. Joseph Grinnell would tell his teams exactly where he wanted them to collect, and it had to be just there. When Walter Taylor wrote from the field that he would collect at another site unless ordered not to, Grinnell replied that Taylor could move on, but that the onus was on him to succeed: "Any way your judgement leads you will be all right, only *get the topotypes*."[60] That is, any failure would be his fault, whether he stayed or gave up and moved on. Collecting to exact specifications was nerve-racking. "This going to a place after topotypes is not always what it's cracked up to be," Ned Hollister wrote his boss at the Field Museum. "I remember my struggles and sleepless nights at Strilacoom until I at last located the *Peromyscus*. I am having the same time here."[61]

Sleepless nights? They were supposed to be an affliction of sedentary office slaves and those who stayed behind to worry about museum politics and money. But in survey collecting, collectors' skill, competence, and reputation were always on the line. And the freedom to depart from itineraries also meant that the blame for failure would rest squarely on the men in the field. This moral tension was no doubt why survey parties worked so hard and endured such difficult conditions to find the animals.

One last little story. In late May 1912 a party led by Harry Swarth arrived in a driving blizzard at the first station of a transect of Kearsarge Pass, in the Sierra Nevada. Just above their camp, drifts were still forty feet deep, and local stockmen warned Swarth that they would not attempt the pass until late June, if then. Swarth wrote Grinnell, proposing that they revise their itinerary, collect somewhere more promising, and return to Kearsarge in August when the birds were sure to be there. He did not want to be a quitter, but the collecting prospects were not good. Swarth's doubts earned him a swift and sharp rebuke: "Fieldwork is in our line of science, and at this stage of the Museum's growth, of preeminent importance," Grinnell lectured his man in the field. "So much so is it, that I think we should take no account of personal preference or comfort (within reason) in discharging this part of our duties. Nothing short of ill-health or accident should terminate an approved itinerary." If Swarth did decide not to stay, Grinnell directed him to return straightaway to Berkeley, so that Grinnell could take his place at Kearsarge—he "would jump at the chance." Miss Alexander supported the museum's fieldwork because she was confident that collectors would do their utmost, and Swarth's backing off a tough task would set a bad precedent. So stay put, Grinnell advised, buy a camp stove and keep the tent warm and dry while skinning; eat well, and "cheer up, Harry." Conditions were bound to improve, and even the worst trip would eventually end. Disagreeable conditions "are incidents apparently to be expected . . . in field work. And fieldwork is the essence of the profession which you and I have elected to follow."[62]

Swarth felt maligned—he was a most able and conscientious field man—but he stayed put, collecting what few birds there were and trying to convince Grinnell that the small bag was nature's fault, not his own.[63] What pained was not the discomfort of spring blizzards, but the doubt cast on his skill and competence. The physical discomforts were common to any outdoor pursuit; the psychic assaults were peculiar to survey collecting.

Fig. 5–5. Harry Swarth, Joseph Grinnell's right-hand man at the Museum of Vertebrate Zoology, 1922. Courtesy Bancroft Library, Berkeley, California, Banc Pic 1973.044-pic, negative no. 033524.

CAREERS

Another difference between ordinary and survey collecting was that survey work was an investment in a career. It was work performed not (just) for pleasure and personal satisfaction, like amateur collecting—or to earn a livelihood, like commercial collecting. Survey collectors exercised their skills in the expectation of upward mobility into careers as curators or professors. Survey work was an investment in a professional future, and collecting was a stage in an occupational life cycle. Skilled collecting earned credibility that could be invested in better jobs and improved working conditions.[64] However, the rules of getting credit and making careers by collecting were changing, as professors and curators made fieldwork a part of their own professional roles.

Initially, natural history surveys seemed to give field-trained naturalists new opportunities as scientists. But this hope proved false, as formal education became the preferred entry into scientific careers. Although a practical apprenticeship remained a necessary part of the training of field workers, it ceased to be a sufficient one.

Why that happened is a complicated story, but the social logic is simple. In a society that gives access to middle-class occupations mainly through formal education, occupations that are carried on without visible training will be less valued, simply because it seems that anyone can do them. Accessible to anyone, they will seem to require no special skill and thus to have no special economic—and hence social—value. Joseph Grinnell saw how this logic made natural history seem a second-rate science:

> There are a lot of travellers (loafers), geologists, etc., around the country, who think it quite a scientific thing to "dash off" a local list every now and then, with the result that our records are lumbered up with lots of rubbish. . . . [Consequently] [t]here is a general notion that to be an ornithologist (with a capital O) requires no brains—and it don't seem to in very many cases. Biologists humiliate me by telling me (and I am unable to defend) that our literature is chiefly the result of the lowest dilettantism; that no training at all is required to bring one into prominence as a "contributor" to the science; that ornithology is now no science at all, but an amateurish recreation.[65]

In reality, some uncredentialed naturalists made greater contributions to science than many academic biologists ever did, but the association of collecting with casual naturalizing and recreation made it easy to disparage field apprenticeship. As academic degrees became more the norm, individual field experience ceased to be a sufficient basis for a museum or academic career. Skill in collecting and fieldwork was valued as a preliminary to academic training, or as an adjunct to it, but not as an alternative point of entry into careers in survey science.

Vernon Bailey's career exemplifies the apprentice type. He hailed from a farming family with New England roots, who migrated from Michigan to the cheap land frontier of Minnesota in 1870. He had little formal education (to his chagrin), but like all his family had a deep respect for learning and aspired to turn his practical interest in wildlife into a career as a professional scientific collector. Through his apprenticeship with Merriam in the late 1880s and association with scientists of his own age, like Leonhard Stejneger, Bailey gradually realized that he too might have the right stuff to become a survey naturalist, even without academic training.[66] Lucky for him, he was uncommonly skilled and worked for a man who put his trust in graduates of the school of hard knocks.

The change from field apprenticeship to credentialing was not gradual but abrupt—and unforeseen. Initially, the burgeoning of surveys and museum expeditions created a demand for field workers that colleges could not supply. So expedition planners turned to the large pool of young men with practical experience as hunters or amateur collectors. In 1891 the botanist John Coulter called for the training up of a new practical profession of field collectors—"a race of field-workers who shall follow their profession as distinctly and scientifically as the race of topographers."[67] For a few years in the early 1890s it seemed that credibility gained in practical fieldwork would lead to careers in survey work. But in the space of a few years around 1900 that road to upward mobility was abruptly closed, as those with formal credentials cornered the market.

This demographic trend can be traced in the staff of the U.S. Biological Survey. (Table 5–1.) Of its seven founding members appointed between 1885 and 1890, three had college-level degrees (B.A., B.S., M.D.), and two more had some years of college work. This cohort came from the small social stratum of educated Easterners who were serious amateur naturalists—"practitioners" in Nathan Reingold's useful social typology—with access to Washington social and political networks.[68] In contrast, only five of the next twelve men appointed to the survey's staff (1891–1902) had college training; the rest had a high school education or less. Most were very young men with lots of experience of outdoor work but little book learning. This was partly necessity and partly choice: there were at the time few college-educated naturalists eager to enter government service, which reinforced Merriam's preference for on-the-job training. But this pattern was short-lived: five of the six men who joined the survey between 1902 and 1916 had bachelor's or higher degrees.

Careers of museum curators display a similar pattern of change. Most first-generation curators lacked higher education and came out of the world of serious amateur collecting. But rather quickly they were succeeded by college-trained biologists. The American Museum appointed eleven men to its curatorial staff between 1871 and 1892, of whom just two had college-level degrees and two more some college work. And even the two that were college-educated were anomalous: one (Henry Fairfield Osborn) had his main appointment at Columbia University, and the other had an M.D. degree, which suggests amateur training as a naturalist. This situation changed abruptly in the late 1890s. Of the nineteen curators appointed between 1895 and 1915, just one lacked a higher degree, and ten had Ph.D. degrees.[69] The pattern is striking: for almost twenty years curators came mainly out of the world of amateur collecting, rarely out of academia. Then in the space of a few years the door to a curatorial career was effectively closed to those who lacked academic credentials.

It was the same again with curators in the U.S. National Museum. Twenty-two were appointed between 1869 and 1894, of whom five had college degrees and three had M.D.s. (Two more had some college work.) In contrast, of the twenty-one appointments to the curatorial staff between 1895 and 1915, only three did not have college degrees, and four had Ph.D.s. Just one had only primary schooling (Ned Hollister), and he was distinctly anomalous.[70] As in the American Museum, the transition from field apprenticeship to academic credentialing was sudden and complete in the mid-1890s. The pattern at the Field Museum in Chicago is less stark but consistent: of the first four appointments (1893–94),

TABLE 5–1.
U.S. Biological Survey staff, by date of joining

Date	Name	Birth	Age	Degree	State	Dates with BBS
1885	Merriam, C. H.	1855	30	MD	NY	1885–1910
1885	Fisher, A. K.	1856	29	coll	NY	1885–1931
1886	Barrows, W. B.	1855	31	BS	Mass	1886–1894
1887	Bailey, V. O.	1864	23	hs	Minn	1887–1933
1889	Palmer, T. S.	1868	21	BA	Cal	1889–1933
1890	Nelson, E. W.	1855	35	coll	NH	1890–1929
1891	Beal, F.E.L.	1840	51	BS	Mass	1891–1916
1891	Loring, J. A.	1871	20	hs	Ohio	1891–1897
1891	Todd, W. E.	1874	16	hs	Ohio	1891–1899
1892	Goldman, E. A.	1873	19	hs	Ill	1892–1944
1892	Preble, E. A.	1871	21	hs	Mass	1892–1935
1895	Howell, A. H.	1872	23	hs?	NY	1895–1940?
1895	Oberholser, H. C.	1870	25	hs?	NY	1895–1941
1897	Osgood, W. H.	1875	22	BA	NH	1897–1909
1897	Starks, E. C.	1867	30	coll	Wis	1897–1899
1899	Oldys, H.	1859	40	BLaw	DC	1899–1912
1901	Lantz, D. E.	1855	46	coll	Pa	1901–1918
1902	Hollister, N.	1876	26	priv	Wis	1902–1909
1904	McAtee, W. L.	1883	21	MA	Ind	1904–?
1905	Henshaw, H. W.	1850	55	hs?	Mass	1905–1916
1909	Dearborn, N.	1865	44	BS	NH	1909–1920
1910	Jackson, H.H.T.	1881	29	PhD	Wis	1910–1936
1910	Wetmore, F. A.	1886	24	BA	Wis	1910–1924
1916	Taylor W. P.	1888	28	PhD	Wis	1916–1935

Sources: U.S. Civil Service lists for 1887–1907, *Annual Reports of the Secretary of Agriculture*, and *American Men of Science*.

one had a higher (M.D.) degree, and another was a college drop-out. Of the twelve staff appointed between 1897 and 1920, only one did not have a higher degree. These of course were the leading museums, and doubtless many smaller ones could not be so choosy in picking staff.[71] However, it is safe to say that by about 1900 anyone who aspired to a curatorial career and lacked academic credentials was likely to be disappointed.

Supply was doubtless one reason for this reversal: universities by then were turning out substantial numbers of trained biologists (though the vast majority were in the laboratory branches of the science). But the more important reason, I think, was that the rules of making careers had changed. Academic credentials had become the social marker of modern, high-status occupations.

By the early 1900s the usual path to a career in survey work was some combination of field experience and academic study. In 1908 Grinnell laid out a typical career track to Walter Taylor, then a promising undergraduate. Finish college with a major in systematic zoology, Grinnell advised him, and spend vacations as a field assistant at the museum, working up to assistant curator. Publish papers and make a reputation as a specialist in western fauna, and learn museum methods on the job as an apprentice. Grinnell insisted that promising students acquire a thorough academic preparation.[72] But he also made sure that graduate students were equally well trained in fieldwork, so that they would not revert to "the purely 'laboratory' type of zoologist."[73]

For field biologists the new rules were a mixed blessing. Because academic departments of biology were biased toward laboratory work, they tended to set lower standards for fieldwork (the logic, I guess, being that field science was irremediably second rate). Grinnell described one such case, of a recent graduate who sought to publish a list of New Mexico birds that was not up to par. She had taken Grinnell's museum course but had no field experience, and her formal training never taught her the standards of good taxonomic work. "She had some ability, and I did my best to inculcate an analytical conscience in regard to identifications and old records," Grinnell reported:

> But there is this trouble, here in this University, as evidently elsewhere, that major Profs. in zoology encourage their graduate students in the belief that any sort of systematic or faunistic treatment can be brot [*sic*] to satisfactory conclusion within the brief one or two years required for a degree. It may be all right for the student and Prof., but it's mighty bad for the literature of the science! The value of experience and consequent judgment is discounted.[74]

Field apprenticeship was slow and did not fit into the faster-paced schedule of academic credentialing: that was one reason why museums remained important sites for training systematic biologists, and why the practical skills of collecting remained scarcer than paper credentials.

That was the point of Harry Swarth's joking compliment to Grinnell on the perfection of his specimens: "What a dandy lot of material you are getting. Everything arrived in perfect condition. It's a shame that you have brains enough to do anything except collect! You ought to stay in the field, and hire a book keeper."[75] Searching for an assistant curator for the Field Museum in 1921, Wilfred Osgood decided to consider only applicants with B.A. degrees, but reluctantly. It was hard enough finding a collector with experience and a flair for fieldwork, he complained, much less one with academic credentials as well.[76]

In fact, field men could with enough talent and luck make careers in survey or museum work, even without academic credentials. For example, when Herbert Stoddard was tapped for a job as research biologist with the U.S. Biological Survey—he was then a junior taxidermist at the Milwaukee Museum—his friends had to reassure him that his lack of advanced schooling was no disqualification, and he went on to a distinguished career as a wildlife biologist. Likewise, when Colin Sanborn, an assistant curator at the Field Museum and a seasoned collector, decided that he ought to take a leave and get an advanced degree, Wilfred Osgood told him not to bother. The best of the older generation of curators were not college trained, he pointed out, and though college training was now usual, it was not absolutely essential for curatorial jobs.[77] Sanborn took the advice, and his career did not falter.

Other field-workers were less sure of their prospects. Harry Swarth reluctantly resigned his position as curator at the Museum of Vertebrate Zoology for a nonacademic post at the Los Angeles Museum because he had never been to college: "It seemed to me rather foolish," he explained to Grinnell, "for a man without any University credentials (whatever his ability might be) to expect much advancement in a University position." Though Swarth was already an excellent systematic zoologist, Grinnell let him go. But craftily he fed his protégé research projects—working up a collection, publishing a revision—knowing that publications would serve as credentials. Swarth never did get a degree, but a paper trail of substantial research papers gave him the confidence and standing to resume his career at Berkeley.[78]

But not everyone was so lucky. For many young field men, the rising value of formal credentials meant that skill and experience in the field ceased to be a sound investment in a career. Collecting

Fig. 5–6. A collector for the American Museum of Natural History (possibly J. H. Batty) with his day's catch of iguanas and burro (Pardo), Mexico, January 1904. Note the proud hunter's pose. Courtesy Department of Library Services, American Museum of Natural History, negative no. 13657.

and logistics remained essential skills for survey work, but they were redefined as a craft occupation and not, as previously, an apprenticeship to a learned profession. Taxidermists suffered the same social demotion about the same time, and for the same reasons.[79] In the early 1900s a class boundary divided what once had been a seamless social space.

The sticking point in the career cycle usually came when a skilled collector and preparator aspired to move on to a more intellectual kind of work, analyzing collections and publishing taxonomic descriptions and revisions. Such "literary" work, as it was often called, required a thorough knowledge of the complex taxonomic literature. Or a collector might aspire to a post as assistant curator, with duties organizing and managing expeditions: work that required both scientific knowledge and social skills in dealing with patrons and the press. These were indoor, managerial skills. At the threshold between field and "literary" work, between occupation and profession, was where careers stalled.

The experience of Thomas Large illustrates the point. Large got his B.A. degree in zoology and was working toward an M.S. degree when Stephen Forbes put him in charge of the collecting side of the survey of Illinois fishes. He labored for three years, but when the fieldwork was finished in 1902 he was unceremoniously replaced by a man of his own age, Robert Richardson, who was "already experienced in manuscript work in ichthyology" and could write up the collection for publication. It was his lack of a higher degree and "literary" experience that proved fatal to Large's career ambitions.[80] He had hoped that his record of achievement in organizing field parties and collecting would count in his quest for a collegiate post, but feared that "some hungry Phd. [*sic*] from Chicago University will make a deeper impression. . . . I wish I had that 'M.S.' Degree." But he did not have it, and it did matter.[81]

So when Large received a copy of *The Fishes of Illinois* by Forbes and Richardson and discovered that his part in the work was barely mentioned, he wrote Forbes an anguished letter telling what that work had meant to him. He had organized the many wagon expeditions to every corner of the state, and was personally in charge of most, with the help at most of one extra man. He had organized high school and college teachers to send in collections from the central and western parts of the state. He had identified and recorded 90 percent of the survey's 200,000 specimens, endured malarial districts in summer and a dirty basement workroom in winter, and organized the Illinois Museum's collection of decaying "alcoholic" fish. He had also made personal sacrifices because he "wanted the reputation that comes from scientific work." He took a cut in salary, exhausted his savings and went into debt, nearly lost his home, and was forced to put his family in boarding houses during his extended field trips. "Now this is what the Illinois Fish Book means to me."[82] Large acknowledged that Richardson had "literary" skills that he did not, but clearly felt that skill, efficiency, and fortitude in the field were no less creditworthy.

Large experienced practical survey work as a career investment, but in an older way: it was his personal commitment and the pain and sacrifice of the work that gave it social value. It was the same social logic that made nineteenth-century heroic explorers and traveler-naturalists authoritative social types. But the moral economy of field apprenticeship no longer obtained. Natural history survey had become an academic discipline, and skilled collecting, though essential, was in itself no longer social capital. Large thought he was making a career investment, only to discover that he was just doing a job for pay. To borrow a concept of Steven Shapin's, he had become a kind of "invisible technician."[83]

What happened to Thomas Large next we do not know; his letter to Forbes was his last. He had been teaching school and was thinking of trying his hand at homestead farming in Idaho, where—he could not have foreseen—a climate cycle was about to turn from wet to dry.

Another of Forbes's field assistants, Adolph Hempel, likewise expected that his expert collecting for the Illinois Natural History Survey would lead to an advanced degree and a career as a college teacher or a museum preparator or collector. Forbes encouraged his ambition, giving him scientific books to read, but in the field he was worked so hard that he had no time for anything scientific. "I was a sort of handy-man," he later complained, "good to take tow [tow net] collections, work the stations, hand seins and dredges, examine the intestines of fish for parasites, clean up dirty laboratory dishes, identify fish, and collect their stomachs for food study, test different plankton nets, etc; and then if there was any time left I could put that upon my definite line of work."[84] And without credit for scientific achievements there was no hope of a scientific career. Collecting per se remained a craft occupation, not a stage for launching a scientific career.

Robert Richardson, in contrast, deftly managed the transition from field to literary work. Following his stint with the Illinois fish survey, he resumed his education at Stanford and Berkeley, where he served as general handyman in the zoology laboratory and museum—"a virtual servant of the whole department," he recalled. But he made it clear that he regarded such work as temporary, and when Forbes asked him to take charge of a second Illinois River survey, Richardson demanded the right to pursue his own research and to publish work independently. A period of fieldwork in the summer season was fine, he allowed, but done year-round it would preclude study and writing up. Richardson never got a Ph.D. but did secure a permanent post with the Illinois Natural History Survey.[85]

A more complicated case is Malcolm Anderson's. He was an experienced and able professional collector who had learned the craft as a teenager with Grinnell in California. After picking up a B.A. degree he spent seven years in the North Pacific and East Asia collecting for the British Museum: an experience that left him wondering if he could ever "live in a civilized way again or not—I haven't tried it for so long."[86] Then at the age of thirty he decided to turn his achievements as a collector into a career as a curator. Hired by Wilfred Osgood to lead an expedition to South America in 1912, Anderson hoped he would "be privileged to assist in working up my collections and be duly credited"—that is, turn practical service to scientific account. "I wish a museum position in America," he explained, "with a chance to rise, and a chance to work up my collections."[87] Osgood was sympathetic but warned him not to expect an easy transition: "As an explorer and field worker you have won your spurs, but of course it would remain to be seen how you would succeed in the more serious side of the work." Still, Osgood saw no reason why Anderson should not "work in the expectation of studying and publishing in a large share of the material you get."[88]

In the event, Anderson was fired immediately upon his return from the field, probably as a cost-cutting measure. (The Field Museum preferred hiring collectors for particular expeditions over keeping them on salary.)[89] Whatever Osgood thought of Anderson's potential, those in charge of the museum apparently regarded collecting and curating as distinctly different—and unequal—occupations. Anderson remained a freelance collector until his untimely death in 1919.

Not everyone was unhappy with the changing rules of career making; some field men preferred freelance collecting to the mixed indoor and outdoor work of a curator. For example, the great collector George Cherrie served as assistant curator at the Field and Brooklyn museums but decided that full-time commercial collecting suited him better.[90] Another famous collector, Rollo Beck, was employed for a time by the California Academy of Sciences and the American Museum, but returned to the work he loved best. Freelance collecting was an arduous and risky occupation, but it afforded variety and independence, and no interludes of desk work.[91] Other collectors who failed to make the transition to curatorial jobs made good careers in wildlife biology, a science that was more practical and less "literary" than systematic biology. A good example here is another of Grinnell's protégés, Joseph Dixon, a college dropout who was happier in the field than the classroom and became a superb collector and expedition man. Annie Alexander hoped he might join the staff of the museum at Berkeley, but Grinnell ruled that out: "The standards of published research have been rising all along," he explained, "and naturally Dixon is handicapped by having discontinued his college work and practice in the literary and scientific fields." Dixon had a fruitful career as a wildlife biologist at Berkeley and later with the National Park Service.[92]

Many amateur collectors must have been disappointed when skill in collecting and fieldwork proved not to be a stepping stone to a career in science. Collecting parties kept bumping into them. In Wrangle, British Columbia, for example, one of Grinnell's parties encountered a Mr. Walker, a local worthy who arranged for transport upriver to Telegraph Creek. He did court and police work for a living, but was a keen naturalist and dreamed of getting a job at the University of Washington.[93] But as Robert Ridgway once wrote to the anxious mother of a young aspiring ornithologist: "The study of natural history affords . . . a very agreeable and instructive *recreation*. . . . As a *means of livelihood*, however, it must, in at least ninety-nine cases out of a hundred, prove a complete failure, success being . . . a question not only of exceptional ability but also of exceptional circumstance."[94] In addition to being skilled in collecting and field craft, one had also to be lucky and in the right place at the right time.

Women in the Field

I do not know of a woman who made it as a professional survey naturalist (though Annie Alexander and Louis Kellogg came close). However, survey parties were not an exclusively male preserve: women did participate, just unofficially. Most were wives of survey biologists, and they were often present, to judge from scattered evidence.[95] And they were not just along for the ride, but contributed to expedition work in varied and important ways. Some managed logistics and supervised camps—extensions of the work of managing a household economy.[96] Others made independent careers as nature writers and used collecting expeditions to gather literary material. A few became expert shooters and collectors. Wives of expedition patrons also appear, some of whom learned the art of skinning and putting up specimens. Prevented by social convention from having scientific careers, married women were re-

Fig. 5–7. Florence Merriam Bailey at work in the field at Queens, New Mexico, 1901. Vernon Baily Papers, box 17, folder "photo album, Texas 1901, 1904." Courtesy American Heritage Center, Laramie, Wyoming, negative no. 14084.

sourceful in adapting existing roles in order to take an active part in expedition work.

For example, Malcolm Anderson's wife was with him on the Field Museum's expedition to Brazil, and presumably others as well. She cooked and "aided me in a hundred ways," Anderson reported. "What insects we have are due to her industry."[97] C. Hart Merriam's wife, Elizabeth, was with him on at least one expedition—to the San Francisco Mountains in 1889—and probably more, as she was an ardent outdoorswoman and a skilled skinner and preparator. She also exercised a civilizing influence on her husband, who sported rough habits when he was with the boys: for example, dining ostentatiously on grilled wildcat and skunk, to his companions' disgust.[98] Florence Merriam Bailey (Hart's sister) became Vernon's regular companion on his field trips, for ex-

Fig. 5–8. Hilda Wood and her husband-to-be, Joseph Grinnell, in the hills near Pasadena, California, May 1906. Courtesy Bancroft Library, Berkeley, California, Banc Pic 1973.044-pic.

ample, in his surveys of New Mexico in 1903 and in the Dakota marshlands in 1908. Florence was never a collector, but worked independently at her own profession of writing nature essays and field guides for amateur birders. She would follow Vernon's itinerary if his collecting sites also gave here the chance to observe animal life; and if they did not, she would go off on her own. The couple liked to signal to each other in the field using bird calls, and Vernon figured as "the mammalogist" in some of Florence's essays.[99] Frank and Fanny Chapman were another working couple, beginning in 1898 on their honeymoon at Ma Lathams' Oak Lodge, where Fanny was instructed in the art of skinning and making specimens. The two worked together in the field throughout their active lives.[100]

Edith and Frederick Clements, who were an inseparable (and childless) couple, were also a working team. In addition to her own literary interests, Edith did the camera work for Frederick's ecological surveys and served as chauffeur and mechanic, as we saw in the last chapter. Hilda Grinnell was another enthusiastic outdoorswoman and was often with Joseph in the field, contributing in various capacities as cook, preparator (she put up fine specimens), and possibly also as collector.[101] Among academic biologists, family collecting was not unheard of. For example, Carl Hubbs conducted his surveys of the fishes of the arid West as a family team, with his wife, Laura, and their three daughters all sharing the hard and often disagreeable labor of wading and seining in murky ponds. Though not a trained biologist herself, Laura became a skilled field-worker and Carl's indispensable assistant, and they coauthored scientific papers.[102]

What we rarely or never see, however, is an unmarried or lone women taking part in survey expeditions. The constraints of gender roles would have made that difficult. Should a lone woman suffer a serious accident or illness in the field, for example, men would have to administer bodily care.[103] There was also the problem of propriety. Any woman at that time who went off into the outback with a group of men could expect to be the object of malicious gossip. Annie Alexander worried about that when putting together her second Alaska expedition in 1908, a party consisting of herself and two men. Of course there was no question of actual impropriety, but Grinnell had to admit that there would be those who would whisper anyway. When Alexander announced that she would take a female companion with her, Grinnell feared that it would be a Sunday naturalist unused to hard expeditionary work:

> You know as well as I that even the best intentions fail, when a little hardship comes along, and then—homesickness, unhappiness, discontent—unless the person in question has the innate love of field-work, and has a definite purpose in view, as *you* have. . . . I can think of no one now whom *I* would risk taking on such a trip. Our Cooper Club ladies are all "parlor naturalists"! Or, at most, "opera-glass observers" who get as far as the Yosemite by stage occasionally![104]

Alexander twitted Grinnell for exhibiting male bias, but if it was that, it was not deeply rooted. As we know, Alexander did find a suitable companion in Louise Kellogg, who had no experience of fieldwork or science (she had majored in classics in college) but liked working with her hands and quickly became an able field worker, winning Grinnell's approbation. As Alexander's biographer observes, "Grinnell based his impressions of Kellogg on her performance in the field and her contributions to the museum's collections. To him, little else mattered."[105] Just so.

For Alexander and Kellogg field collecting provided the freedom to pursue an unconventional lifestyle, do skilled work that they liked and did well, and contribute to science, all the while enjoying outdoor life. Although Grinnell encouraged Alexander to work up her collections herself—to take a step toward career science—she never did. It was not that she doubted her intellectual capacity, I think, or felt constrained by gender roles. Rather, it was that everything she did she had to do in a first-rate way, and she knew that to do first-rate science she would have to submit to formal academic training, which was not to her liking. Besides, as a hunter and collector she already had just about everything she wanted. A letter to Grinnell hints at some such motives: "I haven't mental vigor enough to pursue any study alone," she wrote, "and would have to put myself under someone's guidance and get away from all present connections until I had gotten a footing in something—and it would not be science." Her biographer Barbara Stein puts it well: "Her patronage and amateur status gave her access to a world she loved while allowing her to control the nature and extent of her participation in it."[106] Her life reminds us that values were lost as well as gained when natural history was academized.

Gender bias was not the main impediment to women becoming professional survey biologists, I think; nor was it that the rigors of survey collecting were inherently unappealing to women, or that fieldwork had a macho-male culture that repelled them, as was the case in a prestigious discipline like high-energy physics.[107] As we have seen, Victorian women were encouraged to participate in sport hunting and camping, and did. And the women who did go on expeditions seem universally to have taken much pleasure in the experience. For Evelyn Field, the wife of Marshall Field, Jr., a collecting expedition to South America appears to have been a welcome change from the life of a Chicago society woman, and a chance for productive intellectual work in the company of intelligent and active men.[108] For Edith Clements and Florence Bailey, expeditions were a chance to pursue a literary career while enjoying an intimate and a *working* companionship with their husbands, a relation that was seldom possible in a society that had long since separated the places of domestic and professional work. As Florence Bailey's memorialist put it: "The rich experiences of the outdoors, especially the great Southwest which she loved, the companionship of her husband, and the stimulation of the work they were accomplishing—these were the rewards of the arduous life she chose to pursue."[109] It is a relation that is vividly brought to life in the fictional relation between William Adamson and Matilda Crompton, the hero and heroine of A. S. Byatt's novella *Morpho Eugenia*.[110]

Far from being repelled by a male work culture, women seemed to enjoy a culture in which men and women were expected to share work equally, work to the same standards, and even to compete. For example, when Annie Alexander and Louise Kellogg returned from a season in the field with lots of fine material, Harry Swarth had a hard job convincing them they had done well: "They seem to feel rather discouraged that they did not get more than your [Grinnell's] party!"[111]

It was social structure and custom that limited women who might have liked to make survey science their life's work. Survey had become a career, and careers were understood to be full-time and all-consuming jobs.[112] It was not the work of survey collecting that kept women out, but the occupational structure of middle-class society. Like uncredentialed male field workers, women found themselves on the wrong side of a new social boundary. As the line between recreational and professional collecting hardened, and more white-collar occupations required educational credentials, women were left in the role of hobbyist—just as collectors who lacked academic credentials were left in the role of craftsmen. Career trajectories of survey naturalists, male and female, followed the social contours of work in middle-class society.

Identity

The social logic of careers leads us to the related issue of identity. As I and others have observed, identity was a problem for field scientists generally. Since people in modern industrial societies mainly derive their identities from their occupations, the resemblance of fieldwork to outdoor recreation gave it an ambiguous social meaning. This ambiguity was compounded by the varied sorts of people that field scientists mixed with in the course of their work.[113] Because they hunted, camped, did messy work with their hands, and shared the work of shooters, cooks, and carters, survey naturalists were an especially ambiguous social type.

Survey collectors were regularly mistaken for some other sort of person, often disreputable. Collecting small mammals near Lake Champlain, C. Hart Merriam was mistaken for a bank robber hiding out, and in the mountains of southern Appalachia, for a revenue agent—a potentially fatal mistake in that region. Collecting snails in the Tennessee River basin, Charles C. Adams was taken for a detective, a deadbeat, and a crazy man.[114] Collecting in Chicago's outer suburbs, Frank Daggett was accosted by irate landowners who thought he meant to steal their chickens.[115]

Aven Nelson, a leading expert on western flora, was multiply misidentified in his years of fieldwork. Collecting along the Platte River he was taken for a crazy man who lived in the nearby hills. Shoshone Indians took him for a medicine man. A tourist in Yellowstone Park thought he was a mailman (his collecting bag resembling a mailbag). Others mistook him for a camper (taking his vasculum for a folding bed), and for a fisherman (mistaking dried specimens for bait). Suspicious citizens of a Wyoming town suspected he was an unlicensed peddler or canvasser, and railroad workers assumed he was a tramp. The foreign born were more perceptive: a German hotel cook engaged him in botanical conversation, and a group of immigrant railroad laborers knew at once what he was about.[116] The amateur collector was a more familiar social type in Europe, it seems, than in America's inner frontiers, where they were assumed either to be making a buck, up to no good, or just odd.

And it was not just locals who misread the demeanor of field scientists: they sometimes mistook themselves. Climbing a butte near Scott's Bluff, Nebraska, Edith Clements was panicked "by the presence of a lunatic in the vast and lonely expanse." At least she assumed the man was a lunatic from his strange and obsessive behavior, climbing the bluff with a pickaxe and spending hours tap-tapping at the rock. Perhaps he was just a deluded miner, she thought, but he was "such a dilapidated, wild-looking individual" that she took no chance and hid behind a rise until reassured by the appearance of a whistling tourist with a Kodak.[117] More likely Edith had simply encountered a geologist or paleontologist at his work.

These instances of mistaken identity were not signs of ignorance, but of the genuine difficulties in establishing the identity of strangers in that place and time. The inner frontiers were full of people engaged in varied pursuits that were not all that different from survey collecting. Thus it was not obvious on the face of it whether a stranger with a gun and bag was a poacher, a sportsman, or a scientific collector. And what visual clues distinguished a man doing ecological survey from a timber cruiser, game warden, or tourist with guidebook and Kodak? People unaccustomed to seeing collectors in their fields and woods naturally took them for more familiar sorts.

Scientific collectors and sport hunters were especially alike, as we have seen, and their relations were mixed. Although collecting parties would on occasion hire them, hunters and trappers were more usually competitors—especially when demand for animal pelts was high or stocks declining. In Alaska after bears, Harry Swarth was beaten to the best place by Indian hunters who cleaned up.[118] With predator or "pest" species, in contrast, local residents were usually glad to have collecting parties help them with their problem. For example, a ranch wife encouraged Grinnell to clean out a large colony of yellow-billed magpies that had been devouring her chicks and eggs.[119]

Competition between commercial and scientific hunters was sometimes heated and even violent. Farmers along the Sagamon River in Illinois hated seiners and would hide barbed wire underwater in the good seining places, shredding collectors' nets. At Havana, on the Illinois River, collectors for the Illinois Natural History Survey were more actively discouraged. Their laboratory flatboat was twice scraped by a commercial fishing steamer, and their launch was sunk in a collision, clearly not by accident. At Galena, they suffered the hostility of commercial fishermen who suspecting them of being poachers posing as collectors—not unreasonably, given the prevalence of illegal fishing in the area.[120] The point is that pot hunters, poachers, and scientists did much the same kind of work and could easily be mistaken one for the other.

In fact, the chief practices of survey collecting—series shooting and seining—were strictly illegal when performed by sport hunters and fishers. The differences were in intention and purpose, and these were invisible to casual onlookers. As laws against pot-shooting become more stringent, and as wildlife preservation movements gained strength, it was easy for scientific collectors to be tarred with the same brush as pot-shooters. The annual routine of obtaining hunting licenses afforded survey collectors unwelcome reminders of the resemblance of their own practices to some more ancient and disreputable ones. Most fish and game wardens understood the value of scientific collecting and willingly granted licenses for series collecting. But they also knew that many sport hunters and anglers did not share their tolerant view. To many sportsmen, scientific collectors were simply competitors for a diminishing resource who enjoyed rights that they did not: rights to exceed legal limits, take nongame species, shoot out of season, and use methods that for them were proscribed. There is scattered evidence that fish and game officials became stingier with permits for scientific collecting in the 1910s, not because they thought scientists were abusing the privilege, but because they did not relish explaining to sportsmen why scientists could do what they could not.[121]

The situation was further complicated when sportsmen and amateur collectors realized that by posing as scientists they could get permits more easily and avoid both game laws and the moral ambiguity of recreational killing: it was for science, not for fun. Such abuses of the licensing system were apparently widespread in the early 1900s. In California, collecting bird eggs and nests had become such a fad that wardens ceased to issue permits for out-of-season shooting of all birds, not just game, except by scientific institutions. The game commission wanted to aid scientific work and had been liberal with permits, one warden explained, but "gross abuse" had arisen, so they had to close down on private collectors: "Anything that makes for the furtherance of ornithological science we favor, but private collectors we do not favor."[122] Faced with the ambiguous identities of scientists, private collectors, and pothunters, wardens took the easy road of treating all alike. Survey collectors' ambiguous identity was thus rooted in the everyday activities of the people of the inner frontiers.

Landscapes and social roles were in flux in such places, and as social historians have shown, it is often in times of unusual social change that identity, and anxiety over mistaken identity, become intense concerns.[123] For example, wildlife preservationists were redefining subsistence and market shooters alike as lawbreakers and killers.[124] As extractive economies gave way to recreational ones, displaced professional hunters became tourist guides or scientific collectors, hoping to sell or give their collections to museums in return for a job as resident taxidermist—as one Ernest L. Brown,

"The Minnesota Taxidermist," vainly hoped to do.[125] Amateur collectors were going to college and becoming wildlife biologists; and volunteer deputy wardens were becoming employees of state wildlife agencies. Cases of mistaken identities were symptomatic of these environmental and social dislocations.

Among the urban gentry, dress codes were indicators of social identity—or good cover for those who wished to lose the ones they were born with or acquire new ones. In the field, however, dress was a less reliable indicator, because the requirements of rough living enforced a practical attire that made it difficult to tell one social type from another. Seasoned collectors joked about their wild and disreputable appearance after a sojourn in some outback. After a month in the Mexican bush, Edward Nelson remarked on the group's ragged beards, torn clothes, and down-at-heel look. Roland Harper, on a tramp through the forests of west Florida in his most raggedy clothes, expected to be mistaken for a hobo or some other suspicious character and was surprised (and perhaps a little disappointed?) when he was not.[126]

In Death Valley with Merriam's expedition in 1891, and with a newspaper reporter in tow, Vernon Bailey and his colleagues were careful about their presentation of self: "We tell him all the yarns we can think of but have to place him under bonds not to tell the whole truth," he confessed. "These intrepid and daring sons of science would not show up well [to city readers] in dirty, ragged clothes and mounted on old bony plugs." In a retrospective account of their expedition, Bailey captioned a photograph of the group "a distinguished group, not of frontier bad men, but of scientists."[127] (See figure 4–1.) It was a common trope. In the field in Brazil, Grace Thompson Seton described Karl Schmidt and Colin Sanborn suited up for a nighttime hunt: "[They] looked like two gunmen. Shotguns, pistols, big felt hats upon which are attached acetylene lamps. . . . We bade goodby to the brigands as they clumped away in their hip boots of rubber, the forest is dripping with moisture."[128] These were insider jokes, but they show that appearance and identity were also serious issues. This is perhaps why one of Alexander Ruthven's amateur naturalists wore business attire to collect in the field and always managed to return with garments unstained and high collar unwilted.[129] It was, I surmise, an assertion that collecting was a respectable, middle-class activity.

In arcane occupations like science, outsiders can only judge the authority of knowledge claims by claimants' social reputations and presentations of self. And in a world where white coats are the familiar emblem of scientific authority, rough-clad field workers were disadvantaged. Mistakable identity was funny to those who experienced it, but for the reason that the best jokes are funny: because they get at uneasy issues of social difference and standing.

Identity was not just a personal but an important practical matter for survey collectors, because their work depended on the trust and cooperation of different sorts of people who knew nothing of science. That is why Thomas Large arranged to have "Illinois Natural History Survey" painted on the side of the survey's field wagon: he hoped it would win the respect and trust of farmers who might otherwise run them off as trespassers—which, of course, legally they were. However, Large also worried that advertising his scientific identity might "add a little to expenses sometimes" and provoke unwelcome attention.[130] The label on the wagon was a public advertisement that Large had an institution behind him, and an expense account. The identity of scientist gave survey collectors access and legitimacy, and put them visibly on the right side of the game laws. But anonymity could also be useful, enabling them to go about their work unremarked. Being mistaken for ordinary campers could be useful cover. Creating and manipulating identities, and creating false ones, were useful gambits in the business of scientific collecting, as in many other occupations.

Conclusion

As the natural landscapes of the inner frontiers afforded the venues and the material for survey collecting, so did their social landscapes structure the meanings of the working experience. These meanings were as varied as the kinds of work that people did there: that is Richard White's point when he writes that we know nature through work.[131] He was thinking mainly of practical workers, but those who work at play are no less important: hikers, alpinists, tourists, campers, and sport hunters. After all, recreation and tourism were major industries of the inner frontiers after the extractive phase had passed. And to these we must also add the work of scientific workers—freelance collectors, survey and wildlife biologists, amateur naturalists, and ecologists—who came to collect, map, observe, photograph, and record. Natural history survey, like outdoor recreation, was an activity distinctive of this special phase in the settlement history of North America. It was a pursuit of middle landscapes.

Every outdoor occupation knows nature in its own distinctive way, and many experienced it as some mixture of work and play, including survey collectors. For them work and play, career and recreation, were more equally mixed than most. Because it was outdoor and traveling work, survey collecting of necessity closely resembled recreational camping and hunting; it was just more intensive, sustained, and purposeful. Recreational nature-goers worked at play, but as an alternative and complement to their real work, not as a career. Survey biologists in the field worked to make careers, but experienced their labor in part as recreation. Survey work was an aesthetic and cultural experience, restorative of physical and mental health, as well as a job and a career investment. That mixture of carnival and Lent—Paula Findlen's trope applies well here—was what attracted people to survey work and what gave that vocation its emotional texture and social meanings.[132]

Of all the ways of working and knowing nature, survey collecting may embody most fully the transitional character of the inner frontiers. Here were landscapes moving from wild and imperfectly known to settled and intimately known; economies moving from extraction to service; and rural societies connecting with urban cosmopolitan ones. Survey collecting embodies this moment of transition: its practices, a mix of intensive and intensive; its culture, a mix of economic and recreational; its career structure poised uneasily between craft and learned profession. Survey naturalists knew the liminal nature of the inner frontiers more truly than workers who knew them as either wild or tame, through pure work or unalloyed recreation. Both handwork and headwork, practical occupation and exacting science, serious work and recreation, natural history survey opens a window on a cultural landscape as distinctive as the natural landscapes in which its work was done.

CHAPTER SIX

Knowledge

KNOWLEDGE of biodiversity was the end to which museums were built and surveys organized; expeditions funded and launched; plants picked and pressed; animals killed and transformed into scientific specimens; vast collections lovingly assembled, cataloged, and protected from decay. To know all the plant and animal species with which we share our little blue-green planet—their numbers, natures, places, and relations to their environments and to each other—that is what impelled collectors into the field. To have a passenger list of our communal Ark as complete as one could make it.

But to count the species, as Noah did, means first knowing how properly to distinguish one from another: it means knowing what counts as natural units. The science of biodiversity is above all a science of categories, and of the gaps between natural kinds that identify species as the unitary things that, all together, make up the living world of nature. No units, no diversity. No categories, and nature is an unstructured blur.

Describing and sorting have for some time now been widely regarded as something less than first-rate science—"mere" description and cataloging. They have been taken to be essential but not overly creative preliminaries to the real work of theoretical analysis. In fact, the work of identifying, describing, and categorizing natural kinds is highly exacting and creative. It is not just facts that are thus produced—lists of species and subspecies—but the very categories by means of which we perceive and understand our world.[1] Those who regard categorizing as a mere preliminary implicitly assume that describers simply sort items into preexisting stable categories—but they do not. Categories are created, recreated, and even destroyed as classifiers apply them to material things. The act of sorting and categorizing is as theoretically creative as any scientific act.

The anthropologist Marshall Sahlins has observed that every cultural system puts itself at risk whenever it is exercised in the real world. As he put it: "[Culture] changes precisely because, in admitting the world to full membership in its categories, it admits the probability that the categories will be functionally revalued."[2] Sahlins was writing about social categories and ways of creating distinctions of status and power, but his observation is no less true of scientific categories and classifications. We study taxonomic work to learn what happens when naturalists admit nature into "full membership" in their systems of classification.

Historically it is true that description and inventory did cease to be a high-prestige practice, yielding precedence to more overtly theoretical modes. But to think therefore that the one is a lesser preliminary to the other immodestly assumes that change in science is progressive, and that what comes after is by definition superior to what came before. It is not, or not always: just different. Each generation of scientists finds it useful to highlight their own novelty and worth by contrasting their practices with those whose imperfections and limits history has already revealed. But that rhetorical gambit is simply an artifice of the process of making new knowledge and careers. A generation of biologists fastened the "mere" onto describing and classifying in part for social reasons. Our task here is to show how survey collectors, in assembling and ordering their collections, gave a particular shape to their intellectual landscape.

Modes of collecting favor certain natural categories, and the categories of survey naturalists differed from those of Linnaeans and Humboldtean explorer-naturalists, and differed again from those of later quasi-experimental field workers. In this chapter we will inquire into how collecting practices shaped the basic categories of animal taxonomy, and how those categories in turn shaped field practice.

Systematic biology is a famously, even notoriously elaborate system of categories—species, genus, family, order, class, kingdom, phylum, plus many more intermediate sub- and supercategories—that is constantly changing (to the dismay of its numerous passive users). The species is the most basic and stable category. It is the one and the only natural unit: that is, the one that actually exists in nature. Higher categories are human inventions designed for our own purposes, a cultural rather than a natural order (though not an arbitrary or subjective one). Species alone are not inventions (though, because Linnaean bionomials link a generic to a specific term, their names are in part artificial and change as genera are reshuffled and renamed). But species themselves exist in nature, their separateness enforced by spatial or biological barriers to interbreeding. And though taxonomists differ vigorously about species' identities and names, opinion tends to stabilize over time rather than to cycle in an endless flux.

One reason taxonomists disagree is that their conception of species depends very much on the size and quality of the collections they have to work with. This is a fact well known to practitioners, though less appreciated by historians who have produced a voluminous literature on the conceptual and philosophical aspects of species without paying much attention to taxonomists' actual practices. This is unfortunate, since intellectual changes cannot be understood without reference to collecting and sorting practices: in particular, the slow change from a typological or Linnaean conception of species to a biological or populational one.[3] It is a complex story, but the plot hangs on taxonomists' increasing appreciation of species' internal variability. And that appreciation was in turn directly related to taxonomists' access to large and comprehensive collections, assembled by the methods of survey collecting.

It was changes in field practice, I argue, that drove change in natural categories: the change from type to survey collecting in the late nineteenth century, and from survey to populational collecting in the mid-twentieth. The first created the subspecies as a fundamental unit; the second created the breeding population. Survey collecting was meant to (and did) make species taxonomy a more rigorous science, but in doing so, it ultimately transformed classification from an end in itself to an instrument of evolutionary theory.

Linnaeus and the Linnaeans regarded species as homogeneous entities. Individuals obviously varied from one another, but all were merely more or less imperfect variants of ideal types: types were the real things. This conception reflected the prevalent creationist account of the origin of species—and taxonomists' practice of collecting only representative specimens and relating species definitions to single and unique "type" specimens. This somewhat curious custom was designed to keep order in a community of describers who varied greatly in their scientific and literary abilities:

if a description was doubtful, one could always refer back to the actual thing itself.

These practices of collecting and managing taxonomic data were well suited to the Linnaean project of naming and classifying a huge backlog of natural kinds: it was fast, simple, and relatively undemanding, and could be done with existing collections and without firsthand field experience. The Linnaean binomial system was invented precisely so that naturalists of every degree of training could assist in the great project of taking inventory, naming, and classifying.[4] The initial effect of survey collecting was to make Linnaean taxonomy a more exacting science. With more complete data on the range of individual variation, taxonomists could be more confident in their discriminations between confusingly similar species. And wider geographical knowledge made for fewer redundant names, or synonyms. (Species that wander out of range often appeared as "new" ones to local collectors, who knew neighboring areas less well than their own, and so gave names to phantom species.) Interregional comparison sharpened Occam's razor.

But knowledge of intraspecies variability had unexpectedly subversive effects, undermining creationist accounts of the origin of species and revealing the existence of biological units below the species level. Exact knowledge of variation invited naturalists to regard species not as ideal categories but as biological things. The culmination of this phase was the invention of the subspecies, first as a kind of provisional Linnaean species, then as a distinct natural unit—a species caught in the act of evolving. Subspecies were a creation of survey collecting: they were detectable only in collections of a certain size and scope. They were scientific objects only in the context of a particular mode of collecting practice. But their heyday was brief. As collecting became still more intensive, revealing even greater intraspecies variability, taxonomists began to see species as congeries of local populations, and subspecies lost their standing both as taxonomic and evolutionary entities. Breeding populations, stripped of formal Latin binomials, became the important units of evolutionary biology.

Taxonomic categories thus arose and fell with changing collecting practices. That is the story I want to tell here: a story not of routine fact-gathering giving way to proper scientific analysis, but of one kind of collecting and conceptual ordering succeeding another.

Species and Survey Collecting

To see why categories reflect collecting practices we must look at how taxonomists work, and how their practices evolved as field collecting was intensified and scaled up.

Sorting things is what taxonomists mainly do. Imagine a large table covered with one big pile of unsorted specimens and lots of little piles, which grow and shrink until every specimen is placed in one pile or another. These separate piles represent species (or subspecies), which receive Linnean binomials (or trinomials) and are entered as verbal descriptions into the taxonomic literature. That is the ideal; in reality many specimens may not fit unambiguously in any pile, or could equally well be sorted into several. Too many leftovers are a sign that something may be wrong with the categories; likewise, if different sorters end up with different piles.

Variation is taxonomists' problem, in both meanings of the word: the object of obsessive study, and what complicates their efforts. The object is to pigeonhole nature's diverse forms; the complication is that nature is too infinitely diverse to be easily pigeonholed. The trick of proper sorting is knowing which variations are significant taxonomically and which ones are just random noise. One handbook of taxonomic practice lists seventeen major types of variation, most of them irrelevant except as nuisances and pitfalls to be avoided. These latter include variations due to age, sex, season, habitat, reproductive oddities, genetic polymorphism, disease pathologies, population density, and postmortem damage to specimens.[5] Individual variations have no taxonomic significance, and mistaking a distinctive individual for a definite species or subspecies will reflect badly on a sorter's competence.

Among the traits that genuinely distinguish species are the morphological characters of size, shape, proportions, color, color pattern, and anatomical structures of various kinds, especially those related to feeding and breeding. But these traits have different values as distinguishing characters, and it is never absolutely clear which ones matter most. Variation within species may be as great as variation between them, and some species interbreed, producing hybrids to confound the classifier. Then there are the nonmorphological traits recorded in collectors' field notes: geographical location, habitat, behaviors, and ecological relations. Relatively unimportant before the age of survey, these ecological characters became crucial to modern taxonomic practice. Taxonomists have always to judge which of these varied traits are diagnostic.

We may think that sorting species is entirely a matter of defining things. That, after all, is how species appear in the taxonomic literature: as elaborate word pictures of defining features. But the actual process of sorting is more a matter of recognizing gaps and discontinuities between things. In ordering a world of more or less continuous variation, what counts is the ability to perceive gaps—the negative spaces between species. That is the crucial skill that separates (so to speak) the taxonomic sheep and goats. This was especially so for the collectors of the survey period, who were often dealing with large families of similar species, like the insectivorous mammals, rodents, or bony fish.

It is because its key step is recognizing gaps that taxonomic practice depends so crucially on the size and comprehensiveness of collections. Depending on their depth and scope, collections can make gaps seem to appear where there are in fact none in nature, or can make nature's actual gaps fade into invisibility on the sorting table. There is an optimal density of data that gives gaps maximum visibility, and as it happened, the practices of survey collecting produced data of just that sort.

Consider the situation of Linnaean taxonomists, who liked collections with a few representative specimens of every species. With just a few specimens from widely scattered places, they could not know for sure if observed differences were those of distinct species or meaningless individual variants. With just a few scattered specimens, sorters could have no idea of the normal range of variation within species, so could if they wished (and many did wish) take every visible difference for a species gap, when in fact they were simply the normal differences between the individual members of a single species. Likewise, lack of specimens from neighboring locales, or of closely related forms, invited sorters to see as new species, forms that unbeknownst to them had already been described and named. In either case, the pleasure and prestige that comes from being the first to describe and name a new species will encourage sorters to see gaps where there are in fact none. Synoptic collecting and collections (that is, a pair of every kind) did nothing to discipline collectors' desire to make discoveries: indeed, it gave that desire free rein.

But does it really matter if the taxonomic literature is clotted with phoney or duplicated species? After all, the literature of laboratory science is replete with wrong or redundant data, which disappear soon enough from neglect and overlayering of new facts. "[A]rtefacts and half-born statements which stagnate like a vast cloud of smog," is how Bruno Latour and Steve Woolgar aptly describe the literature of experimental science.[6] But taxonomists' literature is special, perhaps unique, in its respect for its own deep history. Taxonomists work not just for their own kind, but are also the bookkeepers and guardians of nature's order for all scientists, indeed for all the world—Adams, Eves, and Noahs combined. So they cannot just allow names to appear and disappear, or multiply redundantly. They have rules of order: for example, the rule of priority, which says that if a form has been given more than one name, its legitimate name is the one given to it by its first describer. Taxonomists thus can allow no species name, once published, to be forgotten; they list disused names as "synonyms," all the way back to Linnaeus. Unlike experimental scientists, who can sweep bad data quickly and quietly into the dust bin and forget them, taxonomists can forget nothing. So for taxonomists the many "new" species defined on the basis of incomplete and synoptic collections are a problem that cannot be ignored and forgotten but must be remembered and managed forever, however annoying and wasteful of their time. In taxonomy the sins of the fathers are visited not just on their sons and sons' sons, as the proverb says: mistakes and redundancies are an accumulating burden on the whole community.

Collecting representative specimens served Linnaean taxonomists well enough for nearly a century. It enabled them to process species rapidly, especially those already known because they were of practical or aesthetic interest to humankind (birds, predators, game animals). But for unfamiliar faunal groups or those composed of many similar forms, the limitations of synoptic collections threatened the credibility of the Linnaean system. It invited describers to create nonexistent species and synonyms; and when experts disagree and when species identities and names change too fast, biologists may conclude—and have done so from time to time—that species are not a natural category after all, but merely expressions of classifiers' social customs and order, not nature's. With growing knowledge, typological collecting makes nonexistent species appear real, and real ones seem to be cultural artifacts with no fixed identity.

The practices of survey collecting went some way to preventing these foibles of Linnaean taxonomy. Combining comprehensive geographical coverage with local series collecting was crucial: by revealing the full range of intraspecies variability, it enabled sorters to distinguish individual variations from those that demarcated species gaps, and it discouraged them from claiming as new species any forms that were extreme but still within a species' normal range of variation. Likewise, knowing the ranges of variability of neighboring or closely related species enabled sorters to weed out hybrids and place in-between specimens with confidence in one pile or another. In other words, large collections made more visible the gaps that actually existed between species, and less visible those that were the products of wishful thinking. Abundant data disciplined classifiers' unruly appetite for new things.

The quality of taxonomic knowledge thus depended directly on the scale and thoroughness of field collecting. Survey collecting reduced the temptation to excessive splitting that was built into taxonomists' communal practices of valuing new species above all else, and of naming novelties for themselves. Linnaean taxonomy thus became an exacting science, and the species concept more secure.

Taxonomists in the survey period were well aware of the benefits of mass collecting. As Robert Ridgway put it, deciding on species and subspecies "is very much a matter of material, both from a geographical and a numerical point of view." Systematics "enters the exact phase of the science," Edward D. Cope observed, "[as] soon as sufficient material becomes available."[7] Only with large and comprehensive collections, Ernst Mayr observed, could taxonomists feel secure in their conclusions: "The study of such collections has the gratifying feeling that it is unlikely that his finding will be upset by future discoveries." (He was speaking of the great Whitney collection of South Seas birds at the American Museum, the apotheosis of survey collecting, with which he worked out the concept of geographical speciation.)[8] And taxonomists were discouraged from prematurely publishing work on incomplete collections, knowing they were certain to suffer correction. "Only the working taxonomist knows," Mayr reflected, "how many revisions were started and then set aside for lack of adequate material."[9] It was a healthy reticence.

Taxonomists are rarely heard to complain of having too many specimens, but in fact too many could be as much a problem as too few. Robert Ridgway once had this experience when, as curator at the National Museum, he was obliged to sort a series of 129 specimens of a perplexing genus of birds collected from an area just five miles wide. It was "not very pleasant or promising," he reported, because there were so many minor intermediates that did not fit unambiguously in any pile. The task of sorting would have been easier with fewer specimens representing the same overall range of variation, but with some intermediates left out, or if the collecting had been done in a larger geographical area, so that the main forms would be visibly localized. As it was, "the gaps between the extreme forms are completely filled by specimens presenting every intermediate phase of variation."[10]

This unusual little episode reminds us that there is an optimal density of data for making natural units clearly visible: neither too few for an area, nor too many. Collections that were too sparse invited sorters to detect gaps that were in fact just artifacts of incomplete collecting. With collections that are too dense, natural gaps will be obscured by intermediates that make the gaps too narrow for sorters to take seriously. Likewise, for field naturalists there is an optimal balance of extensive and intensive collecting, thorough enough to prevent phoney gaps from appearing, but not so thorough as to make real gaps vanish in a smog of variants.

The practice of biological survey was, as it happened, just in that middle ground—a consequence of the logistical imperatives of collecting that was both extensive and intensive. Surveys covering entire states or regions had limited time in the field and many places to work. Consequently, survey teams could not afford to stop too long at any one locale and were prevented from collecting in the kind of depth that would confound sorters with variations. There were always more stations on their itinerary that had to be worked before the first autumn snows brought field seasons to an end. Series collecting from well-spaced stations produced just about the right density of data for natural gaps between species to be most easily seen.

It is odd to think that an exacting science depends on having limited data to work with, but so it is, and probably is as well for any science that uses qualitative units quantitatively. For example, genetics: Edgar Anderson tried to imagine what mapping genetics would be like if geneticists had to treat the variability of their unit characters biometrically.[11] Probably it would be like Ridgway's experience of sorting a too dense collection of specimens—"not very pleasant and promising." Total knowledge of nature is knowledge of noise, which is no knowledge at all; less than total knowledge is a knowledge of gaps, patterns, and meanings.

In other ways as well, survey collecting helped make Linnaean taxonomy more exacting. The custom of taxonomists themselves collecting in the field, rather than leaving that task to specialists, was crucial, because modern taxonomy depends as much on geographical and ecological knowledge of living creatures as on physical measurements made on preserved specimens. Such knowledge could be gained from field notes, of course—if collectors kept good notebooks—but it was best acquired first hand, in situ. Arnold E. Ortmann, whose revision of Pennsylvania crayfish (*Cambarus*) was a model, observed that for modern taxonomy it was essential that taxonomists do their own collecting, because only the ones who did the sorting knew what specimens from what locales would make for exact discriminations.[12]

In his great study of Andean birds Frank Chapman was likewise impressed by "the great difference in results obtained by modern methods of collection as compared with those employed by natives." Most of the large collections then in museums had been made by local or commercial collectors who kept no field notes, and for modern taxonomy such collections were useless. Chapman tried to work up such older collections but finally gave it up as "simply hopeless without some personal knowledge of the country." With that personal experience, Chapman found that variations that had seemed merely random could be seen to be geographically patterned and crucial clues to how Andean bird species had dispersed and evolved.[13] It was especially difficult to work without field experience in regions where different habitats were packed closely together, as they often are in the tropics. Firsthand knowledge of habitats provided essential clues to judging the value of morphological variation.[14]

Sorting specimens by geographical location was often as revealing as sorting by physical differences, especially where the latter were small. A good example is Edgar Anderson's account of how he and Robert Woodson sorted out the spiderworts (*Tradescantia*), a distressingly varied genus, into which botanists had over the years stuffed numerous and inconsistent species. Anderson and Woodson brought order to this tangle by sorting alternatively by place and form. The plant (a familiar weed) is generally hairless, but many individuals have fine hairs on or near the flowers, some of which are pointed, some knobbed. Anderson and Woodson sorted hairy specimens into pointed and knobbed and found that the former all came from the Great Plains and Rocky Mountains, while the latter all came from the Midwest and Southeast. They then sorted the hairless plants geographically into eastern and western, and observed that those with a few hairs had them in different anatomical locations in the two groups—a pattern that would before have seemed random noise. The chaos of variation began to take coherent shape, and by working back and forth between morphological and geographical sorting, Anderson and Woodson finally showed that there were not many but just two species of *Tradescantia*, both highly variable but occupying distinct regions and varying in ways that could be readily distinguished—once they knew what patterns to look for.[15]

Sorters seldom describe their methods of work as cogently as Anderson did. (I have found no photographs of taxonomists at work: plenty of unpeopled storage rooms, but no sorters at tables of specimens.) But they may occasionally say what they do if provoked by attack, as Robert Ridgway was when a British ornithologist dismissed all subspecies of American game birds as mere individual variations and accused Ridgway of mistaking sexual variations in the mountain partridge (*Oreortyx pictus*) for subspecific ones—a tyro's error. Ridgway revisted his specimens, cross-tabulating males and females separately by their subspecific characters (brown and gray napes) and by geographical locale. The tabulated data left no doubt that he was dealing with natural categories: the brown-naped birds were all from one region, the gray-naped from another, with virtually no overlaps.[16] Although such cross-tabulations were seldom published, it seems likely that Ridgway's method was a standard operating procedure. Of course this diagnostic technique did not work with just any collection, but only with those that were geographically comprehensive and deep—that is, collections assembled by survey methods.

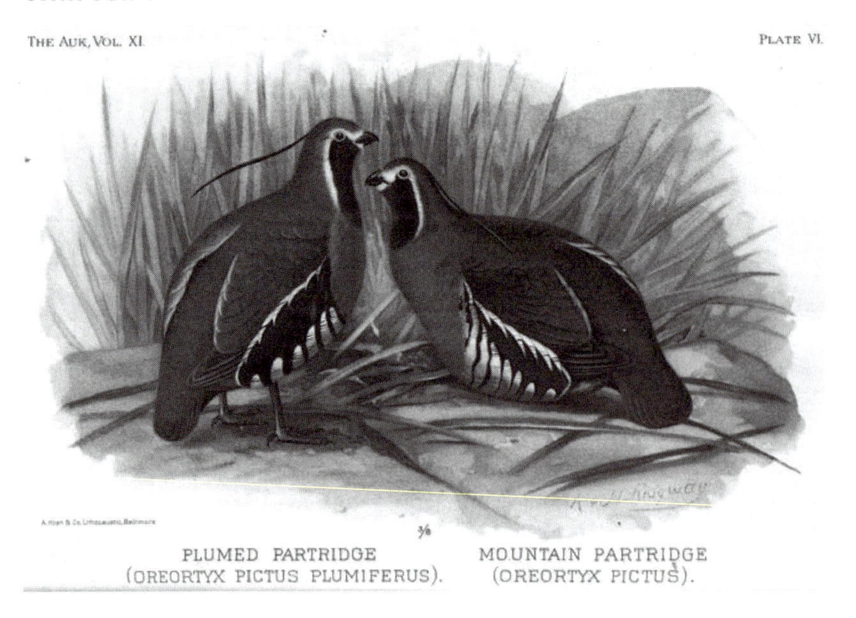

Fig. 6–1. Subspecies of the mountain partridge, used by Robert Ridgway to demonstrate the reality of the subspecies category. In Robert Ridgway, "Geographical, versus sexual, variation in *Oreortyx pictus*," *Auk* 11 (1894): 193–97, p. 193.

Direct experience of species in nature could also give taxonomists a conviction of their reality that was hard to get from preserved specimens. Edgar Anderson, who spent five years crisscrossing North America collecting species of *Iris*—a bewilderingly variable genus—testified to his growing conviction of the stability of species. He had expected to find evidence of transitional forms, but was impressed instead by how much his species, for all their variations, stayed the same. "They were of different fabrics," Anderson wrote: "One might compare them to two old English villages, one in the sandstone region and the other in limestone. In each village there would be no two houses alike but all the houses in one village would be made of limestone, all those in the other made of sandstone." It was the same with species: "The variation *within* could never be compounded into the variation *between*. The two species were made of two different materials."[17]

The entomologist Alfred Kinsey had the same experience collecting gall wasps of the genus *Cynips*—another wildly variable and confusing form. Trying to order specimens indoors, one might doubt that species were anything but arbitrary categories, but such doubts were dispelled by seeing how in nature species varied but always in the same ways, in one place after another over vast areas, like a river, always different but always the same. "[A]fter such field experience," Kinsey reflected, "one comes to feel there is a reality summed up in the word 'species' which is more than a few cabinet specimens or a bottle full of experimental material or a Latin binomial in a textbook. It is an existent, tremendous population of living individuals," each with its own identity, historical development, and geography.[18]

Assurance of the reality of species was vital for a science that dealt in differences that could be barely perceptible or invisible to outsiders poised to dismiss all taxonomy as merely subjective. Survey collecting brought empirical precision, self-confidence, and public credibility to a science that seems always on the defensive. Taxonomists tell stories of colleagues possessing almost magical powers of discrimination, who can see at once differences that others could not see at all. Carl Hubbs had that knack with fish, for example, and Ernst Mayr with birds.[19] Mayr recounted how, as a young medical student, he was invited to do volunteer work at the Berlin Museum and was handed trays of some species of brown creepers that were so alike that even experienced sorters took days to sort them out and then got a third of them wrong. Mayr returned the trays in half an hour with no mistakes. "It turned out that I had realized and recognized the intangible differences in about one glance," he recalled.[20] Such power may seem uncanny, but it is not, any more than is the power that some clinicians have of diagnosing enigmatic ailments. It is simply an ability to see subtle patterns of variation in complex traits, and is a skill acquired by deep and extensive experience with handling things.[21] It is a power that was called forth by working with the deep and comprehensive collections brought home by survey teams.

Taxonomists: A Natural History

Not all biologists were prepared to admit that a collecting science could ever be as rigorous and exact as an experimental one. The ichthylologist Carl Eigenmann took his friend Charles Davenport to task for disparaging taxonomy as a trivial pursuit, when in fact it was in practice not so different from Davenport's own work in experimental heredity:

> You are attempting to get at the transmissability of certain characters. To do this you have to keep an endless number of uninteresting but exact records, describe birds, regulate your incubators and feed your chicks. . . . Now I am laboriously examining a lot of dead smelly fish and writing formulae that enable me and should enable others to distinguish one form from another. Don't imagine for a moment that this is any more edifying or pleasurable than turning eggs or regulating brooders or feeding checks but it is my method of keeping records and it is just as important for me to distinguish between minute differences found in nature as it is for you to distinguish slight differences in the combs you produce.

How would Davenport like it, Eigenmann demanded, if some one who had not seen his chickens announced that his series of combs were all alike? "I think you would be one of the first to recommend a lunacy commission or an entire disregard for such vaporing."[22]

Edgar Anderson, who was trained as a geneticist and came to fieldwork later, also observed how eminently modern taxonomic practice was, with its standardized, international rules of practice. "Taxonomists," he wrote, "though frequently considered hopelessly conservative, were really among the first scientists to seize upon the efficient principle of standardized interchangeable parts, or rather, they invented it for themselves before its advantages had become apparent to the industrial world."[23]

Ernst Mayr likened museum and herbarium collections to experimentalists' precision instruments: "[A]s the cellular biologist needs electron microscopes, Warburg apparatuses, centrifuges, and other expensive types of laboratory equipment and chemicals, so does the systematic biologist need collections."[24] Alexander Ruthven made a similar point about type specimens: "To biologists, types are what original instruments of high precision are to physicists and engineers and are the base of primary description."[25] An even better analogy might be made with standard weights and measures, the basis of all precise empirical science.

Early twentieth-century taxonomists were fond of comparing modern taxonomic practice with experimental genetics. Carl Hubbs boldly declared that modern taxonomy, with its vast data

Fig. 6–2. Examining a tray of specimens from the newly purchased Rothschild collection, American Museum of Natural History, February 1935. Ernst Mayr is second from the left. Courtesy Department of Library Services, American Museum of Natural History, negative no. 314574.

and biometric techniques for measuring variation, was as great an advance on the old taxonomy as modern cytogenetics was on Galton's genetics.[26] Ernst Mayr also observed that taxonomic sorting was as quantitative as genetic mapping: the more abundant the data, the more precise the results.[27] It was, Mayr mused, surprisingly like doing experiments, manipulating variables one at a time, not experimentally but by selecting various geographical situations: "The ornithologist knows his material so well that he can do what the geneticist also does, that is to pick out one particular character and study its fate under the influence of geographical variation."[28] For that to be true, of course, one needed a collection that was deep and complete, like the Whitney collection of South Pacific birds, of which Mayr was in charge.

There was some bravado in these assertions of taxonomic dignity, but also much truth. In fact, taxonomy had become rigorous and exacting, thanks to comprehensive series collecting. Modern taxonomists value exact and abundant data; they count, measure, and compute; they turn geographical comparisons into quasi experiments.[29] And it all begins in the field, with exacting and systematic collecting.

Survey collecting also encouraged taxonomists to shift conceptually from a narrow, typological view of species to a broad and biological one. Taxonomists sort themselves into "lumpers" and "splitters"—that is, those who downplay small differences and sort specimens into a few large piles; and those who value small differences and like a table covered with many small piles. In modern terms, lumpers are a broad-species type, favoring capacious species with much internal variation and many subcategories, whereas splitters are a narrow-species type, who favor species with little internal variation and subdivision. Taxonomists sometimes ascribe these different habits to individual variability: for example, physiological differences in visual acuity, or emotional attachment to particular, beloved kinds of creature.[30] But in fact, lumpers' and splitters' differing habits have less to do with individual eyesight or temperament than with the size and depth of the collections they have to work with.

Periods in which collecting is expanding but not yet comprehensive will favor splitters, because incomplete knowledge of natural variability makes it easy to see a few distinctive individuals as representatives of true species. Also, any increase in the pace of discovery can foster a gold-rush mentality and call forth many phantom species. Such events are not an uncommon occurrence in classifying sciences. In chemistry, for example, the coincident invention of periodic tables and chemical spectroscopy produced a gold rush of "new" elements, many of which proved to be figments.[31] It was the same in the early days of genetic mapping, when recognition of a mutant phenotype triggered "epidemics" of similar mutants.[32] And so, too, with species.

As collections grow and fill in, however, the world becomes a kinder place for lumpers, who downgrade species to subspecies or synonyms, and root out false pretenders from taxonomic lists. Not infrequently in such times lumping gets out of hand, as splitting had previously. There is credit to be earned by taxonomic housecleaning: not perhaps as much as by discovering new species, but credit enough to inspire overzealous purging—species "epidemics" in reverse. Newly enlarged collections made it easy to discredit species identified on the basis of now all too obviously incomplete data. At such moments good species may suffer temporary loss of identity and standing—a kind of virtual mass extinction, until revisers with even more comprehensive data achieve a balance between broad and narrow species concepts.

The sporadic but relentless growth of collections has also produced cycles of confidence and disbelief in the reality of species. With just a few specimens to sort, and happily unaware of the complexity of natural variation, taxonomists will feel confident in their work of creating new species. However, as specimens pile up and the full reality of natural variation begins to be apparent, confidence is succeeded by uncertainty and doubt that such a variable world can be divided into natural units. The nadir of despair will occur when collections are large enough to discredit earlier designations of species, but still too small to reveal the true natural gaps between kinds. Then taxonomic "nihilists" will deny that species are anything more than human conventions, which we invent and name for our own convenience—like the colors we see in the continuous electromagnetic spectrum. Taxonomists have spent years or decades in these uncomfortable middle stages, making piles on the sorting table in the certain knowledge that their judgments will be overturned in the next revision. But as collections become more complete and comprehensive, opinions tend to converge, and sorters can proceed with confidence that their judgments will last.[33]

Taxonomists whose careers overlapped the invention and spread of survey collecting experienced the whole cycle of splitting and lumping, confidence and despair—confident splitting in the aftermath of the collecting rush of the mid-nineteenth century, then growing doubts and episodes of lumping, and returning confidence and balance in the first few decades of the twentieth century. However, each taxonomic specialty—birds, mammals, reptiles, fishes, insects—experienced these cycles at different times, depending on the extent to which practitioners engaged in mass collecting.

Ornithologists were the first to adopt modern taxonomic practices, because birds were the first faunal group to be deeply collected and comprehensively known, even before the age of survey collecting. Mammalogists were the next to go through this cycle, between the 1890s and 1910s, as collections became full and comprehensive. In North America more specimens were collected in just the 1890s than hitherto in the entire world, the number of known forms increasing from 363 in 1885 to about 1,450 in 1901. The American Museum acquired some six thousand specimens of North American mammals in the early 1890s alone.[34]

Ichthyologists entered the cycle still later, in the 1910s to 1930s. Less collectable than birds and mammals, fishes are kept not as dried skins but as whole (and smelly) specimens in jars of alcohol or formalin, and they are expensive to house and unpleasant to work with. If fishes were as aesthetically pleasing and pleasant objects to handle as birds, David Starr Jordan surmised in 1905, the taxonomy of fishes would be as advanced as that of birds. He was probably right. Twenty years later he saw little improvement; however, changes were in fact under way.[35] Reptiles and amphibians posed the same problems for collectors as fishes and were subjected to modern methods of collecting and classifying about the same time. Alexander Ruthven's 1908 revision of garter snakes, based on extensive field collecting, set an example that others gradually followed.[36]

In entomology modern methods were even slower to be adopted, owing to the spectacularly large number of insect species and the difficulty of knowing them all. Insect fanciers' taste for representative collections of entire families (coleoptera, lepidoptera) also hindered the growth of the deep collections that modern taxonomy required.[37] Collectors are not to blame, really; it is just very difficult to apply the exact methods of survey collecting to forms that come in so very many kinds. And without more or less complete collections, modern exact taxonomic methods simply do not work. (Botanists, though they pioneered in applying experimental methods to taxonomy, adopted mass collecting only in the 1920s and 1930s.)[38]

The uneven front of change among taxonomic specialties caused some discomfort to taxonomists who wanted to promote mass collecting and modern biological concepts of species as universal standards, but could not in good conscience do so for fear of alienating colleagues in other specialties who could not use them for lack of proper data. As Wilfred Osgood put it:

> If we knew all the facts about the characters of animals and plants it is probable we could come to something approaching unanimity as to a scheme of nomenclature. . . . But our knowledge is extremely variable, almost complete in some groups and very deficient for others; and we are trying to apply the same system of nomenclature to all of them.[39]

It was tempting to imagine that exacting analysis could compensate for imperfect data, but such confidence was ill placed.

Thirty years later the botanist Willard Camp made the same point about the uneven progress of plant taxonomy: "[C]ertain segments of the plant kingdom are rather well known; other parts are almost unknown. Yet the taxonomic techniques of the "unknown" segment more or less sets the pace for the whole attitude toward the portions which are better known."[40] In the late 1930s Ernst Mayr tried as an experiment to apply the advanced techniques of bird taxonomy to insects, but found it quite impossible, even with a genus (*Cynips*) that had been collected in depth (by Alfred Kinsey).[41]

The ideal of universality in rules of practice was deeply rooted in the community of taxonomists. But universality could be a reality only in a world of uniformly adequate field data. Meanwhile, departures from the ideal could be practically and morally troubling. Although survey collecting made species taxonomy more exact, the gradual and uneven spread of organized collecting meant that disparities between ideals and realities were probably more acute in the age of survey than ever before. Such disparities are no doubt ubiquitous in science; but I think they must be especially chronic in the collecting sciences.

These are rough-and-ready generalizations, which will no doubt suffer modification as historical data improve. The history of taxonomy is a long way from its own survey mode of practice.

Subspecies and Practice

If survey collecting made species taxonomy a more exacting and confident science, it created subspecies taxonomy de novo. That is, subspecies acquired the demonstrable and stable identity of a taxonomic unit only in collections that were both deep and geographically comprehensive. It is an unusually clear instance of the dependence of units and categories on the scale of collections. The species unit is less critically dependent on collecting practice: one can define it with one specimen or one thousand, though not with the same degree of certitude. Not so with subspecies. Below a certain threshold of scale in collecting they cannot be defined at all: they remain invisible. And above a certain density of collecting they again vanish in a blur of data—as we will see later. That is why the "subspecies research program," as the historian Mark Barrow has called it, coincided almost exactly with the era of organized survey collecting.[42] A taxonomy of subspecies and trinomials became a credible and workable system of classification only within a certain range of field collecting.

Not surprisingly, the meaning of a subspecies changed substantially over time. In the mid-nineteenth century subspecies were generally taken to be provisional species: that is, forms that might or might not prove to be real Linnaean species—species on probation, so to speak. This was not nature's uncertainty, but our own. But by the late nineteenth and early twentieth centuries, subspecies were understood to be a distinct taxonomic—and natural—unit. Biologically they were regarded as incipient species, that is, forms partway to becoming new species but not there yet—evolutionary, rather than merely taxonomic, probationers. Subspecies and species were real but different sorts of things, possessing different taxonomic standing.

For taxonomists in the age of survey collecting, subspecies were biogeographical units, defined (unlike species) more by their spatial than their morphological characteristics. Many subspecies can be distinguished by morphological differences, but these are often slight—and some can only be detected by applying statistics to minute differences. However, in nature subspecies always occupy definite and nonoverlapping geographical ranges, which can be seen through inspection by simple mapping. In practice, their spatial character is the most stable and reliable diagnostic feature of subspecies. Species, in contrast, are defined primarily on the basis of degrees of morphological differences (a habit reinforced by the tradition of "type" specimens). Geographical location may help to identify species, especially in the field, but it is seldom a diagnostic character, because species ranges overlap.

The problem with a spatial definition of subspecies, of course, is that the distributional data are laborious to get, so that in practice, diagnoses often had to be made using incomplete biogeographical data. For that reason there has always been a tension between morphological and spatial methods of diagnosis. Morphological differences are used if exact geographical data are lacking, but usually by applying some arbitrary rule of thumb: for example, the rule that a group of specimens can be deemed worthy of a Latin trinomial if a certain fraction of them (say, three-quarters, or seven-

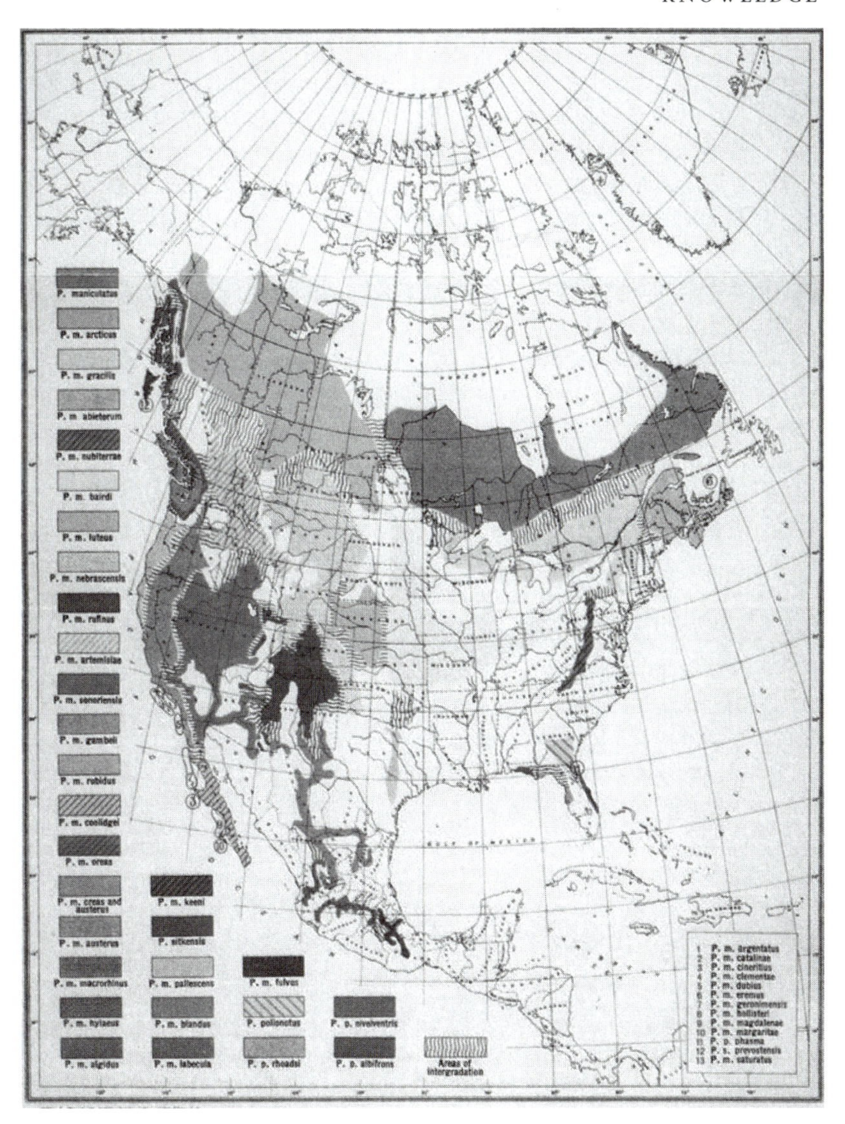

Fig. 6–3. Biogeographical map of the *Peromyscus maniculatus* group of species and subspecies (white-footed mouse). Note the well-defined areas of intergradation, crucial for diagnosing subspecies. In Wilfred Osgood, *Revision of the Mice of the American Genus* Peromyscus (Washington: Government Printing Office, 1909), plate 1.

eighths—the fraction varied) can be identified by sight without knowing their geographical origins. It was hardly an exact method, though it worked well enough. But it was geographical specificity that made subspecies a legitimate taxonomic unit.[43]

Subspecies are natural units, but not as unambiguously as species are, because unlike species they are not reproductively isolated, but interbreed freely where their ranges meet. This is the biological reason why subspecies do not live together "sympatrically" (in the same place) as species do. Species have evolved different habitat preferences, so do not directly compete and can cohabit. Subspecies remain sufficiently alike in their behaviors that they do compete; and in any particular habitat one and only one subspecies will have an adaptive edge sufficient to outreproduce the other and in time achieve sole occupation. In addition, because subspecies interbreed, forms that do not achieve geographical separation blend back into their parent species. That is why definite geographical range and intergradation along boundaries are the strongest empirical criteria of subspecific status. It is also why subspecies have often been regarded as incipient species, or, more precisely, as partly divergent forms that may or may not further separate into distinct species.[44]

It will be evident by now why subspecies taxonomy depended so specifically on the practices of survey collecting. To diagnose subspecies, taxonomists need collections large enough to reveal the full range of individual variation. More important, they require series of specimens from every portion of the relevant geographical area—both from the centers of ranges, where typical forms usually reside, but also and especially from the boundaries between subspecies' ranges, where interbreeding will occur if it does at all. The practices of survey collecting were precisely suited to produce such collections: locally intensive, producing large series; and geographically extensive, covering entire faunal regions. Survey collecting was the only way for taxonomists to be confident that they were not just dignifying with Latin names some undeserving variant or local race.[45]

In principle, such comprehensive collections could have been assembled by individual local collectors, but in practice they were not, because survey and amateur collecting had such different aims and methods. Individuals collecting for themselves aimed at having a few typical examples of all the species of a region, especially the rarer ones. As one of Joseph Grinnell's collectors put it: "The great trouble with the bird men in Cal[ifornia] and most other places, is that they waste their time on getting a general collection of 6[00] or 700 skins and then get weary."[46] In contrast, survey collectors' desire for large numbers of common species would reveal patterns of subspecific variation. Also, individual collectors liked to collect in the centers of species' ranges, where specimens were typical, whereas survey collectors combed species' entire ranges and especially the critical boundaries between ranges where hybrids and intergrades would occur—specimens without value for a synoptic collection. What made a collection attractive to amateur naturalists made it useless for purposes of subspecies taxonomy, and vice versa. In practice, only organized survey collecting could produce the deep and comprehensive collections required for subspecies taxonomy.

It is not surprising that amateur collectors were famously hostile to the trinomial system of nomenclature and to the whole idea of subspecies taxonomy, as Mark Barrow has recounted.[47] Few amateur naturalists were able to distinguish subspecies or cared to; trinomials were an annoying and confusing imposition of distinctions of interest only to museum specialists—indeed, visible in the field only to them. To museum taxonomists, amateurs were looking in the wrong places and in the wrong way.

Besides the data, survey collecting also provided a kind of existential evidence of the reality of subspecies, in the repeated experience of finding particular forms only in certain geographical areas and never in others. This sensible experience of boundary gaps could convince where verbal description and even maps could not. For example, Joseph Grinnell was frustrated when a committee of the American Ornithologists' Union balked at accepting a new subspecies designation for a California bird. He sent them a few specimens with his formal descriptions—the custom in applications for approval of new species—but these did not persuade. As a man of the field, Grinnell was annoyed but not surprised: "Of course you know how unsatisfactory anything but a series is for modern subspecific determination. But series are bulky [to mail]. I would like to exhibit (and *explain*) mine *personally* before the Committee. . . . I should like to show physiographic maps, or, better, take you around to the different habitats!"[48] Experience could dispel doubt where descriptions and data alone could not. Experience of California's complex biotopography made it especially easy to believe that subspecies were real, because there were so many of them packed so closely together, yet separate. (Grinnell identified eleven in one species, within just ninety miles of San Francisco!) But one had to be there, in the field, to get a visceral sense of the reality and constancy of subspecific territories.

Even at height of their popularity, between the 1880s and 1930s, subspecies never had the secure identity that species enjoyed. General biologists largely detested the category, for the same reasons that amateur collectors did: it was too arcane, too confusing, too mutable; their work was better served by a few generic animals. If species taxonomy seemed a necessary evil to laboratory biologists, then subspecies were a quite unnecessary evil. The standard joke was that taxonomists invented subspecies only so that they would have something new to name after themselves when they ran out of species. One taxonomist was told by colleagues—jokingly, he thought, or hoped—that "ornithologists, having examined the bills, feet, and feathers of all of their species, invented the subspecies in order to stall off the day when they would have to investigate some other characters."[49] Since subspecies were of no evident use to experimental or general biologists, they must have been invented for personal and social reasons—that was the logic of the jokes.

One reason for this skepticism was the recurring episodes of nomenclatural churning. When collections passed a threshold of scale, taxonomists would often engage in wholesale demoting of species to subspecies, or of elevating subspecies to species rank. This happened in ornithology in the United States in the late nineteenth century, and again in Europe from the early 1900s to the 1920s, when a world list of almost 19,000 "species" was reduced in short order to less than 9,000, plus a roughly equal number of subspecies.[50] Memories of such massive, and to outsiders seemingly pointless, changes were long-lived.

Another problem was that the criteria for defining subspecies, though conceptually precise and unambiguous, were in practice often applied by imprecise rules of thumb: for example, the three-quarters or seven-eighths rules of morphological difference. Variable rules seemed blatantly subjective. There was also the logical problem of island forms and of forms so widely separated that they had no common boundary. These might intergrade if they were not separated by physical barriers, but the test could not be performed,

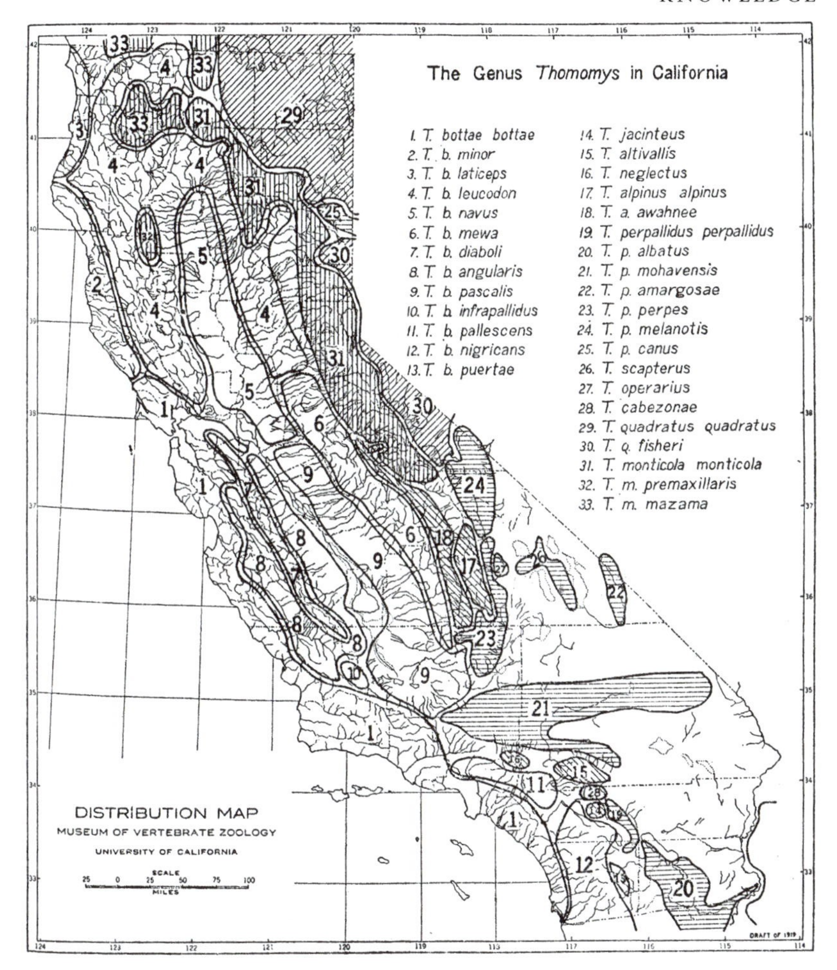

Fig. 6–4. Biogeographical distribution of the California species and subspecies of pocket gopher (*Geomys*), indicating that region's unusually compact and complex ecology. In Joseph Grinnell, "Geography and evolution in the pocket gophers of California," *Annual Report of the Smithsonian Institution* (1926): 343–55, p. 347.

so taxonomists fell back on extrapolations from known instances: a kind of educated guessing.

For example, one could assume that forms separated along environmental boundaries were probably subspecies, since subspecific characters were believed to have arisen as adaptations to different environments. It seemed a plausible shortcut. As Frank Chapman put it: "When . . . the environments, so to speak, intergrade, we may be reasonably sure that the forms to which they have given rise will also merge; and under such conditions it is customary to rank them as subspecies without the confirmation of intergrading specimens." Plausible, but quite unprovable. In cases of island forms, the problem could be sidestepped by arbitrarily classifying all as subspecies—or species, one way or the other—leaving it to future workers to sort out. This rule of thumb had the virtue of being mechanically "objective"—that is, consistent and transparent.[51] But it was biologically indefensible: little better than an admission of ignorance, just less candid.

So taxonomists generally preferred to use degrees of morphological difference to extrapolate from known cases of intergrading to cases in which such evidence was lacking. For example, differences of size and color intensity generally did not distinguish species, but differences of morphological shape and color pattern did, and this rule of thumb could be assumed to apply to subspecies as well.[52] With enough experience, taxonomists could hope that their judgments would be right in more cases than not. But the method was not "objective" in the modern sense, because it depended not on universal rules but on individuals' experience and know-how—on their "judgment and . . . systematic tact," as Ernst Mayr put it: it was "only a matter of taste."[53] That was distasteful for many biologists and even some taxonomists, like C. Hart Merriam, who deplored a system that invited taxonomists who lacked evidence to rely on the "personal equation." Better, he thought, to abandon the intergradation rule and decide exclusively on degrees of morphological difference.[54] But that seemed old-fashioned, and few taxonomists followed his lead.

In fact, taxonomists' rules of thumb did work, because judgment and "systematic tact" were not personal idiosyncracies, but ways of knowing that had been tested and refined by the work of many hands. Such methods are opaque only to those who lack the firsthand experience of the work.[55] Survey collecting made subspecies taxonomy a sound and workable system: not as exacting as species taxonomy, and messy around the edges, but a mode of practice that experienced collectors and sorters could use with confidence. It was knowing nature through work.

Subspecies: The History

Not surprisingly, the concepts and practices of subspecies taxonomy co-evolved with the practice of mass collecting. Something like the idea of subspecies appeared as early as the 1830s, took on its modern meaning in the 1880s and 1890s, as survey collecting became the norm, and was adopted more or less universally by 1910. Although subspecies taxonomy came to be known as the "American system," its first practitioners were Dutch, Swedish, and German ornithologists, who in the 1840s and 1850s used trinomials to designate variants without giving them permanent Linnaean standing—species in limbo. Significantly, these naturalists—Herman Schlegel, Carl Sundevall, C. L. Gloger, A. Thomas von Middendorff—were explorer-naturalists who had extensive field experience as collectors. In contrast, taxonomists who held to the concept of species as invariant types tended to be museum men who worked just with specimens. As Middendorff observed, "Nature appears very different to the travelling naturalist when he daily pursues his researches amidst the richest animal life, impressed by its endless shapes; and very different to the specialist handling a few dry skins in a museum."[56] It appears to be a general rule that firsthand experience of the variability of animals in nature inclines naturalists to value local races as taxonomically significant.

In the United States, Spencer Baird was the first important naturalist to routinely apply trinomials to subspecific forms, in his reports on the western railroad and boundary surveys of the 1850s. These massive expeditions produced collections of an unprecedented size (though they would seem small enough to later collectors). Although these surveys were basically transects designed to locate routes for military roads and transcontinental railroads, they were also intended to gather information about natural areas that might be of use to prospective settlers. So expedition naturalists were able to observe and collect in large swaths of territory, not just narrow rights-of-way. It was Baird's large-scale collecting that gave rise to his interest in subspecies and biogeography.[57] And as the man at the center of the National Museum's transcontinental network of collectors and naturalists, Baird was well placed to put the stamp of legitimacy on new taxonomic practices.[58]

Fig. 6–5. Spencer Baird, 1850. National Museum of Natural History archives, record unit 95, box 2, folder 5, photo no. 2005–3755. Courtesy National Museum, Washington, D.C.

As Jürgen Haffer has observed, it was the experience of having to deal with large numbers of variants that drove both Baird and the Europeans (independently) to adopt trinomials.[59] All the new variants coming in from the field were evidently not distinct species, but neither could they be dismissed as individual variants. As possible species they had to be acknowledged, but without flooding the literature with official species, many of which would clearly fail the test of time. Binomials would thus not do for these ambiguous forms. But lumping variants into existing species was even worse, because it would render forms invisible that might one day be revealed as distinct species. Subspecies and trinomials were a practical compromise, publicly registering variants without endangering the sanctity of the well-policed Linnaean system.

Baird was celebrated for his pioneering efforts to make taxonomy a precise, empirical science: collecting large series; insisting on uniformity and precision in labeling and describing specimens; and attaching to specimens their places and dates of collection. The "Bairdian system" was a practical response to the flood of specimens from the first western surveys. It was a system of data management, and trinomials were a part of that system. Baird believed in evolution but was not driven by theory (indeed, was criticized for his theoretical reticence). Rather, he adopted subspecies and trinomals as practical techniques of recording the facts of variation, leaving it to future taxonomists to decide what the facts meant.[60] As Robert Ridgway put it, trinomials were "handles for facts":

> The most important advantage of trinomials is that they serve as convenient "handles for facts," in providing for the naming of forms which are known not to possess the requirements of true species, but which it is equally evident demand . . . proper recognition. Without trinomials it would be necessary to either name such forms as species, and thus convey an idea of their rank which the person bestowing the name knows to be false, or else ignore them altogether, which would be plainly a dereliction of duty and a positive impediment to the progress of science.[61]

Subspecies were thus adopted as a category well before it was clear exactly what they were biologically. (Varieties or local adaptations? Half-evolved species? Divergent forms that had not been separated by the extinction of intermediate forms? All were possi-

Fig. 6–6. The ornithologist Robert Ridgway, 1873, age twenty-three. National Museum of Natural History archives, record unit 95, box 3, folder 43, negative no. 78–10100. Courtesy National Museum, Washington, D.C.

ble meanings.) This may seem odd, but not if we bear in mind that the principal use of subspecies was the practical one of recording facts of variation at a time when the empirical base of taxonomy was rapidly outgrowing the rules of classifying practice.

We also see the connection between mass collecting and advanced taxonomic practice at Louis Agassiz's Museum of Comparative Zoology. Like his archrival Baird, Agassiz had a policy of deep collecting. He collected actively and kept large series for the museum, not just representative types. Like Baird he aspired to make taxonomy more exact by improving its empirical base: only with large collections, he believed, would taxonomists ever distinguish individual from specific differences. Agassiz recognized the existence of forms restricted to particular geographical locales, though he did not call them subspecies and later disavowed their very existence, fearing they would lend support to Darwin's theory, which he passionately opposed.[62] Mass collecting made subspecies visible even to those who believed for theoretical reasons that there should be no such thing.

Although Agassiz's creationist convictions prevented him from using subspecies as handles for facts, they were embraced by his junior colleague, Joel Asaph Allen. An ardent Darwinian, Allen became one of the first Americans to openly adopt the subspecies as a taxonomic unit. Allen knew of Baird's reports of geographically delimited forms and realized that, if confirmed, they would powerfully support Darwin's theory. Encouraged by Agassiz's "greedy collecting policy," Allen embarked on a program of systematic in-depth collecting. He collected extensively in the Midwest in 1867–1868, and in 1868–1869 in Massachusetts and in Florida. He prepared large series of specimens, measured the ranges of variation, mapped boundaries and areas of intergradation, and correlated morphological variations with climatic gradients.[63]

Allen's was a one-man operation, but unmistakably in the survey mode, and it fully confirmed Baird's findings. In his report on the Florida survey (1871) Allen proposed that ornithologists recognize subspecies as a distinct taxonomic unit defined by geographic range and, most important, by intergradation. This stringent requirement—quite unrealistic at a time when few naturalists collected intergrades—was obviously intended to prevent overeager collectors from turning subspecies taxonomy into a surrogate species-making machine, free of the communal policing of Linnaean nomenclature.

For the same reason, no doubt, Allen declined to adopt trinomials, proposing instead a system of binomials augmented by a brief prose description of each subspecies' variations and geographical range. This clumsy expedient may seem oddly cautious for such a bold thinker. However, Allen knew the risk he was taking: giving a new outlet to naturalists' love of naming things might flood the world with phony subspecies. As the intergradation requirement would raise the bar for naming new forms, so would abstaining from trinomials reduce the psychic rewards that came from naming them for oneself.

However, Allen was overscrupulous. In a review of his Florida monograph, the ornithologist Elliott Coues argued that trinomials were just too convenient to forgo, and within a year Allen quietly adopted them himself.[64] Coues advocated trinomials as a practical—and temporary—expedient for recognizing forms that warranted neither binomials nor oblivion. It was bad enough to dilute the Linnaean roster of species with geographical races, he thought, but far worse to bury distinctions that had taken so much work of collecting, sorting, and recording. Why make future taxonomists do all that work over again?[65] In short, trinomials were a practical response—efficient and relatively harmless—to the problems of a stage in mass collecting when taxonomists could see subspecific units but not yet know exactly what they were. For that, still larger collections would be required.

In the next ten or fifteen years most American ornithologists and mammalogists also concluded that trinomials were a convenient if somewhat risky way of managing an ever expanding mass of data on variation. Their use, as Joel Allen observed, "was forced upon us by conviction of their utility and necessity." For Robert Ridgway, ignoring empirical facts was unscientific and "nothing less than suppression or perversion of an obvious truth."[66] Leonhard Stejneger likewise adopted trinomials (reluctantly) in the early 1880s to ensure that upstart forms would not vanish from memory for lack of proper names:

> There are still many ornithologists who would rather suffer the local races to be extinguished from our books than they would allow them to carry the "sacred" binominals. To them the suspecies are pariahs,

Fig. 6–7. Joel Asaph Allen, 1889. Ernst Mayr Library of the Museum of Comparative Zoology archives, Cambridge, Mass. Courtesy of the Museum archives.

> which must not be admitted to the "rank" of the aristocratic species. I, myself, think better of the poor subspecies, believing that science in time, when they are fully understood, will derive great benefit from their recognition, and consequently I accept the cumbersome trinomials rather than to see them go around without any name at all.

Trinomials were a nuisance, Stejneger thought, but "*a very necessary nuisance.*"[67] In short, handles for facts.

Some nervous naturalists compromised by applying prefixes to the third term of trinomials—like "var." (for "variety") or "subsp."—advertising the fact that trinomials were not equal in standing to "sacred" binomials. As Coues put it, naturalists worked up to trinomialism "by virtue of what may be termed subterfuges."[68] For many taxonomists subspecies were not a different kind of unit from species biologically; they just had a less certain taxonomic standing. As the American Ornithologists' Union put it, the trinomial system "amplifies, increases the effective force of, and lends a new precision to, the old [Linnaean] system."[69] It made binomial nomenclature more flexible and exact by putting provisional species on hold for future determination.

In 1884 the American Ornithologists' Union officially adopted plain unmodulated trinomial names for subspecies, together with Allen's stringent criterion of intergradation. There was so little critical discussion of this move that Coues himself felt obliged to post a warning that plain trinomials might crowd checklists with "an almost indefinite number of very slightly differentiated forms . . . which only the eye trained in that special line is able to . . . discriminate."[70] However, his fears proved unwarranted: describers were surprisingly restrained, and the system of subspecies taxonomy spread, along with the practices of mass collecting.

Ornithologists, of course, were the first to adopt the system en masse. While the number of species of North American birds changed little between 1886 and 1931 (from just under 800 to just over), the number of subspecies increased from 183 to 609.[71] The number of subspecies in North American reptiles and amphibia (both newly described and reduced from species status) likewise increased steadily but modestly at the rate of about six per year in the late nineteenth century and about fifteen per year in the early twentieth.[72]

These were substantial increases, but hardly the deluge that some had feared would be unleashed by trinomials. Nor as time passed were there the wholesale purges that some expected: just the usual promotions and demotions of subspecies to species and vice versa. Most subspecies designations held up well, because they were empirically well founded. Survey collecting prevented abuse: with large and comprehensive collections, subspecies could be diagnosed with some certainty. And if overzealous describers ran amok, they could be assured that it would be seen to be their fault, not the fault of their material—a powerful incentive for restraint. The "American system" of trinomials was adopted by a committee of the Fourth International Congress of Zoologists in 1898 and was ratified by the Fifth in 1901, guaranteeing that it would become a universal practice.[73]

With the rich empirical evidence accumulated by survey collectors, there could be little doubt that subspecies were real. As Wilfred Osgood observed in 1909, if specialists in some animal group doubted the reality of subspecies, it meant only that they had not yet assembled collections big enough to achieve certainty.[74] It was the same with intergradation: those who believed that the criterion did not apply to their groups (as entomologists and ichthyologists still did not in the 1910s) simply lacked adequate data. When their collections were sufficiently large and comprehensive, Osgood predicted, they too would find that intergradation worked perfectly well.[75] He knew what he was talking about: his 1910 revision of the species and subspecies of the genus *Peromyscus* was a paradigm of modern taxonomic practice. Survey collecting made subspecies believable.

The lack of such collections, more than theoretical bias, was why British ornithologists were slow to adopt the "American system." In London in 1884 to sell British ornithologists on trinomials, Elliott Coues found a few hard-line defenders of strict binomialism, but mostly what he called "limited trinomialists"—that is, taxonomists who knew they would eventually adopt trinomials, but not until their collections had become large enough to make the system work. At the time, the Eurasian Palearctic bird fauna was simply not as completely known as the North American, and not even the largest European museums had collections as deep as the best in America. Why adopt a potentially subversive practice until it could actually be used with confidence? Resistance was practical, not

Fig. 6–8. Wilfred H. Osgood in the field, Mount Lassen, California, 1898. Photo courtesy of The Field Museum of Natural History, negative no. Z84063. Copy made from archival print.

ideological. Significantly, the British ornithologist most open to the American system, Henry Seebohm, had extensive field experience in Europe and Siberia and had studied intergradations.[76]

Practical familiarity with diagnosing subspecies also gradually clarified their biological meaning. It was always clear enough that domesticated breeds, hybrids, and sports did not qualify as subspecies. Less clear was the meaning of the variations that correlated not with place but with climatic gradients. For example, animals in colder areas were larger and had shorter extremities than those in warmer areas, and those that lived in wetter climates had darker pelts and plumage that those in drier. These variations—clines, we now call them—are not geographically bounded as subspecies are and cannot be divided into discrete units. Yet taxonomists at first conflated the two types of variation. Elliott Coues, for example, wrote that "the third term is given to climatic or geographical races varying according to known conditions, as latitude, elevation, temperature, moisture and conditions of all sorts."[77]

It was only in the mid-1890s that taxonomists fully grasped the difference between subspecific and clinal variation. Consider, for example, the two definitions of subspecies given by Robert Ridgway, one in 1879, the other in 1901. The first defines subspecies as "species in the process of differentiation, incipient species, where the intermediate forms have not yet died out, but where a series gradually leading from one extreme to the other may be obtained."[78] It is an image of continuous variation. The second, in contrast, depicts subspecies as discrete units occupying "a definite geographic area of more or less marked peculiarities of topography, climate, or other physical features. Such forms are fixed, or 'true,' over territory of uniform physical character, the intergrades coming from the meeting ground of two such areas."[79] Exactly how this change occurred is unclear, but there is no doubt that it followed the adoption of series collecting on a continental scale. Two taxonomists said as much in 1901: it was "chiefly through . . . vastly more extensive series of specimens that the distinction between geographic variation and individual variation has been made."[80] Experience with survey collecting made the difference between clinal and subspecific variation spatially self-evident.

The framers of the AOU trinomial conventions had anticipated in 1884 that survey collecting would made subspecies a workable taxonomic category. "[U]ntil the animals of large areas become well known . . . through extensive suites of specimens," they predicted, "neither the necessity of trinomialism, nor the possibility of putting it to the proper test [of intergradation] is apparent."[81] When European ornithologists collected Old World birds as thoroughly as Americans had the birds of the North American West, Allen had predicted, they too would see the necessity of trinomials.[82] And so they did. Survey collecting distinguished subspecies from individual and clinal variations, and gave taxonomists confidence that applying trinomials to regional variants would not just create verbal thickets for a future generation to sweep away. Subspecies taxonomy and survey collecting flourished productively together until further changes in collecting practices ushered in another cycle of doubt and change.

Subspecies in Crisis

Enthusiasm for subspecies began to fade in the late 1930s, and in the late 1940s and early 1950s their value was being widely questioned. An informal survey of leading systematists and geneticists in 1954 revealed near universal dissatisfaction with the practices of subspecies taxonomy. The number of named subspecies was growing explosively, flooding the taxonomic literature with a deluge of forms and names of dubious scientific worth. The foundations of taxonomy shook, as zoologists witnessed a "rapid disintegration . . . from an orderly regime toward chaotic proliferation!"[83] Ernst Mayr, a vocal advocate of subspecies, acknowledged that dissatisfaction was widespread and that critical reappraisal of the concept was overdue.[84]

In 1953 two young Harvard entomologists, William L. Brown and Edward O. Wilson, published a full-scale theoretical critique, and in 1954 a leading taxonomic society invited its members to consider if subspecies should be replaced by different categories, such as local populations or variation clines.[85] Wilson and Brown pronounced trinomials "misleading, cumbersome, and generally repellent, especially to the uninitiated," and predicted that "we shall soon begin to observe the withering of the trinomial and its cumbersome appurtenances—the types, the tinted labels, the ponderous subspecies lists gravely entered in a thousand catalogues, the awkward labelling of masses of 'intergrade' specimens that so unnecessarily consume the few effective working hours a modern taxonomist has."[86] To the younger generation Linnaean trinomials seemed arbitrary and authoritarian. History seemed to be repeating itself, as the ornithologist Charles Sibley observed: in the 1890s excessive splitting of species had led taxonomists to adopt subspecies and trinomials in the 1890s; now fifty years later it appeared that excessive splitting of subspecies would lead taxonomists to reject that system.[87]

Several trends conspired to set off the binge of subspecies splitting in the 1930s. Museum collections continued to grow, bringing more subspecies to light. Also, the bulge of new species discovered in the era of survey collecting provoked a corrective phase of lumping, and thus a secondary bulge in subspecies, as species were demoted wholesale. In mammalogy, for example, a 1951 revision of the weasels (*Mustela*) reduced the twenty-four species of the previous revision (1911) to three species and sixty subspecies. A revision of the coyotes (*Canis*) reduced fourteen species to one, with nineteen subspecies. In one species of pocket gopher over 150 subspecies had been identified and named, with more to come. "If this continues," one mammalogist warned, "eventually every colony of *Thomomys bottae* will bear a formal name."[88] The conversion of British ornithologists to trinomialism in the 1920s and 1930s resulted in such a mass conversion of "species" to subspecies that American ornithologists felt impelled to protest. While bird species were reduced from over 19,000 to less than 9,000, the number of subspecies soared to over 25,000.[89] Herpetologists also succumbed to splitting fever: over 400 were defined by 1953: about half newly identified, and half by demotion of previously described species.[90] A certain amount of churning is normal and healthy in taxonomic practice, a sign of improvement.[91] But the midcentury pace of subspecies production went so far beyond the normal self-correcting cycle that taxonomists wondered if they would ever be able to sort out the mess of new forms and names.

Some taxonomists blamed the Linnaean rules themselves for the mess: in particular the rule of priority, because it made every splitter's sins a burden on the community forever. Entomologists were especially aware of this danger because of the vast number of insect species to keep track of, and the still vaster numbers undiscovered. "Imagine the confusion which will result," one critic warned, "if future invertebrate zoologists find themselves confronted with millions of 'subspecies' names in addition to the hundreds of thousands of names of distinct species already in their catalogues!"[92] "One suspects," wrote another worried describer, "that under the present Rules the future systematist will be something of an intergrade between a philologist and an attorney. Whether he will have any time for zoological work and whether he will be of any use to society appear highly questionable."[93]

Other critics laid the blame for the uncontrolled proliferation on conceptual flaws in subspecies taxonomy: in particular, on the failure to clearly distinguish between subspecies and local populations. That, Brown and Wilson claimed, was what tempted taxonomists to apply trinomials promiscuously to local populations.[94] Defenders of subspecies, like Ernst Mayr, admitted that subspecies were a subjective (though practical and useful) category, but feared that giving up subspecies would only unleash excessive splitting of species by forcing taxonomists to give better subspecies full species status. Best leave the system alone and work around its imperfections, Mayr counseled.[95] As handles for biological facts, trinomials could be used sparingly and modestly; they became a problem only when taken too seriously as taxonomic entities.

However, none of these critiques explains why a system that had worked well for many decades suddenly failed. What made it fail, I think, was not any inherent conceptual or regulatory flaw, but changes in collecting practice. Survey collecting declined in the 1930s, or rather narrowed. Collecting became less extensive and more local and intensive. Expeditions gave way to individual excursions. And the purpose of collecting shifted from inventory and classifying to research on the micromechanics of speciation in local populations. Studies of breeding populations required collecting that was more local and intensive than survey practice, and such collecting revealed the existence of numerous distinguishable local variants, which by the rules of subspecies taxonomy seemed to demand recognition by Linnaean trinomials. In fact they did not, because the aim of collecting in such cases was not taxonomic but biological. The habit of giving Latin names to local forms simply outlived its usefulness and became not just unnecessary but counterproductive to everyone. Naming distracted biologists' work on the dynamics of populations and fatally complicated the work of the taxonomists, who were stuck with having to deal with a multitude of named forms of no taxonomic interest whatsoever. Taxonomists understood perfectly well the difference between subspecies and local populations; the problem was that the practices and aims of collecting had changed, while taxonomic practices had not.

So were the Cassandras right (if premature), who in the 1880s predicted that adopting subspecies would produce a deluge of phony forms? Not really. In their context of practice the danger was small, because the exacting methods of survey collecting disciplined taxonomists' splitting instincts. But when collecting became local and very intensive, these restraints failed. Overabundant data on local variation caused existing practices of data management to break down.[96] As collectors turned up more and more distinct populations, naming them as if they were Linnaean subspecies was nothing but a nuisance to evolutionary biologists and a lien on future taxonomists' time and energy. But the Cassandras who had warned that trinomials would explode the Linnaean order could not have anticipated this altered state of affairs. The divergence of collecting and taxonomic practices was an unanticipated by-product of the gradual evolution of the species concept from a morphological to a biological and populational one—and of the change in collecting from a survey mode to one focused on microevolutionary dynamics. In the process, the practical virtues of treating subspecies as a taxonomic category subject to Linnaean rules became vices and a cost levied on all.[97]

What happened to subspecies taxonomy at the end of the age of survey is oddly reminiscent of what happened to species taxonomy when collecting was too deep and local—Robert Ridgway's experience. In both cases, data that were too dense made natural units disappear, though species and subspecies became invisible in different ways. In Ridgway's case, gaps between species vanished into a continuous spectrum of variants; subspecies simply vanished in a smog of spurious names. It was not that taxonomists could no longer distinguish subspecies: they could, and did. Rather it was that "subspecies" multiplied so greatly when applied to local populations and clines that the rules that gave the real ones meaning were discredited. In effect, the category lost its identity.

Taxonomists proposed several remedies for this crisis. Some proposed simply to abandon subspecies for concepts better able to deal with continuous variation, like clines.[98] But as Ernst Mayr observed, clines and subspecies were different things; the choice was really between subspecies and no subspecies, and taxonomists could hardly deprive themselves of a device so useful for registering taxonomic difference.[99] Others proposed to limit subspecies strictly to populations that could not under any circumstances intergrade, like island forms. That made the definition mechanically objective—but also arbitrary and biological.[100]

Still others proposed replacing universalistic Latin trinomials with vernacular place names—"Montauk A," "Reelfoot Lake," "Rock Island"—in the same way that population geneticists named populations after their locales. Vernacular names lacked the permanence and dignity of Linnaean trinomials, but for critics like Wilson and Brown, that was an advantage: they could be used for a particular project and then discarded, without obligating taxonomists to manage their names forever. "The geographical vernacular is more broadly communicable [than trinomials]," Wilson and Brown rhapsodized, "more frankly expressive, fully as mnemonic, at least as certain in the long run to be precise, and it cuts the taxonomic red tape to practically nothing." Where Linnaean nomenclature had the force of law and was policed, vernacular names were democratic and "unostentatious"—appealing qualities for young taxonomic rebels.[101] But if vernacular place names were ideal for the purposes of evolutionary biology, as a system of taxonomy they were fatally bad. Taxonomists must be universalistic, enforcing standard rules: it is that or chaos.[102]

The crisis in subspecies taxonomy was symptomatic of the broader shift in systematic biology from inventory and housekeeping to evolutionary biology.[103] To evolutionary biologists species are not unchanging objects, as their formal names and descriptions imply, but variable populations that change continually, and every so often split into reproductively isolated forms. A populational view of species was always latent in the practices of survey collecting and subspecies taxonomy. But that potential became fully manifest only at the end of the age of survey. Then subspecies and trinomials came to be seen as unneeded vestiges of the old way of doing taxonomic business, and a hindrance to the new.

The end of the age of survey was an uneasy transition. Some taxonomists were attracted to a biological approach, while others remained true to the ordering and housekeeping side, or attempted a balance. Both are essential aspects of taxonomy, and individuals were torn between them. As the botanist Willard Camp remarked to Ernst Mayr:

> [W]e are—both of us—in a somewhat similar boat. We would both like to throw the morphological interpretation of the species out the window—but in the long run, neither of us has been able to do it. It just isn't consistent with what all the other taxonomists with whom we have to live are doing. Time and time again you lay out a perfectly cold doctrine—and then, expedience becomes the better part of valor, and the old morphological interpretation somehow slips into the picture. In 1941 and 1942 when [Charles L.] Gilly and I were working on our "Structure and Origin of Species" paper we were confronted so often with the same problem that it got to be a standard joke with us. We used to ask ourselves: "Well, are we going to be biologists or taxonomists at this turn of the road?" Too often, I fear, we retreated from the "biological" standpoint.[104]

Camp feared that by adopting a biological view of plant species, which relied on mass collecting and experiments, he might be going beyond what botanists in less intensively collected areas could use or tolerate, and thus impede the wider acceptance of modern views. So he temporized.

Conclusion

The view that classification is descriptive and subjective (because it depends on individuals' experience and judgment), rather than analytic and objective, is the bias of an age that has been persuaded to think of experiment and causal analysis as the only ways to truth. In fact, taxonomists' categories and classifications are no less real than those of experimental science. Perhaps they seem so because few outsiders know exactly how taxonomists do their work. Or is it perhaps because taxonomists' practices, while intricate and arcane, are also familiar in a way that those of experimental science are not? Defining and ordering categories is a fundamental activity of everyday life: we do it all the time, though unsystematically, and that perhaps is why we undervalue sorting and ordering as a way of knowing. In contrast, we are awed by the work that is done in labs, where we cannot go, by arts that we cannot ourselves perform—quasi-sacred arts—while taking for granted the skills we exercise every day in dividing a diverse world into manageable categories. Thus we fail to appreciate how sophisticated and powerful taxonomic methods are, and how worthy of respect.

In fact, experiment and classification are both in their own ways equally exacting methods and may not be as differently constructed as we like to think. Edgar Anderson shrewdly observed that biological sciences are most productive when they invent qualitative units that can then be studied quantitatively.[105] But defining qualitative units always means ignoring variability. Anderson pointed to the characters of experimental genetics, which are not homogeneous units but families of slightly varying phenotypes. But for the purposes of experimental analysis, geneticists treat these families as one kind of thing, and it is that simplification that enables them to sort and count recombinant forms and construct quantitative chromosome maps. Categories and units are necessary fictions: no unit characters, no experimental genetics. It is the same in systematics. Species and subspecies are heterogeneous populations, but for the purposes of analysis taxonomists take them to be discrete qualitative units. Only then can they sort out natural kinds and reckon degrees of relatedness. Species and subspecies are to systematics what unit characters are to genetics: qualitative units that by suppressing variability enable naturalists to see the order in natural phenomena.

The history of collecting in the last two hundred years is one of continual intensification: from the geographically extensive and serendipitous collecting of the age of exploration and empire; to the extensive and intensive methods of the age of survey; and to local and highly intensive collecting in the age of ecology. And with changes in collecting came changes in the practice of classification. As far-ranging exploration made Linnaean species the most useful taxonomic unit, so did survey methods make the subspecies an essential unit. And so, in turn, did more intensive local collecting make the breeding population the preferred unit of study. What appeared in less intensive modes of collecting as mere noise became visible and meaningful units as collecting intensified. Subspecies became visible when it became possible to empirically map their ranges and zones of intergrading. Still more intensive collecting cast doubt on the reality of subspecies and revealed a world not of discrete units but of fluctuating populations. Different collecting practices make the gaps between natural units visible, or not. So we have learned from our brief excursion into taxonomy and collecting history.

In the final chapter we will see how survey collecting wound down in the 1930s, and glance briefly at what its history suggests might come of current efforts to inventory nature's diversity.

CHAPTER SEVEN

Envoi

FROM OUR VANTAGE in the twenty-first century, it is clear that natural history survey was a passing phase in the Western discovery of biodiversity. Like the preceding Linnaean and "Humboldtian" phases, survey was a dominant mode for about fifty years, then gave way to new modes as the circumstances changed that had made it dominant. It was one of the more remarkable ventures of modern science. Hundreds of expeditions launched to every corner of the world; millions of specimens assembled and lovingly preserved—what made all that happen? It was to answer that question that I undertook this book. As we have seen, the answer lies in no single cause, but rather in a confluence of interlocking trends: in physical and human environments; in the culture and customs of nature-going; and in the science of collecting and classifying. None of these trends alone explains the age of survey, but together they do.

To recap the argument: The mosaic landscape of inner frontiers afforded naturalists physical access to a natural world that until then had been arduous, costly, and even dangerous to work in. Drawn by that opportunity, a generation of naturalists seized the chance not just to explore and collect—that had always been possible at least for a few—but to survey and inventory nature's full diversity.

But if physical access was a necessary condition for action, it was not a sufficient one. Naturalists had to have reasons beyond mere opportunity to collect intensively and far and wide. The scientific work of survey had to borrow value and legitimacy from popular social customs of nature-going. The new customs of middle-class recreation and vacationing provided this cultural rationale, I have argued. The experience of nature-going enlarged the pool of potential recruits to survey science and the pool of sponsors who, though not themselves scientists, were sympathetic to a scientific view of nature and were willing to pay for expeditions. The customs of outdoor recreation gave naturalists cultural (and financial) access to nature as infrastructures of travel and communication gave them physical access.

But here again we have part of an argument, not a whole one, because culture does not explain why naturalists did not simply continue older modes of exploratory, cream-skimming collecting, which was after all what recreational nature-goers did. Why did natural history become an exacting survey science? The answer to that question, I have argued, lies in the science itself. Even before the age of survey, taxonomists were beginning to see that comprehensive, in-depth collecting made taxonomic judgments more secure. And Darwin's theory of evolution gave the empirical study of variation a potent theoretical value. We cannot say that these improvements in practice and theory caused collecting to become both intensive and extensive, because one could argue with equal cogency that survey collecting caused taxonomy to become more exact. What we can say is that science drove a positive feedback cycle that caused expeditionary practice, once started, to feed on itself and get bigger and better. Survey generated scientific achievements—new species, new biogeographical principles—that would otherwise have not been made, and these kept survey naturalists going back to the inner frontiers for more.

Natural history survey got started for reasons that were as much environmental and cultural as scientific. But it was scientific achievement, linked to a social system of professional careers, that made collecting a large-scale and exacting science rather than an ad hoc pursuit of a few highly motivated individuals. Environment, culture, and science in concert produced the age of survey. And, I want now to argue, the same forces together brought that age to a close in the 1930s and 1940s.

From Collecting to Observing

Writing in the early 1950s, Ernst Mayr reflected on the recent changes in collecting practice. Commercial and amateur collectors, once numerous, had become virtually extinct social species, as had large-scale museum expeditions. Curators now made short collecting trips for special research projects, to supplement existing collections. Mayr also observed that making specimens had given way to observing living animals in nature, sometimes in the wild but more often in small nature preserves, which resembled nothing so much as plein air natural history labs.[1]

Although museums still sent out expeditions, it is fair to say that survey had by then been superseded by other natural history practices. On the theoretical side there was the combination of taxonomy and population genetics that we know as the "evolutionary synthesis."[2] On the practical side were the new sciences of wildlife biology and applied ecology that took shape in the 1930s and after.[3] As different as they were, these new areas of field science were alike in crucial ways. In both, problems of cause and mechanism seemed more compelling subjects of research than mapping and inventory. Both pursued their subjects by intensive, small-scale research methods and took highly localized breeding populations as their units of study, not species or subspecies. These new field sciences were based less on objects and collecting than on lab-based methods of genetics, physiology, and ecology. Their devotees were less likely to shoot and gather than to observe and record. They departed from survey practice in similar ways.

The environmental historian Donald Worster has called the second half of the twentieth century an age of ecology, and that term could serve well enough for a period in natural history in which population and ecosystem dynamics replaced inventory as the great issues of the day.[4] More and more, experienced collectors like Herbert Stoddard and Joseph Dixon found the pastures of environmental and wildlife science greener than museum work.[5]

The retreat from survey was gradual and uneven. State natural history surveys were giving way to wildlife management by the 1910s, and the U.S. Biological Survey had largely abandoned survey for economic work by 1920. Museums sidled toward experimental science in the 1930s. Take the U.S. Biological Survey, for example. In the early 1900s it acquired the legal responsibility for administering the Lacey Act, which regulated interstate commerce in game birds, and was elevated to full bureau status. From a small and somewhat anomalous operation getting by on annual appropriations, Merriam's survey became an established agency with a regular budget and political visibility. But establishment upset a fragile status quo. There had never been a compelling economic rationale for faunal surveys, so the best defense against congressional critics had been to work inconspicuously and not make statutory claims on budget lines. Politically, survey was a stealth science. But with important public duties and official bureau status also came political pressure to demonstrate the economic benefits of wildlife survey and to limit the bureau's work to what was officially mandated. There would be no more flying under the radar of budget watchdogs. Power brought increased and secure budgets, but at the cost of rigid accountability and a restricted research agenda.

As a result, survey collecting and biogeography were gradually replaced by projects in fur farming, predator control, and game management. Pioneered in the U.S. Geological Survey and perfected in the agricultural bureaus, the project or "problem approach" was the formula that worked politically in scientific Washington, and it was forcefully applied to the Bureau of Biological Survey.[6] Accustomed to a more freewheeling style and increasingly at odds with Congress, Merriam was forced to retire in 1909, and by 1920 the bureau's main work was predator control—to naturalists' dismay.[7] Some survey and collecting work did continue, but on a small scale and mainly in aid of state surveys. Some survey biologists welcomed the new order, like Edward Nelson; others, like Wilfred Osgood, saw it as the end of a golden age of survey—which it was, however one might value what replaced it.[8]

A similar trend in the politics of patronage also caused state natural history surveys to decline in the 1910s, as state governments organized permanent departments of conservation and wildlife management and more actively regulated their natural resources. In 1913 the director of the Michigan Geological and Natural History Survey observed the "general and rapid awakening to opportunities for practical service." Basic science had been eclipsed by science used to encourage and manage the exploitation of states' natural resources.[9] In 1912 a rising tide of public inquiries about fish and game made a "laboratory" for practical wildlife biology seem a better investment for the University of California than faunal survey.[10] Fish hatcheries and game farming were more direct and visible ways than natural history surveys of serving the public's real interest in wildlife.

Museums, with their cultural aims and private funding, continued expeditionary collecting longer than government agencies. The 1920s were their peak decade, and expeditions continued into the 1930s. One might expect that the stock market crash and Depression would have quickly dried up funds for expeditions, and they doubtless did for smaller museums. But for some, at least, the Depression had surprisingly little effect on the pace of expeditioning.

The Academy of Natural Sciences of Philadelphia especially kept up a brisk pace of expeditions throughout the Depression, leading curators in other museums to wonder how they did it.[11] The reason is probably quite simple: unlike better-organized and better-funded museums, the academy had always improvised and operated on a shoestring; having never gotten used to prosperity, it was better able to cope with hard times. The American Museum also continued a brisk pace of expeditioning up to World War II, a dip in the early 1930s being followed by revived activity as the economy seemed to recover after 1937. Although income from museum endowments virtually disappeared, private funding for expeditions held up remarkably well, especially for expeditions in pursuit of big-game animals. For families that still had money to spend, scientific expeditions were good cover for continuing sporting and vacationing habits at a time when conspicuous displays of wealth would have drawn unwelcome media comment. They were hunting safaris in mufti, with the trophy specimens going into dioramas rather than sportsmens' dens.[12] Only the fuel rationing and travel restrictions of World War II brought expeditioning to a halt.

However, expeditions did change in character in the 1930s, becoming generally smaller, shorter, closer to home, and more closely tied to diorama building. More "expeditions" were in fact short trips by individual curators to study wildlife in situ and collect live material for experiments. There were many little signs of a gathering trend from collecting to observing. For example, in a premier ornithological journal the percentage of papers that involved field collecting declined from 70 percent in the late nineteenth century to about 20 percent in 1924.[13] And whereas older field guides offered detailed information on shooting and preparing specimens, new ones in the 1920s had more to say about cameras, blinds, and methods of observing live birds.[14] In 1917 field-workers in the Bureau of Biological Survey were directed to pay less attention to collecting and more to observing animals' life histories and behaviors. Animals were to be captured alive and their behavior studied.[15]

Popular movements for wildlife preservation reinforced scientific reasons to observe rather than collect, and legal restrictions on wildlife harvesting tarred scientific and commercial collecting with the same brush.[16] Where popular participation in natural history had previously sustained collecting, now conservation-minded nat-

Fig. 7–1. C. Hart Merriam with a (for him) uncharacteristic field instrument, near Pyramid Lake, Nevada, July 1903. Cameras would replace guns and traps as field naturalists' instrument of choice. Courtesy Department of Library Services, American Museum of Natural History, negative no. 245774.

uralists helped to discredit it. Their preferred instruments were no longer the trap or gun, but cameras and binoculars.

This trend did not go uncontested. C. Hart Merriam, for one, thought the "maudlin sentimentalists" who opposed all shooting were doing as much harm to field biology as laboratory zealots (though on at least one occasion Merriam was caught on film taking snapshots—see figure 7–1.)[17] When Joseph Grinnell was asked by a normal-school teacher where she might get bird specimens for her classes, he replied ruefully that she would find it difficult because preservationists had all but put collectors out of business, especially younger ones. "The Audubonites," he wrote, "have instilled into the youth of the land fear of various punishments of said youth indulging in any sort of 'bird collecting.' " Some older ornithologists continued to collect, Grinnell reported, but for the most part ornithology was becoming "ornithophily."[18] A collector friend of Grinnell's living in the Philippines likewise condemned preservationists, declaring angrily that "When the Audubonists come out here I shall move to the interior of Celebes or Timoor."[19] But as rare species declined, even Grinnell found it harder to reconcile his devotion to conservation with the scientific imperative of mass collecting. Where does one draw the line between what was necessary and what was indulgence, he wondered; it was often impossible to say.[20]

For other naturalists the shift from survey to project science was more like a blessing in disguise. Biologists involved in state surveys were not sorry to be quit of the perpetual struggle for small change. Henry Nachtrieb seemed relieved when the fund supporting the Minnesota Natural History Survey finally ran dry in 1909. He was weary of trying to justify the economic value of survey to politicians. Besides, the future of field zoology seemed to him clearly to be in conservation and wildlife management.[21] Alexander Ruthven was also eager to cooperate with state wildlife officials.[22] Besides, practical projects could often be coopted for basic research. Joseph Grinnell's study of game animals for the California Fish and Game Commission in 1912 was indistinguishable from a purely scientific field trip. By the early 1920s he was ready to wind down general survey and focus on specific projects: "[T]his sporadic type of field work, as immediate opportunity and problems in hand seem to demand, is now, perhaps, more justified . . . than consecutive all-round vertebrate collecting," he wrote. However, he did mean to do some general collecting each year.[23] Full-scale survey was still the method of choice for underworked regions, but for the smaller problems that remained, the project mode was best.

Museum biologists also saw some good in the change from survey collecting to a mixture of collecting and life study. As Ernst Mayr observed, a mode of practice that combined collecting with observation and even experiments was much more fruitful scientifically than simple collecting.[24] In fact, ecology and behavior had become vital aids to modern taxonomists, who increasingly focused on problems like variation in local populations, transitional forms and speciation, and sibling species (which look identical in preserved specimens and can only be distinguished by differences in behavior). Even the largest collections of specimens no longer sufficed for the new evolutionary taxonomy, which generally required field observation and often prolonged work at a field station, or experimental work on animals in aviaries or terraria.[25]

Another symptom of change was the appearance of new kinds of places for experimental work in nature. Universities, museums, and zoological societies began to underwrite permanent field stations, ranging from marine stations and large preserves like Barro Colorado Island in the Panama Canal Zone to small relic prairie woodlots where plant and animal life (what remained of it) could be charted and measured.[26] The American Museum established a marine laboratory in the Bahamas in 1946 and dreamed of a comparable facility for work on desert reptiles. It operated several small field stations in New Jersey, acquired an islet in Long Island Sound as a plein air lab, and established stations in Arizona, south-central Florida, and the barrier islands of the Gulf coast.[27]

But the trend to experimental work in museums could also threaten systematic biology. For example, in 1942 a new president of the American Museum, Albert E. Parr, decided to redirect funds from taxonomy and evolution to ecological zoology on the grounds that evolution was an established fact and so further research would add little that was new! Parr (an oceanographer, not a naturalist) abolished the Department of Vertebrate Paleontology and declared that collecting in other departments was to be sharply curtailed—all this at a moment when museum biologists like Ernst Mayr and George Gaylord Simpson were leading the most important advance in evolutionary biology since Darwin.[28] Although collecting and taxonomic work did continue—because curators were able to resist—survey collecting on the grand scale became a thing of the past.[29]

A Changing World

We have seen *that* survey collecting waned, and *how*: but *why* did it wane? I think it was our familiar trio of causes: changes in the natural environment, in cultural habits of nature-going, and in taxonomic science, interacting, as they had previously to cause survey to flourish, but now in reverse.

Change in the physical landscapes of North America was one cause of the waning of survey collecting. From their apogee around the turn of the century, the inner frontiers shrank steadily, especially in the prosperous decades of the 1920s and 1950s (and 1990s), when new big money was available for development. An expanding population and the middle-class love of lawns and single-family homes shrank large islands of nature into fragmented islets. Tourist and vacationing industries were in some ways even more destructive of the inner frontiers than the extractive industries they replaced, as roads and vacation homes laced and dotted wild areas.[30] World War I made all-out resource use patriotic, pushed up agricultural and resource prices, and turned the slow retreat of twilight zones into a rout. Fields were plowed and planted to their edges, destroying the brushy edge habitats in which wildlife sheltered. Railroad rights-of-way were also plowed up, destroying refugia of native prairie ecosystems. State-subsidized projects for draining and diking river floodplains and irrigating desert valleys transformed vast natural areas into monoculture.[31] In the 1930s massive federal projects in the Tennessee Valley and the basins of the Columbia and Colorado Rivers transformed entire watersheds into industrial and agricultural landscapes.[32]

But nothing was more destructive of inner frontiers than the family automobile and the concomitant public investment in paved roads. Suburbs exploded outward, invading rural countrysides at a pace not possible with mere rail and trolley.[33] Autos and roads opened up to recreation or development wild areas that had before been safely beyond easy reach. Campaigns for good roads and rural improvement promoted "clean up" campaigns that destroyed the vestigial habitats of native plants and animals. In 1920 Robert Ridgway lamented the zeal with which road officials in his home town of Olney, Illinois, cleared away "brush" and trees from roadsides and fence lines.[34] Untidy and casual modes of land use, which had inadvertently protected wildlife habitats, became morally and aesthetically displeasing. Automobility and biotic cleansing thus turned twilight zones into conurbations and monocultures. The transportation technologies and messy habits that had created the inner frontiers also gradually destroyed them.

Along with changes in the environment came changes in the customs of nature-going that had once sustained systematic natural history survey. Also in decline were the local institutions that once had nurtured a broad interest in the natural world. Local natural history societies had been slowly waning since the late nineteenth century, but by 1930 they were effectively dead, and their demise diminished the cultural space in which a living interest in nature could develop into a scientific one. Amateur networks, which a generation earlier had quickened popular interest and participation in natural history survey, gradually disappeared.[35]

The habit of outdoor vacationing became ever more pervasive in the interwar and postwar periods, as we have seen.[36] But at the same time the symbiosis between recreation and scientific work seemed to attenuate. It was perhaps to be expected: as recreation became everyone's pursuit, it had less need of moral justification. Middle-class people secure in their social identity had less need to rationalize their pleasures as self-improvement. No need to work at play: they could just play. Sport hunting and fishing lost their connection with taxonomy and natural history and became merely healthy and sociable outdoor sports or a way of stocking the family freezer. Outdoor recreation, along with other forms of popular culture, was assimilated to the culture of mass commercialized entertainment. Nature for many Americans became a kind of theme park.[37] Even as more Americans camped, tramped, and toured national parks, they had fewer opportunities to participate actively or virtually in the work of natural history.

In other words, the special circumstances—the inner frontiers, with their blend of culture and nature, and the cultural habits of working at play—no longer existed to sustain systematic natural history survey. The distinctive experiences that had increased the pool of recruits to survey science, and the social meanings that gave survey transscientific value, no longer operated, or operated less intensely. But we also need the third element, science, to explain why the momentum of a flourishing mode of practice did not just carry it along. In fact, changes in the science of survey collecting were probably more important in its demise than in its origins. Survey collecting may have declined mainly because of diminishing returns: as inventories became more nearly complete, it was no longer worthwhile collecting in that way. Survey succeeded so well that it put itself out of business.

Inventories of animal species (vertebrate species, at lest) were becoming nearly complete throughout the survey period. Descriptions of new vertebrate species declined (somewhat bumpily) with each succeeding decade of the twentieth century. Although there were little spurts of discovery in some faunal groups in the 1960s, by 1940 the point was reached in most groups where an estimated nine-tenths or more of existing species were known. Nine-tenths of mammal species accepted as of 1985 were named and described by about 1940. Some groups arrived at the point of diminishing returns earlier, others a bit later. Birds were early: nine-tenths of species accepted in 1985 were known just after 1900, plus or minus a decade depending on the family.[38] For North American reptiles and amphibians, most orders passed the nine-tenths mark (of species accepted in 1953) well before 1940: *Serpentia* about 1900; *Chelonia* (turtles) and *Sauria* (lizards) about 1910; *Salienta* (frogs and toads) about 1920; and *Caudata* (salamanders) in the early 1940s.[39] Among the mammals, carnivores and primates were early—nine-tenths described by 1900; the nocturnal and secretive insectivores and chiroptera (bats) much later, in the 1950s.[40] By 1940 the pace of discovery of new vertebrate species had diminished to just a few per year—hardly enough to sustain programs of survey collecting. It was in the 1930s that the picture of diminishing returns became clear and that systematic biologists began to look for other ways to exercise their skills and make careers.

As I suggested in chapter 1, survey collecting was a practice of the descending phase of a discovery curve, when total inventories were a realistic aim, but when the pace of discovery of new species was still brisk. Earlier, when inventories were very incomplete and pickings easy, reconnaissance and cream-skimming were the methods of choice; and later, when species inventories were nearly complete, survey methods reached the point of diminishing returns, and discoveries were better left to serendipity than to systematic pursuit. Because it was efficient and thorough, survey was inherently self-limiting. It could temporarily reverse the downward trend of discovery, producing new peaks, as we have seen. But in the long run it moved naturalists faster down the discovery curve toward the point where survey methods would no longer pay.

We do not usually think of science as cyclic or self-limiting. We have an image of it as endlessly abundant and unfinished, an ever receding horizon—"endless frontier" was the phrase to conjure with in the postwar world.[41] Perhaps that is because experiment appears to be a limitless cornucopia of new facts (or because modern society harbors an illusion of endless growth in everything). In any case, the historical trajectory of natural history survey certainly was limited, because nature's stock of species is finite. And cycles of discovery may be more general than we think—in other inventory and classifying sciences (for example, particle physics), and perhaps even in experimental sciences. Of course there is no limit to what can be done with inventories. The work of classification never ends because research agendas evolve and new methods of measuring relatedness appear from time to time (for example, molecular measures) and force revisions of taxonomic method. But the task of collecting and taking inventory is finite.

In the larger view, however, it was the combination of scientific with environmental and cultural factors that caused systematic survey collecting to wane, as the same factors operating differently had caused it to become a dominant mode. The scientific rewards of systematic collecting diminished at the same time that its environmental and cultural supports weakened. One might predict that if some kind of natural history survey should again become predominant, it would do so, as before, because of some conjunction of environment, culture, and science.

Biodiversity Revisited

In fact, the invention of cheap mechanized methods of DNA fingerprinting of species has revived survey collecting, though in a new way: not of living specimens in nature, but of bits of DNA from animal and plant specimens collected long ago. (Early techniques of DNA hybridizing required material from freshly killed animals or eggs, but recombinant-DNA techniques generally work well with old museum specimens.) One could say that museum and herbarium specimens are enjoying a second, molecular scientific life. Molecular taxonomists are recapitulating the field collecting of past centuries, but without going afield: their expeditions are to museum storerooms.

Such work is not the armchair collecting of earlier centuries, exactly.[42] It is more like laboratory or computer-room collecting, or data mining. Biologists now inventory banks of data, as survey collectors once inventoried the living ark of nature. Genome-mapping projects could be seen as a kind of survey collecting, the difference being that genome mappers quietly ignore natural variability, whereas field naturalists make it the object of their study. Something of this sort has also occurred with ethnographic collections, which though quite out of favor for ethnographic research are finding a new use as quantitative indicators of cultural encounters.[43]

There have also been renewed calls by naturalists for a total inventory of world species, beginning with E. O. Wilson's in the mid-1980s.[44] More precisely, there have been calls to complete the inventories that the age of survey left unfinished: namely, of the still largely unknown world of invertebrates. A century ago it was rodents, bats, reptiles, and suchlike creatures that drew expeditioners afield; now it is the multitudinous insects and arthropods of tropical soils and forest canopies.

This is a daunting project. Even in the better-known parts of the world, inventories of invertebrate groups are wildly uneven. Take insects: in the United States and Canada, some 76 percent of an estimated 133,000 living species were known in 1983, and 57 percent described. For Australian insects the comparable figures were 61 and 45 percent of an estimated 108,000 species. (British insects, in contrast, numbered some 23,500 species and were 94 percent known and 90 percent described by 1983.)[45] Of the invertebrate fauna of the tropical forests of South America, Africa, Southeast Asia, and the Pacific archipelagos—those great engines of biodiversity—it can only be said that most species remain unknown.[46] Shake or dust a patch of forest canopy, or sift a few square meters of forest soil, and thousands of unknown species come to light. In these vast microrealms the task of natural history survey has barely begun.

This revived impulse to collect, classify, and preserve in some ways resembles the earlier episodes of Linnaean, Humboldtian, and survey collecting. Like these, it is in part a salvage operation, made urgent by the growth and global spread of human populations, especially into equatorial forests. It is also associated with cultural movements that promote public interest in the natural world and its fragility: the green and preservationist movements that abound in the age of ecology. As before, it is now field naturalists who are most vocal in promoting species inventories as a grand design for scientific discovery.

Calls for species surveys often point to the state-sponsored "big science" projects of our time. Like the various genome projects: why, biologists ask, should we blithely spend billions of dollars on total inventories of human, fruit fly, or nematode genes, but regard an inventory of living species as an unaffordable luxury? Why do we pay for inventories of stars and galaxies, and subatomic particles, but not of living kinds? And spend billions exploring extra-planetary space but balk at the cost of exploring our own earthly environment?

Various rationales for a big biodiversity project have been urged, mostly practical, as they were for natural history survey a century ago: new medicinals from plants, gene banks of useful species, understanding of ecosystems on which humans unwittingly depend, and so on. But in the end the strongest and most honest rationale may be simply the satisfaction, even obligation, of knowing the plants and animals with which we cohabit—a custom as ancient as humankind itself.

However, if we have learned anything from the history of natural history survey, it is that scientific ambitions alone are not what put collectors in the field, but rather conjunctures of environmental, cultural, and scientific trends that occur from time to time but unpredictably. In future, as in the past, it is likely to be changes in natural environments, as well as in cultural attitudes to nature and customs of nature-going, that may enable naturalists to realize their dreams of total knowledge of nature's diversity—these, together with trends in science, that resonate with public values.

Whether the conditions now exist that might support another large-scale stocktaking is hard to say. On the one hand, there is widespread popular concern over the extinction of species and habitats, and green politics has real clout in some countries. The areas of the world that most need to be inventoried have become accessible (and threatened) by the expansion of transport infrastructure and settlement as the inner frontiers were a century ago. And there are new forms of cultural consumption of nature—nature films and TV docudramas, and ecotourism—with mass appeal. On the other hand, tiny insects and invertebrates lack the human appeal that caused earlier generations to build dioramas and underwrite collecting expeditions. Nor is it clear that social customs exist on which surveys of small invertebrates might take a free ride, as collecting bats and rodents once did on the middle-class taste for outdoor recreation. And there is the sheer scale of the scientific problem: it is hard to imagine a total inventory of millions of tiny and nearly indistinguishable creatures, many confined to minute habitats. Who would collect, sort, describe, and name them; and how? And to what end?

Most likely, future surveys will not be the comprehensive inventories of the age of survey, but some sort of sampling of species in small plots scattered across large regions like the Amazon or Indonesian forests. Future survey practice is likely to be some mix of expeditionary and project work, and a future science of biodiversity, some combination of taxonomy, ecology, and bioinformatics. Or surveys may be designed for purposes other than complete inventory: for example, to understand how ecosystems survive or fail, as models for our own survival. That after all is what most concerns us. But a complete stocktaking of all the world's invertebrate species in the world? That may be beyond practical reach. If so, then the project of an inventory of our celestial ark, which began with Carl von Linné about 250 years ago and reached its apogee in the age of survey, may indeed be drawing to a close. Biologists will continue to collect, sort, name, map, and classify—but in new ways and to new ends.

Abbreviations

AA	Annie Alexander Papers, Bancroft Library, Berkeley, California
AGR	Alexander G. Ruthven Papers, Butler Library, Ann Arbor, Michigan
AGV	Arthur G. Vestal Papers, University of Illinois Archives, Urbana-Champaign, Illinois
ANS	Academy of Natural Sciences Archives, Philadelphia Pennsylvania
CBD-Gen	Charles B. Davenport Papers, general correspondence, American Philosophical Society Library, Philadelphia, Pennsylvania
CHM	Clinton Hart Merriam Papers, Bancroft Library, Berkeley, California (microfilm edition)
CIW	Carnegie Institution of Washington Archives, Washington, D.C.
CLH	Carl L. Hubbs Papers, Scripps Institution for Oceanography Archives, La Jolla, California
EM-HU	Ernst Mayr Papers, Harvard University Archives, Cambridge, Massachusetts
f., ff.	folder, folders
FEC	Frederic E. Clements Papers, American Heritage Center, Laramie, Wyoming
FMNH	Field Museum of Natural History Archives, Chicago, Illinois
HFN	Henry F. Nachtrieb Papers, University of Minnesota Archives, Minneapolis, Minnesota
INHS	Illinois Natural History Survey, University of Illinois Archives, Champaign, Illinois
JG-MVZ	Joseph Grinnell Papers, Museum of Vertebrate Zoology, Berkeley, California
JG-UC	Joseph Grinnell Papers, Series C-B995, Bancroft Library, Berkeley, California

KPS	Karl P. Schmidt Papers, Field Museum of Natural History, Chicago, Illinois
MMZ	University of Michigan Museum of Zoology Archives, Bentley Historical Library, Ann Arbor, Michigan
PLE	Paul L. Errington Papers, Iowa State University Archives, Ames, Iowa
UM	University of Minnesota Archives, Minneapolis, Minnesota
UWLim	University of Wisconsin Limnology Papers, University of Wisconsin Archives, Madison, Wisconsin
VOB	Vernon O. Bailey Papers, American Heritage Center, Laramie, Wyoming
WER	William E. Ritter Papers, Bancroft Library, Berkeley, California
WHO	Wilfred H. Osgood Papers, Field Museum of Natural History Archives, Chicago, Illinois
WMW	William Morton Wheeler Papers, Harvard University Archives, Cambridge, Massachusetts

Notes

Chapter 1: Nature

1. Edward O. Wilson, "The current state of biological diversity," in *Biodiversity*, ed. Wilson, (Washington: National Academy Press, 1988), 3–18; R. M. May, "How many species are there on earth?" *Science* 241 (1988): 1441–49.

2. Wilson, ibid., 4–5, table 1–1.

3. Ibid., 12–13.

4. Some readers may wonder why I use "taxonomy" instead of "systematics," which connotes a biological and analytical science rather than a merely descriptive and cataloging one. I do so to dignify the practical, craft side of the discipline, which tends to be underrated. Also, the invidious distinction appears to be of recent origin. For example, in 1941 Ernst Mayr and Edgar Anderson agreed that the two terms were essentially synonyms. "Taxonomist" was the preferred American usage, and "systematist" the European (British naturalists used both). Edgar E. Anderson to Ernst Mayr, 9 Jan. 1941; Mayr to Anderson, 27 Dec. 1940, 6 Jan. 1941; all in EM-HU, box 2.

5. Lisbet Koerner, *Linnaeus: Nature and Nation* (Cambridge: Harvard University Press, 1999).

6. Martin Rudwick, review of Simon J. Knell, *The Culture of English Geology, 1815–1851* (Aldershot: Ashgate, 2000), in the *Times Literary Supplement* (4 May 2001): 27. On the history of collecting, in addition to Knell's book: David E. Allen, *The Naturalist in Britain: A Social History* (London: Allen Lane, 1976; reprinted, Princeton University Press, 1994); Mark V. Barrow, Jr., *A Passion for Birds: American Ornithology after Audubon* (Princeton: Princeton University Press, 1998), chaps. 1, 2; Anne L. Larsen, "Not Since Noah: The English Scientific Zoologists and the Craft of Collecting, 1800–1840," Ph.D. diss., Princeton University, 1993; and Anke te Heesen and E. C. Spary, eds., *Sammeln als Wissen: Das Sammeln und seine wissenschaftsgeschichtliche Bedeutung* (Göttingen: Wallenstein Verlag, 2001).

7. Robert E. Kohler, "Place and practice in field biology," *History of Science* 40 (2002): 189–210.

8. Edgar Anderson, *Plants, Man and Life* (Cambridge: Cambridge University Press, 1952; reprinted Berkeley: University of California Press 1969), 39–40.

9. Richard White, *The Organic Machine: The Remaking of the Columbia River* (New York: Hill and Wang, 1995).

10. For example: the contributions to Henrika Kuklick and Robert E. Kohler, eds., "Science in the Field," *Osiris* 11 (1996); Larsen, "Not Since Noah" (cit. n. 6); Jonathan Weiner, *The Beak of the Finch: A Story of Evolution in Our Time* (New York: Knopf, 1994); and Robert E. Kohler, *Landcapes and Labscapes: Exploring the Lab-Field Border in Biology* (Chicago: University of Chicago Press, 2002).

11. Susan Faye Cannon, "Humboldtian science," in *Science in Culture: The Early Victorian Period* (New York: Science History Publications, 1978), 73–110.

12. Data are from Charles G. Sibley and Burt L. Monroe, Jr., *Distribution and Taxonomy of Birds of the World* (New Haven: Yale University Press, 1990). Dates are for first-published descriptions of species accepted as of 1990. The date specimens are described and published is a lagging, though still useful, indicator of collecting activity.

13. Data are from James H. Honacki, Kenneth E. Kinman, and James W. Koeppl, *Mammal Species of the World: A Taxonomic and Geographic Reference* (Lawrence: Association of Systematics Collections, and Allen Press, 1982).

14. Data are from Karl P. Schmidt, *A Check List of North American Amphibians and Reptiles* (6th ed., Chicago: University of Chicago Press, 1953). These figures include species subsequently reduced to subspecies. For data on world amphibia: Darrel R. Frost, "Amphibian Species of the World: An Online Reference," American Museum of Natural History, ongoing (http://research.amnh.org/herpetology/amphibia/index.html). There is apparently no world checklist of reptiles.

15. For example: Richard H. Grove, *Green Imperialism: Colonial Expansion, Tropical Island Edens, and the Origin of Environmentalism, 1600–1860* (New York: Cambridge University Press, 1995); Rob Iliffe, "Science and voyages of discovery," in *The Cambridge History of Science*, ed. Roy Porter, vol. 4, *Eighteenth-Century Science* (Cambridge: Cambridge University Press, 2003), 618–45; and Larry Stewart, "Global pillage: science, commerce, and empire," in Porter, ibid., 825–44.

16. Koerner, *Linnaeus* (cit. n. 5), chaps. 4–6.

17. For example: Lucile H. Brockway, *Science and Colonial Expansion: The Role of the British Royal Botanic Gardens* (New York: Academic Press, 1979); Robert A. Stafford, *Scientist of Empire: Sir Roderick Murchison, Scientific Exploration and Victorian Imperialism* (Cambridge: Cambridge University Press, 1989); David P. Miller and P. Reill, eds., *Visions of Empire: Voyages, Botany, and Representations of Nature* (Cambridge: Cambridge University Press, 1996); John Gascoigne, *Science in the Service of Empire: Joseph Banks, the British State and the Uses of Science in the Age of Revolution* (Cambridge: Cambridge University Press, 1998); Michael A. Osborne, *Nature, the Exotic, and the Science of French Colonialism* (Bloomington: Indiana University Press, 1994); Janet Browne, "Biogeography and empire," in *Cultures of Natural History*, ed. Nicholas Jardine, James A. Secord, and E. C. Sparry (Cambridge: Cambridge University Press, 1996), 305–21; and Nathan Reingold and Marc Rothenberg, eds., *Scientific Colonialism: A Cross-Cultural Comparison* (Washington: Smithsonian Institution Press, 1987).

18. On western surveys: A. Hunter Dupree, *Science in the Federal Government: A History of Policies and Activities to 1940* (Cambridge: Harvard University Press, 1957), chaps. 3, 5, 10; William H. Goetzmann, *Exploration and Empire: The Explorer and the Scientist in the Winning of the American West* (New York: Knopf, 1966); Goetzmann, *Army Exploration in the American West, 1803–1863* (New Haven: Yale University Press, 1959); and Richard A. Bartlett, *Great Surveys of the American West* (Norman: University of Oklahoma Press, 1962).

19. We know little of European museums' collecting practices. Some received specimens from state expeditions (Berlin, Hamburg), and the Paris museum sent out expeditions; the British Museum preferred to purchase specimens on the open market. Gordon McOuat and Lynn Nyhart, personal communications. The need for comparative study is plain enough.

20. Wilfred H. Osgood, "Clinton Hart Merriam, 1855–1942," *Journal of Mammalogy* 24 (1943): 421–36, p. 428. Botanists had the same experience—for example [Editor], "Botany," *Popular Science Monthly* 57 (1900): 328–29, p. 329.

21. Beau Riffenburgh, *The Myth of the Explorer: The Press, Sensationalism, and Geographical Discovery* (New York: Oxford University Press, 1994); William H. Goetzmann, "Exploration and the culture of science: the long good-bye of the twentieth century," in *Making America: The*

Society and Culture of the United States, ed. Luther S. Leudtke (Chapel Hill: University of North Carolina Press, 1992), 413–31.

22. For example: Riffenburgh, ibid., chaps. 8–9; William H. Goetzmann and Kay Sloan, *Looking Far North: The Harriman Expedition to Alaska, 1899* (New York: Viking, 1982).

23. Is exploration a distinctively European and American social category? Other world societies pursued commercial and political expansion and cartography, obviously; but they have seemed less curious about the natural world for its own sake than has the West.

24. See Alex Soojung-Kim Pang, "The social event of the season: solar eclipse expeditions and Victorian culture," *Isis* 84 (1993): 252–77.

25. Frederick Jackson Turner, "The significance of the frontier in American history," in *The Frontier in American History* (New York: Henry Holt, 1920), 1–38, p. 1.

26. David W. Meinig, *The Shaping of America: A Geographical Perspective on 500 Years of History. Volume 3, Transcontinental America 1850–1915* (New Haven: Yale University Press, 1998), part II.

27. For example, William Cronon, "Revisiting the vanishing frontier: the legacy of Frederick Jackson Turner," *Western Historical Quarterly* 18 (1987): 157–76.

28. For example: William H. TeBrake, *Medieval Frontier: Culture and Ecology in Rijnland* (College Station: Texas A&M University Press, 1985); and Robert M. Netting, *Balancing on an Alp: Ecological Change and Continuity in a Swiss Mountain Community* (New York: Cambridge University Press, 1981).

29. Meinig, *Shaping of America Vol. 3* (cit. n. 26), part III. Also Richard White, *"It's Your Misfortune and None of My Own": A History of the American West* (Norman: University of Oklahoma Press, 1991), part 2.

30. Carville Earle, "Place your bets: rates of frontier expansion in American history, 1650–1890," in *Cultural Encounters with the Environment: Enduring and Evolving Geographic Themes*, ed. Alexander B. Murphy and Douglas L. Johnson (Lanham, Md.: Rowman and Littlefield, 2000), 79–105.

31. Ibid., 96–101.

32. Hildegard B. Johnson, *Order Upon the Land: The U.S. Rectangular Land Survey and the Upper Mississippi Country* (New York: Oxford University Press, 1976); White, *Your Misfortune* (cit. n. 29), chap. 6; and Vernon Carstensen, ed., *The Public Lands: Studies in the History of the Public Domain* (Madison: University of Wisconsin Press, 1968).

33. J. Ronald Engel, *Sacred Sands: The Struggle for Community in the Indiana Dunes* (Middletown, Conn.: Wesleyan University Press, 1983).

34. Henry A. Gleason, "The vegetation of the inland sand deposits of Illinois," *Bulletin of the Illinois State Laboratory of Natural History* 9 (1910): 23–174; and Charles A. Hart and Gleason, "On the biology of the sand areas of Illinois," ibid., 7 (1907): 137–272.

35. One naturalist reported that he could find his way anywhere in the United States with only a map, except the pine barrens. There he needed a local guide. James A. G. Kuhn to Alexander Ruthven, 22 Jan. 1906, AGR, box 53, f. 9.

36. J. M. Holzinger, "Some interesting cases of plant distribution," *Plant World* 4 (1901): 185–87.

37. Philip G. Terrie, *Contested Terrain: A New History of Nature and People in the Adirondacks* (Syracuse: Syracuse University Press, 1997); Paul Schneider, *The Adirondacks: A History of America's First Wilderness* (Boston: Henry Holt, 1997).

38. Alexander G. Ruthven, "An ecological survey in the Porcupine Mountains and Isle Royale, Michigan," *Report of the Board of the Geological Survey of Michigan* (1905), 17–55, pp. 17–20; and Charles C. Adams, "Isle Royale as a biotic environment," *Report of the Board of the Geological Survey of Michigan* (1908): 1–56, pp. ix–x, 1–2, 5–6.

39. Christopher F. Meindle, "Past perceptions of the great American wetland: Florida's Everglades during the early twentieth century," *Environmental History* 5 (2000): 378–95; Joseph V. Siry, *Marshes of the Ocean Shore: Development of an Ecological Ethic* (College Station: Texas A&M University Press, 1984); and Hugh Prince, *Wetlands of the American Midwest: A Historical Geography of Changing Attitudes* (Chicago: University of Chicago Press, 1997).

40. Ben Palmer, "Swamp drainage with special reference to Minnesota," *University of Minnesota Studies in Social Science* 5 (1915), p. 17, and chap. 3; Prince, ibid.; and Mary R. McCorvie and Christopher L. Lant, "Drainage district formation and the loss of midwestern wetlands, 1850–1930," *Agricultural History* 67:4 (Fall 1993): 13–39.

41. Joseph Grinnell to Annie Alexander, 13 Feb. 1911, 16 Oct. 1922, AA, box 2; Grinnell to Alexander, 4 March 1918, AA CU-120, box 1, f. 3.

42. Carl H. Eigenmann, "Turkey Lake as a unit of environment, and the variation of its inhabitants," *Indiana Academy of Sciences Proceedings* 11 (1895): 204–17; and William A. McBeth, "The Tippecanoe, an infantile drainage system," ibid. 44 (1910): 341–43.

43. Lewis H. Weld, "A survey of the Huron River valley, II: a peat bog and morainal lake," *Botanical Gazette* 37 (1904): 36–52; Howard S. Reed, "A survey of the Huron River valley, I: the ecology of a glacial lake," ibid., 34 (1902): 125–139.

44. Ray P. Teele, *The Economics of Land Reclamation in the United States* (Chicago: A. W. Shaw, 1927), 46–48; and "Swamp Lands of the United States," Senate Document 443 (60th Congress first session), 5265.

45. Fourteenth Census of the United States, 1920, vol. VII "Irrigation and Drainage" (Washington: Government Printing Office, 1922), 348.

46. Daniel W. Schneider, "Enclosing the floodplain: resource conflict on the Illinois River, 1880–1920," *Environmental History* 1:2 (1996): 70–96; and Thomas G. Scott, "Wildlife research," *Bulletin of the Illinois Natural History Survey* 27 (1958): 179–301, pp. 190–94.

47. Joseph Grinnell to Annie Alexander, 4 March 1918, AA CU-120, box 1, f. 3.

48. Dan Flores, *Caprock Canyonlands: Journeys into the Heart of the Southern Plains* (Austin: University of Texas Press, 1990).

49. Jane Maienschein, "Pattern and process in early studies of Arizona's San Francisco Peaks," *BioScience* 44 (1994): 479–85; and Frederick V. Coville and Daniel T. MacDougal, *Desert Botanical Laboratory* (Washington: Carnegie Institution Publication no. 6, 1903), 12–13.

50. Michael Williams, *Americans and Their Forests: A Historical Geography* (New York: Cambridge University Press, 1989), chap. 8; Robert H. Whittaker, "Vegetation of the Great Smoky Mountains," *Ecological Monographs* 26 (1956): 1–80, pp. 2–4.

51. Williams, ibid., 430–40, fig. 13.1 (436–37).

52. William Cronon, *Nature's Metropolis: Chicago and the Great West* (New York: Norton, 1991); John C. Hudson, "Towns of the western railroads," *Great Plains Quarterly* 2 (1982): 41–54; and Meinig, *Shaping of America* (cit. n. 26), part 1, chap. 1, and part 3, chap. 3.

53. Harvey Thomson, "An excursion to the Platte," *Botanical Gazette* 12 (1887): 219–21.

54. Williams, *Americans and Their Forests* (cit. n. 50), chap. 7.

55. A. M. Holmquist to Warder C. Allee, 16 Oct. 1933, WCA, box 18, f. 10.

56. Cindy S. Aron, *Working at Play: A History of Vacations in the United States* (New York: Oxford University Press, 1999), 49–54; Hal K. Rothman, *Devil's Bargain: Tourism in the Twentieth-Century West* (Lawrence, Kans.: Kansas University Press, 1998), chaps. 3, 4; Alfred Runte, *National Parks: The American Experience* (2nd ed., Lincoln: Uni-

versity of Nebraska Press, 1987), 91–94; and Eugene T. Petersen, "The History of Wild Life Conservation in Michigan, 1859–1921," Ph.D. diss., University of Michigan, 1953, 8–14.

57. Editorial [F. H. Knowleton?], *Plant World* 4 (1901): 179.

58. Harlan I. Smith to Alexander G. Ruthven, 16 Nov. 1911, AGR, box 54, f. 4.

59. Stephen A. Forbes, untitled report enclosed in C. S. Deneen to Forbes, 19 Feb. 1909, INHS series 43/1/1, box 9; Forbes to Dear Sir, 19 March 1914, INHS series 43/1/5, box 2, f. conference; Forbes and Robert E. Richardson, "Studies on the biology of the upper Illinois River," *Bulletin of the Illinois Laboratory of Natural History* 9 (1913): 481–574; Forbes and Richardson, "Some recent changes in Illinois River biology," ibid., 13 (1919): 134–66, pp. 146, 152–53 (reclamation data); and Schneider, "Enclosing the floodplain" (cit. n. 46).

60. John Stilgoe, *Borderland: Origins of the American Suburb, 1820–1939* (New Haven: Yale University Press, 1988), 58–64, chaps. 11–13; Sam B. Warner, Jr., *Streetcar Suburbs: The Process of Growth in Boston, 1870–1900* (Cambridge: Harvard University Press, 1962); and Adam Rome, *The Bulldozer in the Countryside: Suburban Sprawl and the Rise of American Environmentalism* (New York: Cambridge University Press, 2001).

61. Frederic E. Clements, "The relict method in dynamic ecology," *Journal of Ecology* 22 (1934): 39–68, pp. 50–53; Arthur G. Vestal to Frederic E. Clements, 26 Sept. [1923], FEC, box 23; and Henry A. Gleason, "Some unsolved problems of the prairies," *Bulletin of the Torrey Botanical Club* 36 (1909): 265–71, pp. 265–66.

62. Stilgoe, *Borderland* (cit. n. 60), 175–76.

63. John Burroughs, "Spring at the capital," in *Wake-Robin* (1871), vol. I of *The Writings of John Burroughs* (Boston: Houghton, Mifflin, 1905), 139–70, 141–42.

64. Burroughs, "Spring at the capital," ibid., 155–65 (quote, p. 158).

65. Frank M. Chapman, *Autobiography of a Bird-Lover* (New York: Appleton-Century, 1933), 33–34, photo at p. 30.

66. Editorials, *Botanical Gazette* 12 (1887): 112, 163–65; Henry R. Linville to Charles B. Davenport, 5 April, 4 June 1899, CBD-Gen.

67. Stilgoe, *Borderland* (cit. n. 60), chap. 12.

68. Frank S. Daggett to Joseph Grinnell, 15 May 1904, JG, box 5.

69. Frank S. Daggett to Grinnell, 15 May 1904, 19 April 1907, both in JG, box 5; Harry S. Swarth to Grinnell, 13 Sept., 4 June, 4 Sept. 1905,

25 Oct. 1905, all in JG, box 18; Tadpole [pseud.], "Opening of the season at Fox Lake," *Chicago Field* 7 (1877): 213.

70. Joseph Grinnell to Annie Alexander, 8 May 1908, 13 Feb., 9 March, 15 April 1911, all in AA, box 1.

71. Frank M. Chapman to Joel A. Allen, 19 Jan. 1889, in *Frank M. Chapman in Florida: His Journals and Letters*, ed. Elizabeth S. Austin (Gainesville: University of Florida Press, 1967), 46–48 (quote, 46–47).

72. Williams, *Americans and Their Forests* (cit. n. 50), chap. 12, esp. fig 12.1, p. 408.

73. Terrie, *Contested Terrain* (cit. n. 37); Schneider, *Adirondacks* (cit. n. 37); Karl Jacoby, *Crimes Against Nature: Squatters, Poachers, Thieves, and the Hidden History of American Conservation* (Berkeley: University of California Press, 2001), chaps. 1–3; and W. Storrs Leed, *The Green Mountains of Vermont* (New York: Henry Holt, 1955), 200–202.

74. Richard W. Sellers, *Preserving Nature in the National Parks: A History* (New Haven: Yale University Press, 1997); Runte, *National Parks* (cit. n. 56); and Beatrice Ward Nelson, *State Recreation Parks, Forests, and Game Preserves* (Washington: National Conference on State Parks, 1928).

75. Thomas S. Palmer, "Private Game Preserves and Their Future in the United States," *Circular of the Bureau of Biological Survey* 72 (1910). Also Ira N. Gabrielson, *Wildlife Refuges* (New York: Macmillan, 1943).

76. Clements, "Relict method" (cit. n. 61), 50–54.

77. Charles C. Adams, "An ecological survey of prairie and forest invertebrates," *Bulletin of the Illiois State Laboratory of Natural History* 11:2 (1916): 33–280, pp. 34–38. Also Henry A. Gleason to Arthur G. Vestal, 25 Sept. 1910, AGV, box 2.

78. James D. Miller, *South by Southwest: Planter Emigration and Identity in the Slave South* (Charlottesville: University of Virginia Press, 2002).

79. Williams, *Americans and Their Forests* (cit. n. 50), 468–77, chap. 14; and Lawrence M. Vaughan, "Abandoned farm areas in New York," *Cornell University Agricultural Experiment Station Bulletin* 490 (1929): 1–285, pp. 15–18.

80. Harold F. Wilson, *The Hill Country of Northern New England: Its Social and Economic History 1790–1930* (New York: Columbia University Press, 1936), chaps. 5–6; and John R. Stilgoe, "The wildering of rural New England 1850 to 1950," *New England and St. Lawrence Valley Geographical Society Proceedings* 10 (1980): 1–6.

81. Vaughan, "Abandoned farm areas" (cit. n. 79), 7–13.

82. Williams, *Americans and Their Forests* (cit. n. 50), 472–77, fig. 14.3, p. 475.

83. Wilson, *Hill Country* (cit. n. 80), chaps. 9–11; and Eric E. Lampard, *The Rise of the Dairy Industry in Wisconsin: A Study in Agricultural Change 1820–1920* (Madison: State Historical Society of Wisconsin, 1963), chaps. 1–4.

84. Donald Worster, *Dust Bowl: The Southern Plains in the 1930s* (New York: Oxford University Press, 1979); and Jonathan Raban, *Bad Land: An American Romance* (New York: Pantheon, 1996).

85. Frederic Clements to Secretary Lane, 15 Nov. 1918, CIW, f. Ecology projects.

86. Garrett Hardin, "The tragedy of the commons," *Science* 162 (1968): 1243–48. As many critics have observed, Hardin did not distinguish unregulated from regulated commons, which can be used sustainably, for example: Fikret Berkes, ed., *Common Property Resources: Ecology and Community-Based Sustainable Development* (London: Belhaven Press, 1989); and Netting (cit. n. 28).

87. Williams, *Americans and Their Forests* (cit. n. 50), chaps. 8–9; Robert Gough, *Farming the Cutover: A Social History of Northern Wisconsin, 1900–1940* (Lawrence: University Press of Kansas, 1997); Arthur F. McEvoy, *The Fisherman's Problem: Ecology and Law in the California Fisheries, 1850–1980* (New York: Cambridge University Press, 1986); and Grove K. Gilbert, *Hydraulic-Mining Debris in the Sierra Nevada*, U.S. Geological Survey Professional Paper no. 105 (Washington: Government Printing Office, 1917).

88. Will O' the Wisp [pseud.], "Fresh fields for sport," *Chicago Field* 7 (1877): 423.

89. Will Wildwood [pseud.], "A week in the woods: or the fish and fishing of northern Wisconsin," *Chicago Field* 8 (1877): 38, 90–91, 108.

90. Ingomar [pseud.], "The game sections of the Northwest," *Chicago Field* 5 (1876): 182; also Dakota [pseud.], "Should aould acquaintance be forgot, and ne'er brought to mind?" *Chicago Field* 7 (1877): 87.

91. Francis B. Sumner, *The Life History of an American Naturalist* (Lancaster, Pa.: Jaques Cattell Press, 1945), 1–5, 32–34, 38, 42–44, 46–48.

92. Quoted from Henry Gleason diary by Bassett Maguire, "Henry Allen Gleason," *Bulletin of the Torrey Botanical Club* 102 (1975): 274–77, p. 274.

93. Jay N. "Ding" Darling to Paul L. Errington, 4 Jan. 1958, PLE, box 2, f. 1; also R. M. Anderson to Aldo Leopold, 5 Oct. 1931, PLE, box 3, f. 5.

94. Paul L. Errington to Otto Koehler, 9 Jan. 1948, PLE, box 9, f. 1.

95. Joseph Grinnell to Annie Alexander, 27 Sept. 1918, AA, box 2.

96. Ibid., 10 Oct. 1925; also 13 Jan., 5 Nov. 1925; all in AA, box 2.

97. Homer L. Shantz, "Plant succession on abandoned roads in eastern Colorado," *Journal of Ecology* 5 (1917): 19–42.

98. Wilfred B. Shaw, *The University of Michigan: An Encyclopedic Survey* (Ann Arbor: University of Michigan Press, 1942–1958), vol. II, 761–69, quote, p. 762; and Henry A. Gleason, "The biological station of the University of Michigan," *School Science and Math* 13 (1913), 411–15, pp. 411–12.

99. Ulysses O. Cox, "The Zoological Survey of Minnesota," *Indiana Academy of Science Proceedings* 17 (1901): 89–03.

100. John Burroughs, "The return of the birds," in *Wake-Robin* (cit. n. 63), 3–39, p. 23.

101. Vaughan, "Abandoned farm areas" (cit. n. 79), pp. 20–36; quote p. 26.

102. Leo Marx, *The Machine in the Garden: Technology and the Pastoral Ideal in America* (New York: Oxford University Press, 1964); chap. 3; 142 (on Turner), 220–26.

103. Ibid., 245–46.

104. Marx, *Machine in the Garden* (cit. n. 102), 230–36. Also Ralph Waldo Emerson, "Nature" (1836), in *Nature: Addresses and Lectures*, vol. 1 of *Complete Works of Ralph Waldo Emerson* (Boston: Houghton Mifflin, 1903), 1–78.

105. For example: David W. Meinig, ed., *The Interpretation of Ordinary Landscapes: Geographical Essays* (New York: Oxford University Press, 1979); David Lowenthal, "Geography, experience, and imagination: towards a geographical epistemology," *Annals of the Association of American Geographers* 51 (1961): 241–60; and Stilgoe, *Borderland* (cit. n. 60).

106. Thomas J. Schlereth, "Chautauqua: a middle landscape of the middle class," *Old Northwest* 12 (1986): 265–78.

107. Susanna S. Zetzel, "The garden in the machine: the construction of nature in Olmsted's Central Park," *Prospects* 14 (1989): 291–339, pp. 293, 295 (quote); and Daniel M. Bluestone, "From promenade to park:

the gregarious origins of Brooklyn's park movement," *American Quarterly* 39 (1987): 529–50.

108. Paul V. Turner, *Campus: An American Planning Tradition* (Cambridge: MIT Press, 1984), 129–57; and Anne W. Spirn, "Constructing nature: the legacy of Frederick Law Olmsted," in William Cronon, ed., *Uncommon Ground: Toward Reinventing Nature* (New York: W. W. Norton, 1995), 91–113.

Chapter 2: Culture

1. Hans Huth, *Nature and the American: Three Centuries of Changing Attitudes* (Berkeley: University of California Press, 1957); Roderick Nash, *Wilderness and the American Mind* (New Haven: Yale University Press, 1967); Peter J. Schmitt, *Back to Nature: The Arcadian Myth in Urban America* (Oxford: Oxford University Press, 1969); and John Higham, "The reorientation of American culture in the 1890s," in *Writing American History: Essays on Modern Scholarship* (Bloomington: University of Indiana Press, 1970), 73–102.

2. For example: Kathryn Grover, ed., *Hard at Play: Leisure in America, 1840–1940* (Amherst: University of Massachusetts Press, 1992); Steven M. Gelber, "Working at playing: the culture of the work place and the rise of baseball," *Journal of Social History* 16:4 (June 1983): 3–20; Gary A. Tobin, "The bicycle boom of the 1890s: the development of private transportation and the birth of the modern tourist," *Journal of Popular Culture* 7 (1974): 838–49; and Laura Waterman and Guy Waterman, *Forest and Crag: A History of Hiking, Trail Blazing, and Adventure in the Northeast Mountains* (Boston: Appalachian Mountain Club, 1989).

3. Andrew C. Isenberg, *The Destruction of the Bison: An Environmental History, 1750–1920* (New York: Cambridge University Press, 2000), chap. 6; Robin W. Doughty, *Feather Fashions and Bird Preservation: A Study in Nature Protection* (Berkeley: University of California Press, 1975); James B. Trefethen, *An American Crusade for Wildlife* (New York: Winchester Press, 1975); and Mark V. Barrow, Jr., *A Passion for Birds: American Ornithology after Audubon* (Princeton: Princeton University Press, 1998), chaps. 6–7.

4. Sally G. Kohlstedt, *The Nature Study Movement: Science in Public Schooling, 1890–1932* (unpublished book manuscript, 2004). My thanks to Professor Kohlstedt for a copy of her manuscript. Also Dora Otis

Mitchell, "A history of nature-study," *Nature-Study Review* 19 (1923): 258–74, 295–321; and Schmitt, *Back to Nature* (cit. n. 1), chap. 7.

5. Richard White, *The Organic Machine: The Remaking of the Columbia River* (New York: Hill and Wang, 1995).

6. William A. Koelsch, "Antebellum Harvard students and the recreational exploration of the New England landscape," *Journal of Historical Geography* 8 (1982): 362–72; David Strauss, "Toward a consumer culture: 'Adirondack Murray' and the wilderness vacation," *American Quarterly* 39 (1987): 270–86; and Cindy Aron, *Working at Play: A History of Vacations in the United States* (New York: Oxford University Press, 1999), chap. 6. Also Orvar Löfgren, *On Holiday: A History of Vacationing* (Berkeley: University of California Press, 1999).

7. John B. Bachelder, *Popular Resorts and How to Reach Them* (Boston: The author, 1874), 9–11.

8. Aron, *Working at Play* (cit. n. 6), chap. 6; Harold F. Wilson, *The Hill Country of Northern New England: Its Social and Economic History, 1790–1930* (New York: AMS Press, 1967), chap. 14.

9. Aron, ibid., chap. 6, and (on women and family) 57–58, 158, 166, 176.

10. Porter Sargent, *A Handbook of Summer Camps: An Annual Survey* (12th ed., Boston: Porter Sargent, 1935); earlier editions have much useful historical material. Walter S. Ufford, "Fresh Air Charity in the United States," Ph.D. diss., Columbia University, 1897; and Schmitt, *Back to Nature* (cit. n. 1), chap. 10 (on scouting).

11. For example: Dona Brown, *Inventing New England: Regional Tourism in the Nineteenth Century* (Washington, D.C.: Smithsonian Institution Press, 1995), chaps. 2 and 5; Earl Pomeroy, *In Search of the Golden West: The Tourist in Western America* (Lincoln: University of Nebraska Press, 1957); Kerwin L. Klein, "Frontier produces: tourism, consumerism, and the Southwestern public lands, 1890–1990," *Pacific History Review* 62 (1993): 39–71.

12. Waterman and Waterman, *Forest and Crag* (cit. n. 2), 149, 151–59, 183–98; 152–55.

13. Ibid., 153, 156, 159, 184–85, 192–93.

14. Ibid., 154–59 (quote, p. 158).

15. Ibid., 79–91, also chaps. 9–11, 16.

16. Ibid., 145–50; 18–20, table 20.2, p. 204 (trail-making).

17. John F. Reiger, *American Sportsmen and the Origins of Conservation* (New York: Winchester Press, 1975); Louis Warren, *The Hunter's*

Game: Poachers and Conservationists in Twentieth-Century America (New Haven: Yale University Press, 1997).

18. Charles Hallock, *A Sportsman's Directory to the Principle Resorts for Game and Fish in North America*, in Hallock, *The Sportsman's Gazetteer and General Guide* (New York: Forest and Stream Publications, 1877).

19. For example: Thomas G. Scott, "Wildlife research," *Bulletin of the Illinois Natural History Survey* 27 (1958): 179–201, on p. 186; James J. Dinsmore, *A Country So Full of Game: The Story of Wildlife in Iowa* (Iowa City: University of Iowa Press, 1994); and A. W. Schorger, *The Passenger Pigeon: Its Natural History and Extinction* (Norman: Oklahoma University Press, 1955).

20. Ouiskinsin [pseud.], "The Wisconsin Association," *Chicago Field* 12 (1879–80): 49.

21. Marvin W. Kranz, "Pioneering in Conservation: A History of the Conservation Movement in New York State," Ph.D. diss., Syracuse University, 1951, 44–45, 102–105, 122–23; and H. K. Kohlsaat, "The greatest game market in the world," *Saturday Evening Post* 196 (26 April 1924): 72.

22. R. W. Palmer, Jr., "Game Commissions and Wardens: Their Appointment, Powers, and Duties," *Biological Survey Bulletin* 28 (1907), 9–10.

23. "Vacation," *Oxford English Dictionary*; Neil Harris, "On vacation," in *Resorts of the Catskills* (New York: St. Martin's Press, 1979), 101–108, p. 103; Betsy Blackman, "Going to the mountains: a social history," ibid., 71–99; Aron, *Working at Play* (cit. n. 6), 15–16, 32–33, chap. 1; and Huth, *Nature and the American* (cit. n. 1), 105–28, esp. 106.

24. Peter B. Bulkley, "Identifying the White Mountain tourist, 1853–1854: origin, occupation, and wealth as a definition of the early hotel trade," *Historical New Hampshire* 35 (1980): 106–62.

25. William James, "Vacations," *Nation* 17 (1873): 90–91.

26. Bachelder, *Popular Resorts* (cit. n. 7); Hallock, *Sportsman's Directory* (cit. n. 18); Charles Hallock, *The Fishing Tourist: Angler's Guide and Reference Book* (New York: Harper, 1873). Hunters' guides are rich sources on the infrastructure of tourism.

27. Melvil Dewey, "Co-operation in vacations," *Outlook* 52 (1895): 135–36.

28. Harris, "On vacation" (cit. n. 23), 104–106.

29. [Editor], "American sports," *Chicago Field* 10 (1878): 232.

30. Edward Hungerford, "Our summer migration," *Century* 42 (1891): 569–76; and Aron, *Working at Play* (cit. n. 6), chap. 2, also chap. 4 (religious resorts).

31. Hungerford, ibid., 576, 569.

32. Daniel T. Rodgers, *The Work Ethic in Industrial America 1850–1920* (Chicago: University of Chicago Press, 1974), chap. 4; Peter Bailey, *Leisure and Class in Victorian England: Rational Recreation and the Contest for Control, 1830–1885* (London: Routledge & Kegan Paul, 1978), chap. 3; Elizabeth B. Keeney, *The Botanizers: Amateur Scientists in Nineteenth-Century America* (Chapel Hill: University of North Carolina Press, 1992), chaps. 3, 6; and especially Aron, *Working at Play* (cit. n. 6), intro., chaps. 6, 7, esp. 175–77 and 230–36.

33. Aron, ibid., chap. 3; Harris, "On vacation" (cit. n. 23), 102–104.

34. Bailey, *Leisure and Class* (cit. n. 32), 63–70 (quote, p. 64). Bailey writes of the British middle class, but the social dynamic in America was much the same.

35. Aron, *Working at Play* (cit. n. 6), 7–10, 175–77, 230–36; Rodgers, *Work Ethic* (cit. n. 32), 108–10; and Bailey, *Leisure and Class* (cit. n. 32), 65–70, and chap. 6.

36. Rodgers, ibid., 102–105; Aron, ibid., chaps. 1, 3; and Huth, *Nature and the American* (cit. n. 1), 106–107.

37. Bailey, *Leisure and Class* (cit. n. 32), 68–73.

38. Brace [pseud.], "Our national health: less work and more play," *Chicago Field* 2 (1874–75): 573–74; also "Sport vs. work," ibid, p. 295.

39. Bailey, *Leisure and Class* (cit. n. 32), 69–71.

40. On middle-class self-fashioning: Stuart M. Blumin, "Hypothesis of middle-class formation in nineteenth-century America: a critique and some proposals," *American Historical Review* 90 (1985): 299–338; and Karen Halttunen, *Confidence Men and Painted Women: A Study of Middle-Class Culture in America, 1830–1870* (New Haven: Yale University Press, 1982).

41. Herbert Spencer, "The gospel of recreation," *Popular Science Monthly* 22 (1883): 354–59, 358; also Bailey, *Leisure and Class* (cit. n. 32), 66–67.

42. Glenn Uminowicz, "Recreation in a Christian America: Ocean Grove and Asbury Park, New Jersey, 1869–1914," in Grover, ed. *Hard at Play* (cit. n. 2), 8–38 (quote, p. 23); Hungerford, "Our summer migration" (cit. n. 30); Clarence Deming, "City boarders and the farm," *Nation* 53 (1891): 115–16; and E. L. Godkin's many editorials in *The Nation*: for example, "Cottagers and boarders," 35 (1882): 196–87; "The evolu-

tion of the summer resort," 37 (1883): 47–48; and "Boarders' rights on the sea-shore," 37 (1883): 111. Also C. P., "The boarder and the cottager," *Nation* 37 (1883): 97.

43. Jennifer Koop, "Randolph, New Hampshire, a special community: founded by farmers, transformed by trailmakers," *Historical New Hampshire* 49 (1994): 133–56, on pp. 135–36, 139. There is apparently no history of clerical vacationing (Professor Robert Prichard, personal communication).

44. See Nina Lübbren, *Rural Artists' Colonies in Europe 1870–1910* (New Brunswick: Rutgers University Press, 2001).

45. [Editor], "Summer rest," *Nation* 37 (1883): 157.

46. Huth, *Nature and the American* (cit. n. 1), 124–26. The quotations are from William Dean Howell's vacationing novel, *The Landlord of Lion's Head* (1897).

47. Aron, *Working at Play* (cit. n. 6), chaps. 2–3; Hungerford, "Our summer migration" (cit. n. 30); and Julian Ralph, "The spread of outdoor life," *Harper's Weekly* 36 (1892): 830–31.

48. Hungerford, ibid., 569–70; and Blackman, "Going to the mountains" (cit. n. 23), 78–80. Also [Editor], "Changes in summer migration," *Nation* 53 (1891): 210–11; Brown, *Inventing New England* (cit. n. 11), chap. 5; Huth, *Nature and the American* (cit. n. 1), 106–108; and Aron, ibid., 232–35.

49. Schmitt, *Back to Nature* (cit. n. 1), 172.

50. [Editor], "Cottagers and boarders," *Nation* 35 (1882): 196–97; [Editor], "The evolution of the summer resort," ibid., 37 (1883): 47–48; C. P., "The boarder and the cottager," ibid., 97; and [Editor], "Boarders' rights on the sea-shore," ibid., 111.

51. Brown, *Inventing New England* (cit. n. 11), chap. 5., esp. 135–57; William H. Bishop, "Hunting an abandoned farm in upper New England," *Century* 48 (1894): 30–43; Bishop, "Hunting an abandoned farm in Connecticut," ibid., 47 (1894): 915–24; [J. E. Learned], "A suggestion for summer," *Nation* 50 (1890): 195–96; [E. P. Clark], "Summer boarding and rural health," ibid. 52 (1891): 253–54; [E. D. Mead], review of Horace G. Wadlin, "Abandoned farms in Massachusetts," *New England Magazine* 4 (1891): 675–77; and Allen Chamberlain, "The ideal abandoned farm," ibid., 16 (1897): 473–78. John R. Stilgoe, "The wildering of rural New England 1850 to 1950," *New England and Saint Lawrence Valley Geographical Society Proceedings* 10 (1980): 1–6.

52. www.taylorfarmvermont.com

53. Olivier Zunz, *Making America Corporate, 1870–1920* (Chicago: University of Chicago Press, 1990), chap. 5.

54. On the symbiosis of outdoor sport and science: David Robbins, "Sport, hegemony and the middle class: the Victorian mountaineers," *Theory, Culture and Society* 4 (1987): 579–601; and Bruce Hevly, "The heroic science of glacier motion," *Osiris* 11 (1996): 66–86.

55. Bachelder, *Popular Resorts* (cit. n. 7), 9–11 (quote, p. 10).

56. Aron, *Working at Play* (cit. n. 6), 157, 175 (quote).

57. Mark Barrow, "Birds and Boundaries: Community, Practice, and Conservation in North American Ornithology, 1865–1935." Ph.D. diss., Princeton University, 1992, p. 105 (quote); also Barrow, *Passion for Birds* (cit. n. 3), chap. 2.

58. [Editor], "Changes in summer migration," *Nation* 53 (1891): 210–11.

59. James L. Peters, "Collections of birds in the United States and Canada: study collections," in *Fifty Years' Progress of American Ornithology 1883–1933* (Lancaster, Pa.: American Ornithologists' Union, 1933), 131–41.

60. Waterman and Waterman, *Forest and Crag* (cit. n. 2), 184, 189–92, chaps. 13, 17.

61. Reiger, *American Sportsmen* (cit. n. 17), chap. 1; and Thomas L. Altherr, "The American hunter-naturalist and the development of the code of sportsmanship," *Journal of Sport History* 5 (1978): 7–22. Also David K. Wiggins, "Work, leisure, and sport in America: the British travellers image, 1830–1860," *Canadian Journal of the History of Sport* 13 (1982): 28–60.

62. Barrow, *Passion for Birds* (cit. n. 3), 30–33 (quotes, p. 30).

63. Reiger, *American Sportsmen* (cit. n. 17), 25–41. Also Huth, *Nature and the American* (cit. n. 1), 54–57.

64. John M. Mackenzie, *The Empire of Nature: Hunting, Conservation and British Imperialism* (Manchester: Manchester University Press, 1988), 39–41, 50–51 (quote, p. 39).

65. Reiger, *American Sportsmen* (cit. n. 17), 65–69; and Altherr, "American hunter-naturalist" (cit. n. 61).

66. Frank M. Chapman, *Autobiography of a Bird-Lover* (New York: Appleton-Century, 1933), chap. 2.

67. Ibid., 32.

68. For example: Elliott Coues, "On the habits of the moose," *Chicago Field* 8 (1877–78): 27; David Starr Jordan, "On the distribution of the fresh water fishes," ibid., 8 (1877–78): 219; Fred Mather, "Nomencla-

ture," ibid., 8 (1877–78): 209; and "Archer's reply to Dr. Coues," ibid., 12 (1879–80): 299–300 (disagreement over nomenclature).

69. For example, W. H. Ballou, "Natural history," *Chicago Field* 14 (1880): 116–17; and [Editor], "Smithsonian men," ibid., 14 (1880): 155–56.

70. [Editor], "American sports," *Chicago Field* 10 (1878): 232.

71. [Editor], "Field sports for women," *Chicago Field* 1 (1874): 88.

72. For example, [Editor], "Amateur and professional," *Chicago Field* 7 (1877): 422; [Editor], "The violations of the game law," ibid., 8 (1877–78): 104; and Guido [pseud.], "Annual Prairie Club hunt in Arkansas," ibid., 8 (1877–78): 137.

73. [Editor], "The Cuvier Club," *Chicago Field* 15 (1881): 74, 264; [Anonymous], "The Lake Pepin Club." ibid., 5 (1876): 90; and [Anonymous], "Sporting notes from South Bend, Ind.," ibid., 3 (1875): 227.

74. Charles C. Adams to Alexander G. Ruthven, 8 Dec. 1906, AGR, box 51, f. 1.

75. Philip J. Pauly, "Summer resort and scientific discipline: Woods Hole and the structure of American biology, 1882–1925," in *The American Development of Biology*, ed. Ronald Rainger, Keith R. Benson, and Jane Maienschein (Philadelphia: University of Pennsylvania Press, 1988), 121–50; and Robert E. Kohler, "Labscapes: naturalizing the lab," *History of Science* 40 (2002): 473–501.

76. Henry A. Gleason, "The biological station of the University of Michigan," *School Science and Mathematics* 13 (1913): 411–15, 411 (quote), 414–15. Also Francis Ramaley and W. W. Robbins, "A summer laboratory for mountain botany," *Plant World* 12 (1909): 105–10.

77. "The University of Montana Biological Station," in Morton Elrod to Stephen Forbes, 26 April 1913, INHS 43/1/5, box 1, f. State Lab.

78. Huth, *Nature and the American* (cit. n. 1), 87–104; and Schmitt, *Back to Nature* (cit. n. 1), chaps. 2, 4, 11, 12.

79. Francis W. Halsey, "The rise of the nature writers," *Review of Reviews* 26 (1902): 567–71.

80. Ralph H. Lutts, *The Nature Fakers: Wildlife Science and Sentiment* (Golden, Colo.: Fulcrum, 1990), 28–29; and Higham, "Reorientation of American culture" (cit. n. 1), 81.

81. Barrows, *Passion for Birds* (cit. n. 3), 13–17, fig. 1 (p. 15); also Barrow, "Birds and Boundaries" (cit. n. 57), 48–62.

82. Schmitt, *Back to Nature* (cit. n. 1), chap. 12. Also Jonathan Raban, *Bad Land: An American Romance* (New York: Pantheon, 1996), 38–42 (on Porter).

83. Schmitt, *Back to Nature* (cit. n. 1), chap. 11. Gene Straton Porter, *A Girl of the Limberlost* (New York: Grosset and Dunlap, 1909).

84. John Burroughs, "Introduction to Riverside edition [1895]," in *Wake-Robin*, vol. I of *The Writings of John Burroughs* (Boston: Houghton, Mifflin, 1905), xi–xvi, pp. xii–xiii.

85. Halsey, "Rise of the nature writers" (cit. n. 79), 571.

86. Frank Chapman to Witmer Stone, 22 March 1933, ANS 679.

87. Karen Wonders, *Habitat Dioramas: Illusions of Wilderness in Museums of Natural History* (Stockholm: Almqvist and Wiksell, 1993), chaps. 2, 5. Also Lynn K. Nyhart, "Science, art, and authenticity in natural history displays," in Soroya de Chadarevian and Nick Hopwood, eds., *Models: The Third Dimension of Science* (Stanford: Stanford University Press, 2004), 307–35.

88. Much—too much, I think—has been made of moral values implicit in dioramas. See Donn@ Harraway, "Teddy bear patriarchy: taxidermy in the Garden of Eden, New York City, 1908–1930," in *Primate Visions: Gender, Race, and Nature in the World of Modern Science* (New York: Routledge, 1989), 26–58. Telling critiques of Harraway's reading are: Michael Schudson, "Paper tigers: a sociologist follows cultural studies into the wilderness," *Lingua Franca* 7:6 (Aug. 1997): 49–56; and Wonders, *Habitat Dioramas* (cit. n. 87), 223–25.

89. Wonders, *Habitat Diorama* (cit. n. 87), 35–36, 113–14.

90. Wonders, *Habitat Dioramas* (cit. n. 87), 112–25.

91. Charles B. Cory to Frederick Skiff, 29 May 1895; 23 June, 10 Aug., 12 Oct., 5, 9 Nov. 1896; Cory to H. Mintorn and Mrs. Mogridge, 9 Nov. 1896; all in FMNH, Directors Papers, ff. Cory.

92. Wonders, *Habitat Dioramas* (cit. n. 87), 126–36; Chapman, *Autobiography* (cit. n. 66), 164–66 (Cobb's Island group). Field Museum *Annual Report of the Director* (1901–1902): 162; (1910): 100; (1911): 133; (1912): 218 (bird groups); and ibid., 188 (bear group). The bear group is an interesting transitional form, with a background scene painted on the back side of a glass box with no effort to merge painting and group.

93. Dean C. Worcester, untitled report, *Proceedings of the Board of Regents, University of Michigan* (1891–96): 45–53 (quote, p. 545).

94. For an analysis of how the visual illusions of the habitat diorama worked: Wonders, *Habitat Dioramas* (cit. n. 87), chap. 5.

95. Ibid., 126 (quotes).

96. Ibid., 118–19, 120–21, 126–29. Also, John M. Kennedy, "Philanthropy and Science in New York City: The American Museum of Natural History, 1868–1968," Ph.D. diss., Yale University, 1968, chap. 3.

97. Burroughs, "Introduction" (cit. n. 84), xv–xvi.

98. Edward J. Renehan, Jr., *John Burroughs: An American Naturalist* (Post Mills, Vt.: Chelsea Green, 1902), 96 (quote). Also Huth, *Nature and the American* (cit. n. 1), 87–104.

99. For example: Burroughs, "Spring at the capital" (1868), in *Wake-Robin* (cit. n. 84), 139–70, pp. 163–64.

100. For example: Burroughs, "A sharp lookout," in *Signs and Seasons. Writings.* (cit. n. 84), vol. VII; and "Eye-beams," in *Riverby. Writings*, vol. IX, 119–42.

101. Lutts, *Nature Fakers* (cit. n. 80); Schmitt, *Back to Nature* (cit. n. 1), 46–49; Renehan, *John Burroughs* (cit. n. 98), 229–50. This revival of the sentimental nature fable may be symptomatic of the mass commercialization of vacationing.

102. For example, John Burroughs, "Birch browsings," in *Wake-Robin* (cit. n. 84), 171–203.

103. John Burroughs, "The return of the birds" (1863), ibid., 3–39, p. 23; "The Adirondacks," ibid., 77–102, pp. 77–79, 94–99 (quote, p. 97); and "Spring at the capital," ibid., 139–70, pp. 157–65. Also Burroughs, "The heart of the southern Catskills," in *Riverby* (cit. n. 100), 37–66.

104. Steven Shapin, "Pump and circumstance: Robert Boyle's literary technology," *Social Studies of Science* 14 (1984): 481–520 (virtual witnessing). On the uses of particularity in field science: Robert E. Kohler, *Landscapes and Labscapes: Exploring the Lab-Field Border in Biology* (Chicago: University of Chicago Press, 2002); and Kohler, "Place and practice in field biology," *History of Science* 40 (2002): 189–210.

105. Burroughs, "A sharp lookout" (cit. n. 100), 25–26.

106. Wonders, *Habitat Dioramas* (cit. n. 87), 118–19. On the reception of the *Cobb's Island* diorama: Chapman, *Autobiography* (cit. n. 66), 165–66. Naturalists in German museums were equally resistant to diorama displays. Nyhart, "Science, art, and authenticity" (cit. n. 86).

107. Wonders, ibid., 118, also 34–45. Nyhart, "Science, art, and authenticity" (cit. n. 87), 313–15.

108. Wonders, ibid., 109–10, 119–20.

109. Susan Leigh Starr, "Craft vs. commodity, mess vs. transcendence: how the right tool became the wrong one in the case of taxidermy and natural history," In *The Right Tools for the Job: At Work in Twentieth-Century Life Sciences*, eds. Adele E. Clarke and Joan H. Fujimura (Princeton: Princeton University Press, 1992), 257–86.

110. For example: James Buzard, *The Beaten Track: European Tourism, Literature, and the Ways to Culture 1800–1918* (Oxford: Clarendon

Press, 1993); John F. Sears, *Sacred Places: American Tourist Attractions in the Nineteenth Century* (New York: Oxford University Press, 1989); and Joan-Pau Rubies, "Instructions for travellers: teaching the eye to see," *History and Anthropology* 9 (1996): 139–90.

111. Steven Shapin, "The house of experiment," *Isis* 79 (1988): 373–404, p. 383.

Chapter 3: Patrons

1. Robert R. Miller to Carl L. Hubbs, 29 July 1947, CLH, box 24, f. 77. It is not clear when family members were excluded from government expeditions, or if it was ever an official rule. Pamela Henson, personal communication.

2. Robert E. Kohler, *Partners in Science: Foundations and Natural Scientists 1900–1945* (Chicago: University of Chicago Press, 1991).

3. "Proposed explorations and investigations on a large scale," *Carnegie Institution Year Book* 1 (1902): Appendix B, 239–69; Frederic V. Coville, "A plan for botanical research," to Charles D. Walcott, 28 June 1902, CIW, f. Advisory Committee Botany.

4. John Shaw Billings to Nathaniel L. Britton, 29 June 1903, CIW, f. Advisory Committee Botany.

5. Mary P. Winsor, *Reading the Shape of Nature: Comparative Zoology at the Agassiz Museum* (Chicago: University of Chicago Press, 1991).

6. A. Hunter Dupree, *Science in the Federal Government: A History of Policies and Activities to 1940* (Cambridge: Harvard University Press, 1957), chap. 10; Thomas G. Manning, *Government in Science: The U.S. Geological Survey 1867–1894* (Lexington: University of Kentucky Press, 1967).

7. Manning, ibid., 42–43. State geological surveys often included some natural history at first but dropped it when they became permanent. Frederick C. Newcombe, "The scope and method of state natural history surveys," *Science* 37 (1913): 615–22, pp. 615–16.

8. Keir Sterling, *Last of the Naturalists: The Career of C. Hart Merriam* (New York: Arno, 1977), 56–64; and *Report of the Secretary of Agriculture* (1885): 210–11; (1886): 227–59, 463. Also, Jenks Cameron, *The Bureau of Biological Survey: Its History, Activities, and Organization* (Baltimore: Johns Hopkins University Press, 1929).

9. *Report of the Secretary of Agriculture* (1889): 364–66; (1890): 278; (1891): 267–70; (1898): 37–43.

10. Wilfred Osgood, "Clinton Hart Merriam—1855–1942," *Journal of Mammalogy* 24 (1943): 421–36, pp. 422–26; C. Hart Merriam to Vernon Bailey, 13 Jan. 1886, p. 4, VOB, box 1 (on Merriam's personal collectors).

11. *Report of the Secretary of Agriculture* (1887): 399–400; (1888): 477–84; (1889): 363–65; (1898): 37–43; Sterling, *Last of the Naturalists* (cit. n. 8), 122–34; and Cameron, *Bureau of Biological Survey* (cit. n. 8), 20–36.

12. Other agencies empowered to send out expeditions were the Geological Survey, Bureau of American Ethnology, and Fish Commission.

13. Osgood, "Clinton Hart Merriam" (cit. n. 10), 432–35.

14. Joseph Grinnell to Wilfred Osgood, 20 Oct. 1930; Osgood to Grinnell, 18 Sept. 1930; both in WHO, box 4.

15. John C. Coulter, "The future of systematic botany," *Botanical Gazette* 16 (1891): 243–54, pp. 246–48. [Coulter?], editorials, *Botanical Gazette* 12 (1887): 197–98; 15 (1890): 267–68; and 22 (1896): 57.

16. Clarence M. Weed, "The biological work of American experiment stations," *American Naturalist* 25 (1891): 230–36.

17. Thomas R. Dunlap, *Saving America's Wildlife* (Princeton: Princeton University Press, 1988).

18. Newcombe, "Scope and method" (cit. n. 7).

19. Stephen A. Forbes, "State Laboratory of Natural History," *University of Illinois Report of the Trustees* 14 (1887–88): 185–93, p. 186.

20. Henry Nachtrieb to Charles B. Davenport, 28 Jan. 1901, HFN, box 1.

21. Alexander Ruthven, memorandum to Board of Scientific Advisors" [Dec. 1913], AGR, box 54, f. 6.

22. Conway Macmillan to John S. Pillsbury, 1 Dec. 1898, UM Comptrollers, f. 318.

23. P. L. Hatch to Henry Nachtrieb, 2 Jan., 13 June 1893, both in HFN, box 1. Hatch was a physician and amateur ornithologist who rashly agreed to survey state birds, with minimal support and relying on local amateurs; the scheme ended badly, in mutual recriminations.

24. G. M. Schwartz, *History of the Minnesota Geological Survey* (Minneapolis: Minnesota Geological Survey Special Publication no. 1, [1964], 3–11. Harold L. Lyon to Regents, 4 Oct. 1906; Henry F. Nachtrieb to Regents, 29 May 1890; Committee on Geological and Natural History Survey to Regents, n.d. [1896–97]; Conway Macmillan to John S. Pillsbury, 1 Dec. 1898; all in UM Comptroller, f. 317. C. Otto Rosendahl to William Trelease, 7 Jan. 1918, UM Botany, box 1, f. 6.

Newton H. Winchell to Henry Nachtrieb, 18 Jan. 1893; P. L. Hatch to Nachtrieb, 2 Jan., 13 June 1893; all in HFN, box 1.

25. Alexander Ruthven to Bryant Walker, 10 Nov. 1909, AGR, box 52, f. 17; Ruthven to Rolland C. Allen, 17 April 1912, AGR, box 54, f. 4; Ruthven to Charles C. Adams, 29 Jan. 1909, and Alfred Lane to Ruthven, 5 May 1909, both in AGR, box 54, f. 2; Ruthven, "Report of the curator of the University of Michigan Museum to the Board of Regents" (1908–1909): 7–11, p. 10; Rolland C. Allen, "Annual report of the director to the Board of Scientific Advisors to the Michigan Geological and Biological Survey," Dec. 1913.

26. Stephen A. Forbes, "The Illinois State Laboratory of Natural History and the Illinois State Entomologist's Office," *Illinois State Academy of Science Transactions* 2 (1909): 54–67; Forbes, "Report of the State Laboratory of Natural History," *University of Illinois Report of the Trustees* 14 (1887–88): 185; Various authors, "A Century of Biological Research," *Illinois Natural History Survey Bulletin* 27:2 (1958): 85–234; and Robert G. Hays, *State Science in Illinois: The Scientific Surveys, 1850–1978* (Carbondale: Southern Illinois University Press, 1980), chaps. 2, 3.

27. Forbes, "Illinois State Laboratory," ibid., 54–56.

28. *Bulletin of the Washburn Laboratory of Natural History* 1 (1884): 2–3; and "The zoology class," *Washburn Argo* 1:1 (Dec. 1885).

29. Lucien M. Underwood, "Report of the botanical division of the Indiana State Biological Survey," *Proceedings of the Indiana Academy of Sciences* 9 (1893): 13–19; John M. Coulter, "Suggestions for the Biological Survey," ibid. 9 (1893): 191–93; and Underwood, "Report . . . for 1894," ibid., 10 (1894): 144–47.

30. Thomas Hankinson to Alexander Ruthven, 30 May 1914, AGR, box 51, f. 24.

31. Roscoe Pound and Frederic E. Clements, *The Phyto-Geography of Nebraska, I: General Survey* (Lincoln: University of Nebraska Botanical Seminar, 1897; 2nd ed., 1900), 3; and Charles E. Bessey, Roscoe Pound, and Frederic E. Clements, "Report on recent collections," *Botanical Survey of Nebraska* 5 (1901): 4. For what little is known of George Holdridge, my thanks to Jeremy Vetter.

32. Ronald C. Tobey, *Saving the Prairie: The Life Cycle of the Founding School of American Plant Ecology, 1895–1955* (Berkeley: University of California Press, 1981), 20, chap. 1.

33. Pound and Clements, *Phytogeography of Nebraska* (cit. n. 31), 3, 21–22; Roscoe Pound and Frederic E. Clements, "A method of determin-

ing the abundance of secondary species," *Minnesota Geological and Natural History Survey, Botanical Series* 4 (1898): 19–24; and Robert E. Kohler, *Landscapes and Labscapes: Exploring the Lab-Field Border in Biology* (Chicago: University of Chicago Press, 2002), chap. 4.

34. C. W. Hayes, "The state geological surveys of the United States," U.S. Geological Survey *Bulletin* 465 (Washington, D.C.: Government Printing Office, 1911); Howard S. Reed, "A survey of the Huron River valley, I: the ecology of a glacial lake," *Botanical Gazette* 34 (1902): 125–39, p. 125; Charles W. Dodge, "A proposed biological survey of New York State," *School Science and Mathematics* 6 (1906): 371–78; and Theodore Cockerell to William M. Wheeler, 27 Nov. 1903, WMW, box 6, f. Formicidae.

35. Newcombe, "Scope and method" (cit. n. 7), 616–17; Arthur I. Ortenburger to Karl Schmidt, 4 May, 21 June, 16 Sept. 1925, all in KPS, box 15, f. 3.

36. Osgood, "Merriam" (cit. n. 10), 424; and Sterling, *Last of the Naturalists* (cit. n. 8), 58–65.

37. Gerrit Miller to Witmer Stone, 21 Jan. 1899, ANS 955; Dodge, "Proposed biological survey" (cit. n. 34); and Charles C. Adams to Alexander Ruthven, 10 March 1908, AGR, box 51, f. 2.

38. Coulter, "Future of systematic botany" (cit. n. 15), 248.

39. There is apparently no history of state topographical mapping projects, but see J. L. van Ornum, "Topographical surveys, their methods and value," *Bulletin of the University of Wisconsin, Engineering Series* 1 (1896): 331–69. Also Alex Chekovitch, "Mapping the American Way: Geographical Knowledge and the Development of the United States, 1890–1950," Ph.D. diss., University of Pennsylvania, 2004 (forest, soil, and land-use maps).

40. Louis Agassiz's Museum of Comparative Zoology was an exception. Agassiz pursued a "greedy collecting policy," believing that knowledge of variation would make taxonomy more exact. Winsor, *Reading the Shape of Nature* (cit. n. 5), 67–78 (quote, 77); and Mark V. Barrow, Jr., *A Passion for Birds: American Ornithology after Audubon* (Princeton: Princeton University Press, 1998), 77–79.

41. Charles C. Adams to Alexander Ruthven, 22 May 1909, AGR, box 51, f. 2.

42. Robert Ridgway to Witmer Stone, 29 June 1893 and 27 Oct. 1894, both in ANS 681.

43. Winsor, *Reading the Shape of Nature* (cit. n. 5), 144–46. Academy of Natural Sciences *Proceedings* 41 (1889): 430–33; also 36 (1884): 330–

31; 39 (1887): 413–16; 42 (1890): 475–78; 45 (1893): 568–69; 47 (1895): 20–22.

44. Joel A. Allen to Witmer Stone, 24 Oct. 1894, ANS 658.

45. Academy of Natural Science Annual Report, *Proceedings* 53 (1901): 745–47.

46. Wilfred Osgood to Malcolm P. Anderson, 9 Dec. 1913, WHO, box 1.

47. Daniel G. Elliott to Frederick J. V. Skiff, 15 April 1901, FMNH, Directors Papers, f. Elliott 1.

48. Charles B. Cory to Frederick J. V. Skiff, 18 June 1909, WHO, box 9, ff. Skiff.

49. Karen Wonders, *Habitat Dioramas: Illusions of Wilderness in Museums of Natural History* (Stockholm: Almqvist and Wiksell, 1993), 126–31, 148.

50. For example, Wilfred Osgood to Charles B. Cory, 29 Jan. 1917, WHO, box 8, f. Osgood.

51. Ruthven, "Report" (cit. n. 25), 7–11, 8. Also Leonhard Stejneger and Gerrit Miller, "Plan for a biological survey," *Carnegie Institution Year Book* 1 (1902): 244.

52. James L. Peters, "Collections of birds in the United States and Canada: study collections," in *Fifty Years' Progress of American Ornithology 1883–1933* (Lancaster: American Ornithologists' Union, 1933), 131–41, pp. 131–32.

53. Frederick J. H. Merrill, "Natural History Museums of the United States and Canada," *New York State Museum Bulletin* 62 (Albany: University of the State of New York, 1903).

54. For example, Charles B. Cory to Frederick Skiff, 15 Feb., 16 Dec. 1911, both in WHO, box 9, ff. Skiff.

55. Daniel G. Elliott to Frederick J. V. Skiff, [Sept. 1900?] (quote); Elliott to Skiff, 13 Sept. 1898; both in FMNH, Directors, ff. Elliott.

56. Daniel G. Elliott to Frederick Skiff, 1 Oct. 1905, FMNH, Directors, f. Elliott. Cory was still fighting in 1907 to allow curators to go afield: Charles B. Cory to Skiff, 19 Nov. 1907, WHO, box 9, ff. Skiff.

57. Charles C. Adams, "Report of curator of Museum to Board of Regents of the University of Michigan," (1905–1906): 7.

58. John M. Kennedy, "Philanthropy and Science in New York City: The American Museum of Natural History, 1868–1968," Ph.D. diss., Yale University, 1968, 91–94, 108–11, 148, 156–57.

59. American Museum of Natural History *Annual Report* 34 (1902): 14–16; 35 (1903): 14–16; 36 (1904): 15; 38 (1906): 16–17; Frank M.

Chapman, *Autobiography of a Bird-Lover* (New York: Appleton-Century, 1933), 164–65.

60. Robert C. Murphy to Wilfred Osgood, 21 Oct. 1920, WHO, box 7.

61. Reports on the Whitney South Seas Expeditions can be found in the American Museum's annual reports. The project is voluminously documented in the museum's archive and photo collections.

62. *American Museum of Natural History Annual Report* 61 (1929), p. 10; 62 (1930), p. 21.

63. Data are from "Expedition File," American Museum of Natural History archives, New York City, corrected and supplemented by data from annual reports. (This list counts multiple-year expeditions only in their initial years, and excludes informal excursions.)

64. Data are from the annual reports of the Field Museum.

65. Data are from a card file and list of expeditions in the Academy archives, but are doubtless incomplete.

66. Data are from the National Museum's annual reports.

67. American Museum *Annual Report* 19 (1887–88): 8, 12–13 (quote, p. 12); Wonders *Habitat Diorama* (cit. n. 49), 120–23.

68. American Museum *Annual Report* 27 (1895): 21–22 (quote); 28 (1896): 15–16; 29 (1897): 14–15; 32 (1900): 12–15.

69. Academy of Natural Sciences *Proceedings* 41 (1889): 430–33, (quote p. 431); also ibid., 42 (1890): 475–78; 45 (1893): 568–69; 57 (1895): 20–22; 53 (1901): 745–47.

70. For example, Daniel G. Elliott to Frederick Skiff, 16, 26 Feb. 1898, both in FMNH, Directors, ff. Elliott.

71. American Museum *Annual Report* 22 (1890–91): 7; 23 (1891–92): 12; 24 (1892): 10–11; 26 (1894): 9–10; 27 (1895): 21–22; 28 (1896): 15–16; 29 (1897): 14–15; 35 (1903): 14–16; 39 (1907): 29–30; 42 (1910): 36–41. Museums also continued to use professional collectors, whose relations with curators were generally cordial.

72. Harry Swarth to Joseph Grinnell, 5 Sept. 1907; also 12 Dec. 1906 and 19 Nov. 1907; all in JG-MVZ, box 18.

73. Correspondence in FMNH, Directors, ff. Elliott.

74. Harry Swarth to Joseph Grinnell, 3, 7, 28 May, 21 Aug., 11 Oct. 1913; 1 Jan., 14 Feb., 8 April, 1 July 1914; all in JG-MVZ.

75. Alexander Ruthven, untitled history of the Michigan Museum [early 1930s], MMZ, box 1, f. histories.

76. Wilfred Osgood to Carl Akeley, 7 June 1921, WHO, box 1.

77. Wilfred Osgood to Oldfield Thomas, 27 Aug. 1924 (quote), 16 March 1925, both in WHO box 9.

78. Ibid., 11 March 1926, WHO, box 9.

79. Frank Chapman to Osgood, 31 Oct. 1928, WHO, box 1.

80. Joseph Grinnell to Annie Alexander, 29 May 1916, AA, box 2.

81. Ibid., 26 May 1921, 5 June, 12 Oct 1928; all in AA, box 2.

82. For example, Charles B. Cory to Frederick Skiff, 29 Feb. 1912, 26 June 1914, 24 March 1916, and other correspondance, all in WHO, box 9, ff. Skiff.

83. Ruthven received his Ph.D. and joined the Michigan faculty in 1906.

84. Wilfred B. Shaw, *The University of Michigan: An Encyclopedic Survey* (Ann Arbor: University of Michigan Press, 1942–1958), vol. 4, 1502–19, pp. 1504–1505; Alexander Ruthven to George Shiras, 20 Jan. 1911, AGR, box 52, f. 8.

85. Ruthven to Bryant Walker, 9 Feb. 1909, AGR, box 52, f. 17.

86. Charles C. Adams, Report of curator of University Museum to Board of Regents of the University of Michigan (1904–1905): 9–10; (1905–1906): 9–11.

87. Alexander Ruthven to William B. Mershon, 23 Dec. 1909 [Jan. 1910]; Mershon to Ruthven, 28, 30 Dec. 1909, 4 Feb. 1910; all in UMM, box 1, f. Charity Islands 1910.

88. Ruthven to George Shiras, 20 Jan., 9 Dec. 1911; Shiras to Ruthven, 25 Jan., 21 Dec. 1911; all in AGR, box 52, f. 8.

89. George Shiras to Alexander Ruthven, 17 April, 7 Dec. 1915, 2 Feb. 1917; Ruthven to Shiras, 12 Feb. 1917; all in AGR, box 52, f. 8.

90. Alexander Ruthven to William W. Newcomb, 17 June 1912, AGR, box 52, f. 2.

91. William W. Newcomb to Alexander Ruthven, 20 June 1912; also Ruthven to Newcomb, 27 Jan., 2 April 1913, 9 June 1922, 8 March 1923; all in AGR, box 52, f. 2.

92. For example, Mario Biagioli, *Galileo Courtier: The Practice of Science in the Age of Absolutism* (Chicago: University of Chicago Press, 1993), chap. 1.

93. Alexander Ruthven to Bryant Walker, 1 April 1906, 5 June 1907, 28 May and 9 June 1908, 8 Dec. 1909, 10 April 1916; Walker to Ruthven, 31 Jan. 1915; all in AGR, box 52, ff. 16–20.

94. Alexander Ruthven to Charles C. Adams, 18 Nov. 1910, AGR, box 51, f. 2.

95. Alexander Ruthven to Bryant Walker, 16, 20 and 26 April 1909, 10 March, 20 Oct. 1910, 7 Feb., 30 May 1911; Walker to Ruthven, 24 Oct. 1910; all in AGR, box 52, f. 17.

96. Alexander Ruthven to Bryant Walker, 26 April 1910, and Walker to Ruthven, 30 May 1910, both in AGR, box 52, f. 17; Ruthven to Walker, 20 June 1912, AGR, box 52, f. 18.

97. Barbara R. Stein, *On Her Own Terms: Annie Montague Alexander and the Rise of Science in the American West* (Berkeley: University of California Press, 2002).

98. Ibid., 48, 54–57. Annie Alexander to C. Hart Merriam, 22 Feb. 1906 (quote), CHM, reel 28.

99. Stein, ibid., 61–65; Joseph Grinnell to Annie Alexander, 3 Nov. 1907, AA C-B1003, box 1; Alexander to C. Hart Merriam, 11 March 1932, CHM, reel 28; Merriam to Grinnell, 19 Dec. 1907, JG, box 13.

100. Stein, ibid., 86–87; Annie Alexander to Joseph Grinnell, 23 Nov. 1907, AA 67/121, box 1; Alexander to Grinnell, 6 Jan. 1911 (losing interest), AA CU-120, box 1, f. 2; Grinnell to Alexander, 20 Nov., 6 Dec. 1907, both in AA C-B1003, box 1. Also Grinnell to William E. Ritter, 20 May, 10, 23 Nov. 1907, all in WER, box 10.

101. Edmund Heller to Joseph Grinnell, 9 Jan. 1907, JG, box 10.

102. Joseph Grinnell to Annie Alexander, 14 (quote) and 20 Nov. 1907, both in AA C-B1003; Grinnell to C. Hart Merriam, 27 Nov. 1931, CHM, reel 58.

103. Joseph Grinnell to Annie Alexander, 2 Nov. 1907, AA C-B1003.

104. Ibid., 28, 31 Jan., 16 Feb. (quote), 1 March 1908; all in AA C-B1003, box 1.

105. Stein, *On Her Own Terms* (cit. n. 97), 786–87 (quote p. 81); Alexander to Benjamin I. Wheeler, 30 Jan., 19 Feb. 1908; Alexander to Joseph Grinnell, 14 Dec. 1908; all in AA 67/121, box 1.

106. Joseph Grinnell to Annie Alexander, 29 Dec. 1913, AA CU-120, box 1, f. 2.

107. Edmund Heller to Joseph Grinnell, 9 Jan. 1907, JG, box 10.

108. Annie Alexander to Joseph Grinnell, 2 Feb. 1907, AA 67/121, box 1.

109. Joseph Grinnell to Annie Alexander, 20 Nov. 1907, AA C-B1003, box 1; Stein, *On Her Own Terms* (cit. n. 97), 92.

110. Joseph Grinnell to Annie Alexander, 2 Nov. 1907 and 4 April 1908; 7 Feb. 1910; 31 May, 11 June, 11 Aug. 1910; all in AA C-B1003, box 1.

111. Annie Alexander to Joseph Grinnell, 11 Dec. 1910, AA CU-120, box 1, f. 1; Grinnell to Alexander, 14 Dec. 1910, AA C-B1003, box 1.

112. Harry Swarth to Joseph Grinnell, 27 June 1910 (quote), 5 April 1910, both in JG-MVZ.

113. For example, Joseph Grinnell to Annie Alexander, 15 Sept., 4 Oct. 1913, both in AA C-B1003, box 1.

114. Stein, *On Her Own Terms* (cit. n. 97), 120–37, 181–89; Joseph Grinnell to Annie Alexander, 2 Oct. 1916, AA CU-120, box 1, f. 3.

115. Joseph Grinnell to Annie Alexander, 7 Oct. 1918, AA C-B1003, box 2.

116. Stein, *On Her Own Terms* (cit. n. 97), 224–52.

117. Joseph Grinnell to Annie Alexander, 8 Dec. 1926, AA C-B1003, box 2.

118. Stein, *On Her Own Terms* (cit. n. 97), 88–96, p. 117 (quoting Alexander to Grinnell, 6 Jan. 1911). On Alexander's endomment of the MVZ, see Stein, ibid., 155–64.

119. Joseph Grinnell to Annie Alexander, 20 Sept. 1922, AA C-B1003, box 2.

120. Annie Alexander to Joseph Grinnell, 28 Feb. 1909, AA 67/121, box 1.

121. Joseph Grinnell to Annie Alexander, 11 July 1908, AA C-B1003, box 1.

122. William W. Newcomb to Alexander Ruthven, 11 Nov. 1912 (quote); Ruthven to Newcomb, 14 Nov. 1912; both in AGR, box 52, f. 2.

123. Bruce Hevly, "The heroic science of glacier motion," *Osiris* 11 (1996): 66–86; Naomi Oreskes, "Objectivity or heroism? on the invisibility of women in science," *Osiris* 11 (1996): 87–116; and Henrika Kuklick and Robert E. Kohler, "Science in the Field," *Osiris* 11 (1996): 1–14, p. 6.

Chapter 4: Expedition

1. Richard Sorrenson, "The ship as a scientific instrument in the eighteenth century," *Osiris* 11 (1996): 221–36.

2. Mary P. Winsor, *Reading the Shape of Nature: Comparative Zoology at the Agassiz Museum* (Chicago: University of Chicago Press, 1991), 198, 273–74.

3. Sorrenson, "Ship as scientific instrument" (cit. n. 1), 226–33.

4. Vernon Bailey to family, 25 Jan. 1891, VOB, box 1.

5. Vernon Bailey to family, 1 Sept. 1889; also 25 Aug. and 6 Oct.; all in VOB, box 1. Keir Sterling, *Last of the Naturalists: The Career of C. Hart Merriam* (New York: Arno, 1974), 270–79.

6. C. Hart Merriam, "Plan for a biological survey of South and Central America," *Carnegie Institution Year Book* 1 (1902): 267–69.

7. Daniel G. Elliott to Malcolm P. Anderson, 11 Dec. 1912, WHO, box 1.

8. Joseph Grinnell to Annie Alexander, 2, 20 Nov. 1907, both in AA C-B1003, box 1.

9. For example, Charles B. Cory to Frederick Skiff, 4 Nov. 1912, WHO, box 9, f. Skiff.

10. C. Hart Merriam to Vernon Bailey, 22 Oct. and 1 Nov. 1888, both in VOB, box 1.

11. Daniel G. Elliott to Malcolm P. Anderson, 11 Dec. 1912, WHO, box 1, f. Anderson; Wilfred Osgood to Robert H. Becker, 10 May 1913, WHO, box 2.

12. For example, Wilfred Osgood to Karl Schmidt, 5 July 1938, KPS, box 16, f. 2.

13. Wilfred Osgood to Charles B. Cory, 30 Apr. 1912, WHO, box 8, f. Osgood; Joseph Grinnell to Annie Alexander, 18 Aug. 1911, AA C-B1003, box 1.

14. "Melbourne A. Carriker, Jr. to Charles Cadwalader, 14 Aug. 1933 (quote), 31 May 1933, and 16 June 1933, all in ANS 900.

15. Grace Thompson Seton, diary, July 22, 24, 1926; Karl Schmidt to Mr. Gerhard, 16 Aug., 1926; all in FMNH, Expeditions, box 11, no. 88. Schmidt to D. C. Davies, 13 Oct. 1926; Colin Sanborn to Davies, 22 Dec. 1926; both in FMNH, Expeditions, box 12, no. 88.

16. Charles C. Adams, "Instructions to the field party" [1904], MMZ, box 1, f. Porcupine Mountains.

17. For example: Karl P. Schmidt to Wilfred Osgood, 20 Dec. 1926, FMNH Expeditions box 12, no. 90. Daniel G. Elliott to Malcolm P. Anderson, 17 Oct. 1913; Wilfred Osgood to Anderson, 9 Dec. 1913, 1 Jan. 1914; all in WHO, box 1. Osgood to John T. Zimmer, 25 Aug. 1922, WHO, box 10. Osgood to Oldfield Thomas, 5 March 1913, WHO, box 9.

18. D. C. Davies to Herbert L. Stoddard, 28 Jan. 1926, WHO, box 9.

19. Alfred M. Collins to Wilfred Osgood, 2 Dec. 1914, WHO, box 3.

20. Frank Chapman to Wilfred Osgood, 3, 16 Dec. 1914, both in WHO, box 2. Alfred M. Collins to Field Museum, 28 April 1915, WHO, box 3.

21. Frank Chapman to Wilfred Osgood, 4 Oct. 1923, WHO, box 1.

22. Wilfred Osgood to Karl Schmidt, 14 June 1929, KPS, box 15, f. 5.

23. Karl Schmidt to D. C. Davies, 16 Aug. 1926, and subsequent correspondence in FMNH Expeditions, boxes 11 and 12, ff. Brazil 1926–1927.

24. [Alexander Ruthven], "Instructions to the members of the field party to work in the region of Fort Davis, Texas . . . 1914," MMZ, box 1, f. Texas Expedition.

25. Robert E. Kohler, *Lords of the Fly: Drosophila Genetics and the Experimental Life* (Chicago: University of Chicago Press, 1994), chaps. 4–5.

26. Daniel G. Elliott to Frederick Skiff, 13 Sept. 1898, FMNH Directors, ff. Elliott.

27. Wilfred Osgood to Malcolm P. Anderson, 26 Sept. 1912, WHO, box 1.

28. Wilfred Osgood to Robert H. Becker, 23 June 1913, WHO, box 2.

29. Ibid., 7 Jan., 25 Feb. 1914; Becker to Osgood, 10 March 1914; all in WHO, box 2.

30. Frederic E. Clements to Fernandus Payne, 28 Feb. 1933, FEC, box 70.

31. Wilfred Osgood to Oldfield Thomas, 5 March 1913, WHO, box 9.

32. Wilfred Osgood to Robert H. Becker, 23 June 1913, WHO, box 2.

33. Joseph Grinnell to Annie Alexander, 8 May 1908, 13 Feb., 9 March, 15 April 1911, all in AA C-B1003, box 1.

34. Edmund Heller to Joseph Grinnell, 5 Nov. 1906, JG, box 10.

35. Harry Swarth to Joseph Grinnell, 2 June; also 12 Dec. 1906, 5 Sept., 19 Nov. 1907; all in JG, box 18.

36. Joseph Grinnell to Annie Alexander, 15 April (quote) and 29 April 1911, both in AA C-B1003, box 1.

37. Ibid., 13 Feb. 1911 (quote), AA C-B1003, box 1. Grinnell to Alexander, 22 Dec. 1920, 16 Oct. 1922; Grinnell, "Proposed five-year program," 20 Dec. 1920; all in AA C-B1003, box 2.

38. Joseph Grinnell to Annie Alexander, 7 July 1923; Grinnell to Charles Sheldon, 12 Feb. 1924; both in AA C-B1003, box 2. Joseph Grin-

nell, "Museum of Vertebrate Zoology," in University of California *Report of the President* (1908–1910): 117–24, p. 122.

39. Kohler, *Lords of the Fly* (cit. n. 25), chap. 3.

40. Joseph Grinnell to Annie Alexander, 8 Dec. 1926, AA C-B1003, box 2.

41. Harry Swarth to Joseph Grinnell, 10 July 1911, JG-MV.

42. Wilfred Osgood, "Clinton Hart Merriam, 1855–1942," *Journal of Mammalogy* 24 (1943): 421–57, pp. 429, 435–36; and Sterling, *Last of the Naturalists* (cit. n. 5), 431–32.

43. Susan Leigh Starr and James Griesemer, "Institutional ecology: 'translations' and boundary objects: amateurs and professionals in Berkeley's Museum of Vertebrate Zoology, 1907–39," *Social Studies of Science* 19 (1989): 387–420.

44. Joseph Grinnell to Lansing K. Tevis, 11 Sept. 1913, JG-MVZ.

45. Joseph Grinnell to Annie Alexander, 22 Jan. 1917, 11 Feb. 1923, both in AA C-B1003, box 2; 18 Aug. 1911, ibid., box 1.

46. Joseph Grinnell to Annie Alexander, 18, 26 March 1908, both in AA C-B1003, box 1.

47. For example, C. Hart Merriam to Vernon Bailey, 5 March 1887, 12 Dec. 1890, both in VOB, box 1. Alexander Ruthven to Vernon Bailey, 27 Nov. 1912, AGR, box 51, f. 9.

48. Joseph Grinnell to Annie Alexander, 18 Feb., 8 April 1908, both in AA C-B1003, box 1.

49. Charles C. Adams, "Instructions" (cit. n. 16).

50. Charles C. Adams to Arthur G. Vestal, 24 June, 11 July 1910, both in AGV, box 2.

51. [Alexander Ruthven], "Instructions" (cit. n. 24).

52. Vernon Bailey, "Field reports," n.d. [June 1932?], VOB, box 13, f. Field Reports.

53. For example, Alexander Ruthven to A. W. Andrews, 10 March 1910, 8 June, 7 Nov. 1914; Andrews to Ruthven, 1 Nov. 1914; all in AGR, box 51, f. 5. Ruthven to Vernon Bailey, 27 Nov. 1912, AGR, box 51, f. 9.

54. Joseph Grinnell to Annie Alexander, 18 Feb., also 8 April 1908, both in AA C-B1003, box 1.

55. Vernon Bailey, "Field reports," n.d. [June 1932?], VOB, box 13, f. Field Reports.

56. Joseph Grinnell to Annie Alexander, 15 Sept. 1913, AA C-B1003, box 1.

57. Seton, field diary, 6 July 1926 (cit. n. 15).

58. C. Hart Merriam to Vernon Bailey, 6 Nov. 1888; Bailey to family, 18 Nov. 1888; both in VOB, box 1; Joseph Grinnell to Annie Alexander, 29 Sept. 1914, AA C-B1003, box 1; Grinnell to Harry Swarth, 7 Sept. 1916, JG-MVZ.

59. For example, C. Hart Merriam to Vernon Bailey, 5 March 1887, VOB, box 1.

60. For example, Vernon Bailey to family, 21 Aug. 1887; C. Hart Merriam to Bailey, 28 June 1887, 2 Dec. 1889; all in VOB, box 1.

61. Joseph Grinnell to Annie Alexander, 20 Sept. 1910, AA C-B1003, box 1.

62. C. Hart Merriam to Vernon Bailey, 12 June 1888; 10 June, 3 Nov. 1889; 20 May, 4 June, 2 and 10 July, 3 Dec. 1890; 20 April 1892; 26 May 1893; all in VOB, box 1.

63. Vernon Bailey to family, 1 Dec. 1889 (quote); also C. Hart Merriam to Baily, 2, 5, 8, 11 Dec. 1889; all in VOB, box 1.

64. For example, Joseph Grinnell to Annie Alexander, 4 Oct. 1913, AA C-B1003, box 1; Harry Swarth to Joseph Grinnell, 5 April 1910, JG-MVZ.

65. Vernon Bailey to Anna Bailey, n.d. [May 1888]; also Bailey to family, 1, 9 Oct. 1887; all in VOB, box 1.

66. Vernon Bailey to C. Hart Merriam, 5 Sept. 1887, VOB, box 13.

67. Vernon Bailey to family, 20 April 1890; 10 June 1888; April 27, 1890; all in VOB, box 1.

68. Ibid., 26 Aug. 1888, 26 May 1889, 18 Nov. 1888; C. Hart Merriam to Bailey, 21 June 1890; Bailey to family, 15 May 1887; Merriam to Bailey, 18 May 1887; all in VOB, box 1.

69. Frank Stephens to Joseph Grinnell, 5 May, 3 June, 17 Sept. 1905, 24 Aug. 1906, all in JG-UC, box 17.

70. Joseph Grinnell to Walter P. Taylor, 24 June 1912, JG-MVZ.

71. Harry Swarth to Joseph Grinnell, 21 Aug. 1908, JG-MVZ.

72. Grinnell to Swarth, 23 March (quote), 15 May 1911; Swarth to Grinnell, 20 May 1911; all in JG-MVZ.

73. Frank M. Chapman to Joel A. Allen, 23 Jan, 1889, in *Frank M. Chapman in Florida: His Journals and Letters*, ed. Elizabeth S. Austin (Gainesville: University of Florida Press, 1967), 48–50 (quote, p. 48).

74. Robert E. Richardson to Stephen A. Forbes, 26 Sept., 3 Oct. 1909, both in INHS 43/1/1, box 10.

75. Harry Swarth to Joseph Grinnell, 14 Aug. 1910, JG-MVZ.

76. For example: Grinnell to Annie Alexander, 16 May 1918, AA C-B1003, box 2.

77. Joseph Grinnell to Annie Alexander, 2 Dec. 1914, AA C-B1003, box 1; Harry Swarth to Grinnell, 15 June 1909, JG-MVZ; Alexander to Grinnell, 27 Feb. 1911, AA CU-120, box 1, f. 2.

78. Ulysses O. Cox to Henry Nachtrieb, 21 May 1900, HFN, box 4, f. 48. Alfred C. Weed to D. C. Davies, 20 Aug. 1924, WHO, box 10.

79. Harry Swarth to Joseph Grinnell, 21 March 1911; also Grinnell to Swarth, 23 March 1911; both in JG-MVZ.

80. Charles Kofoid to Stephen A. Forbes, 28 April 1899, INHS 43/1/1, box 10.

81. Joseph Grinnell to Walter P. Taylor, 24 June 1912; Taylor to Grinnell, 21 June 1912; both in JG-MVZ.

82. See also Stuart McCook, " 'It may be truth, but it is not evidence': Paul du Chaillu and the legitimation of evidence in the field science," *Osiris* 11 (1996): 177–200.

83. Annie Alexander to Joseph Grinnell, 27 Feb. 1911, AA CU-120, box 1, f. 2.

84. C. Hart Merriam to Vernon Bailey, 9 June 1892, VOB, box 1.

85. For example: C. Hart Merriam to Vernon Bailey, 7 April, 25 June 1890; Bailey to family, 27 April 1890; all in VOB, box 1; Merriam to Bailey, 24 Aug. 1895, VOB, box 2. Alexander Ruthven to Bryant Walker, 12, 30 Oct. 1904, both in AGR, box 52, f. 16. Wilfred Osgood to Robert H. Becker, 21 April 1913, WHO, box 2.

86. Alex Soojung-Kim Pang, "The social event of the season: solar eclipse expeditions and Victorian culture," *Isis* 84 (1993): 252–77; Pang, *Empire of the Sun: Victorian Solar Eclipse Expeditions* (Stanford: Stanford University Press, 2002); and Jane Camerini, "Wallace in the field," *Osiris* 11 (1996): 44–65.

87. Charles C. Adams to Charles B. Davenport, 8 July 1899, CBD-Gen.

88. Edith Clements, "Biographical data ESC," Feb. 1947, p. 4, FEC, box 109, f. Edith.

89. Joseph Grinnell to Annie Alexander, 15 April 1911, AA C-B1003, box 1.

90. Herbert H. Ross, "Faunistic surveys," Illinois Natural History Survey *Bulletin* 27 (1958): 127–214, pp. 131–32.

91. C. Hart Merriam to Vernon Bailey, 7 April 1890, VOB, box 2.

92. John C. Hudson, "Towns of the western railroads," *Great Plains Quarterly* 2:1 (1982): 41–54.

93. C. Hart Merriam to Vernon Bailey, 1, 9 July 1897, VOB, box 2.

94. For example: Seton field diary for July 1926, in Brazil (cit. n. 15); D. C. Davies to Marshall Field, 25 April 1926, FMNH, Expeditions, box 11, no. 89; Karl Schmidt to Mr. Gerhard, 16 Aug. 26, FMNH, Expeditions, box 11, no. 88.

95. Joseph Grinnell to Annie Alexander, 28 June 1911, AA C-B1003, box 1; also Walter P. Taylor to Grinnell, 11 June 1911, JG-MVZ.

96. Vernon Bailey to family, 19 June 1887, VOB, box 1.

97. Harry Swarth to Joseph Grinnell, 25, 30 June 1919, both in JG-MVZ.

98. Joseph Grinnell to Annie Alexander, 12 March, 17 April 1917, both in AA C-B1003, box 2.

99. For example, Angelo Heilprin, "Report," *Academy of Natural Sciences Proceedings* 42 (1890): 478–80.

100. Alexander Ruthven, "A biological survey of the sand dune region on the south shore of Saginaw Bay, Michigan," Michigan Geological and Biological Survey, Publication no. 4 (1911): 13–34, pp. 13, 15.

101. Alexander Ruthven to Light House Board, 8 Feb. 1910; Ruthven to A. W. Andrews, 4 June 1910; both in MMZ, box 1, f. Charity Islands. Charles C. Adams, ed., *An Ecological Survey of Isle Royale, Lake Superior* (Lansing: State Printers, 1909), ix–x, 1, 16.

102. H. K. Harring to Edward A. Birge, 4 Aug. 1917, 15 Aug. 1916, both in UWLim, box 5.

103. Joseph Grinnell to Annie Alexander, 11 July 1908; 29 Sept., 2 Dec. 1914, 23 June 1915; all in AA C-B1003, box 1.

104. James A. G. Kuhn to Alexander Ruthven, 22 Jan. 1906, AGR, box 53, f. 9.

105. Francis Harper to Charles Cadwalader, 23 Aug. 1929; Harper to Witmer Stone, 4 Aug. 1929; both in ANS 113, box 3, f. 58.

106. For example: "Jack" to "Shamrock" [Ruthven], 6 Dec. 1909, MMZ, box 1, f. Mexico 1910; Karl Schmidt to D. C. Davies, 12 Oct. 1926, FMNH, Expeditions, box 12, no. 90; Seton field diary 4 July 1926 (cit. n. 15).

107. Melbourne A. Carriker, Jr., "Experiences of an Ornithologist along the Highways and Byways of Bolivia," 56–58, ANS 900F.

108. J. E. Hallinen to Stephen A. Forbes, 7 Oct.; also 6 Dec. 1892; both in INHS 43/1/1, box 5.

109. Harry Swarth to Joseph Grinnell, 10 June 1919, JG-MVZ.

110. Seton, diary, 4 July 1926 (cit. n. 15). John T. Zimmer to Wilfred Osgood, 1 Aug. 1922, 7 Jan. 1923, both in WHO, box 10.

111. Joseph Grinnell to Annie Alexander, 5 Nov. 1925 (quote); also 13 Jan. and 10 Oct. 1925; all in AA C-B1003, box 2.

112. C. Hart Merriam to Vernon Bailey, 20 July 1887 (quote); Bailey to Merriam, 5 May 1889; both in VOB, box 1.

113. Alexander Ruthven to Hermon C. Bumpus, 19 March 1906, AGR, box 52, f. 9. Ruthven to Bryant Walker, 6 Aug. 1915, 31 May 1917, both in AGR, box 52, ff. 19, 20.

114. For example: J. E. Hallinin to Stephen A. Forbes, 7 Oct. 1892, INHS 43/1/1, box 5; Thomas Large to Forbes, 13 Oct. 1900, 1 and 11 June 1901; and Robert E. Richardson to Forbes, 3 July 1903; all in ibid., box 10. The one woman in the party lodged in town.

115. Edmund Heller to Joseph Grinnell, 1 March 1908, JG-MVZ, box 10.

116. Alexander Ruthven to Bryant Walker, 2 July 1914, AGR, box 52, box 19.

117. Annie Alexander to Joseph Grinnell, 1 May 1910, AA C-B1003, box 1, f. 1.

118. Vernon Bailey to family, 12 May 1889; also 5 May 1889; both in VOB, box 1.

119. Ibid., 16 Sept. 1888 (quote); also 23 Sept. 1888; both in VOB, box 1.

120. Ibid., 8 July 1888 (quote); also 24 June 1888; both in VOB, box 1.

121. Joseph Grinnell to Annie Alexander, 18 Aug. 1911, AA C-B1003, box 1.

122. Harry Swarth to Joseph Grinnell, 23 Aug., 9 Sept., 5 Oct. 1916; Grinnell to Swarth, 28 Aug. 1916; all in JG-MVZ.

123. Joseph Ewan, "San Francisco as a mecca for nineteenth-century naturalists," in *A Century of Progress in Natural Sciences 1853–1953* (San Francisco: California Academy of Sciences, 1953), 1–64, p. 36.

124. George B. West to William M. Wheeler, 8 Dec. 1895 (quote); also 16 Feb. 1896; both in WMW, box 36, f. W. Also West to Wheeler, 10 May 1896, WMW, box 36 f. W entomology.

125. Frank M. Chapman, *Autobiography of a Bird-Lover* (New York: Appleton-Century, 1933), 88–90 (quote, p. 89). On Oak Lodge: Austin, *Chapman in Florida* (cit. n. 73), 56–7; 88–90.

126. Henry B. Ward, "The fresh-water biological stations of the world," *Science* 9 (1899): 497–508.

127. Charles F. Baker to William M. Wheeler, 20 March 1914, WMW, box 5.

128. James A. G. Kuhn to Alexander Ruthven, 22 Jan. 1906, AGR, box 53, f. 9.

129. Wilfred Osgood to Robert H. Becker, 13 Aug. 1913 (quote), WHO, box 2.

130. C. Hart Merriam to Vernon Bailey, 5 March 1887, 1 May 1889 (quote); Norman J. Coleman to Vernon Bailey, 4 March 1887; all in VOB, box 1.

131. Ernst Mayr, E. G. Linsley, and R. L. Usinger, *Methods and Principles of Systematic Zoology* (New York: McGraw-Hill, 1953), 64–65, chap. 4.

132. Joseph Grinnell to Annie Alexander, 1 Nov. 1918, AA C-B1003, box 2. Also Edward A. Goldman to C. Hart Merriam, 24 July, 3 Aug. 1916, both in CHM, reel 57.

133. Alexander Ruthven to Bryant Walker, 2 July 1914 (quote), AGR, box 52, f. 19; Ruthven to Walker, 2 Jan. 1911, AGR, box 52, f. 17; Ruthven to Arthur S. Pearse, 21 Feb. 1913, AGR, box 52, f. 5; Ruthven to Melbourne A. Carriker, 11 Feb. 1913 and Carriker to Ruthven, 14 March 1913, both in MMZ, box 1, f. Expeditions Colombia.

134. Vernon Bailey to family, 12 June 1887, VOB, box 1.

135. C. Hart Merriam to Vernon Bailey, 10 Jan. 1889; 12 Feb. 1889; both in VOB, box 1.

136. Ibid., 5 Nov. 1890, VOB, box 1.

137. Vernon Bailey to family, 2 Nov. 1890, VOB, box 1.

138. A. R. Witson to Edward A. Birge, 1 May 1913, UW-Lim, box 7; H. K. Harring to Birge, 25, 27 July 1916, both in UW-Lim, box 5.

139. Joseph Grinnell to Annie Alexander, 17 April 1917, AA C-B1003, box 2.

140. Lee Dice to Frederic E. Clements, 7 Nov. 1923, FEC, box 63. Alfred C. Kinsey, "The Gall Wasp genus *Cynips*: A study in the origins of species," *Indiana University Studies* 16 (1929), 11–15.

141. Roland M. Harper to Arthur G. Vestal, 31 Aug. 1924, AGV, box 5; 27 Oct. 1912; also 31 May 1914, and 19 July 1917; all in AGV, box 2.

142. [Victor Shelford?], "Reserves for research and instruction in universities and colleges," n.d. [c. 1931], WCA, box 22, f. 2; Warder C. Allee to Victor Shelford, 6 March 1937, WCA, box 22, f. 3.

143. Daniel T. MacDougal to Robert S. Woodward, 17 May 1910 (quote), CIW, f. Desert Lab reports. Also MacDougal to Woodward, 1 Feb. 1910, CIW, f. MacDougal Mexico travel; 17 June 1910, CIW, f. Desert Lab reports.

144. Kinsey, "Gall Wasp" (cit. n. 140), 14.

145. For example: Daniel T. MacDougal to Robert S. Woodward, 2 Aug. 1909, CIW, Desert Lab employees.

146. Joseph Grinnell to Annie Alexander, 22 Jan., 12 March, 17 April 1917, 16 Oct. 1922, 16 Oct. 1922, all in AA, box 2; Grinnell to Harry Swarth, 11, 15 Sept. 1922; Swarth to Grinnell, 19 Oct. 1922; all in JG-MVZ.

147. Osgood, "Clinton Hart Merriam" (cit. n. 42), 431. Frank Chapman to Merriam, 18 Aug. 1913, CHM, reel 40.

148. Edith Clements, "Biography of Frederic E. Clements," on pp. 2–3, FEC, box 109. Edith Clements diary, 5 June 1918 (quote), also 22 June–1 July 1914, 17 Aug. 1915, 29 Nov. 1933, all in FEC, box 114.

149. Arthur G. Vestal to Frederic E. Clements, 16 Sept. 1923, 4 Aug. 1924, both in FEC, box 64.

150. Witmer Stone, "American ornithological literature 1883–1923," in *Fifty Years' Progress of American Ornithology 1883–1933* (Lancaster: American Ornithologists' Union, 1933), 29–41, p. 41 (birdflight); Kohler, *Lords of the Fly* (cit. n. 25), 204 (fungus gnats).

151. Carl L. Hubbs and Boyd W. Walker, "Abundance of desert animals indicated by capture in fresh road tar," *Ecology* 28 (1947): 464–66.

Chapter 5: Work

1. Richard White, *The Organic Machine: The Remaking of the Columbia River* (New York: Hill and Wang, 1995).

2. For example: Sharon Traweek, *Beamtimes and Lifetimes: The World of High Energy Physicists* (Cambridge: Harvard University Press, 1988), chap. 3; Lorraine Daston, "The moral economy of science," Osiris 10 (1995): 3–24.

3. Steven Shapin, " 'A scholar and a gentleman': the problematic identity of the scientific practitioner in early modern England," *History of Science* 29 (1991): 279–327; Shapin, "Who was Robert Hooke?" In Robert Hooke: *New Studies*, ed. Michael Hunter and Simon Schaffer (Woodbridge, U.K.: Boydell Press, 1989), 253–86; and Mario Biagioli, *Galileo Courtier: The Practice of Science in the Age of Absolutism* (Chicago: University of Chicago Press, 1993).

4. For example: John T. Zimmer to Wilfred H. Osgood, 14 Jan. 1922, WHO, box 10; Rollo Beck to Osgood, 15 Nov. 1911, WHO, box 2; Edward W. Nelson to Osgood, 1 Dec. 1911, WHO, box 7; Charles C. Adams to Alexander Ruthven, 16 July 1906, AGR, box 51, f. 1.

5. A. C. Best to Witmer Stone, 2 Feb. 1929, ANS 113, box 3, f. 58.

6. Vernon L. Bailey to C. Hart Merriam, 27 Jan. 1886, CHM, reel 35; and Keir B. Sterling, *Last of the Naturalists: The Career of C. Hart Merriam* (New York: Arno Press, 1974), 107. Wilfred Osgood, "Clinton Hart Merriam, 1855–1942," *Journal of Mammalogy* 24 (1943): 421–57, p. 425; and Colin C. Sanborn, "Wilfred Hudson Osgood: 1875–1947," *Journal of Mammalogy* 29 (1948): 95–112, p. 104.

7. Vernon Bailey, "Hiram Bailey," n.d. [1928], 37–39, VOB, box 11, f. biographical notes.

8. Herbert L. Stoddard to Paul L. Errington, 31 Dec. 1937, PLE, box 5, f. 5. On Errington's early life as a trapper, see Errington, *The Red Gods Call* (Ames: Iowa State University Press, 1973).

9. William M. Wheeler to Henry H. Donaldson, 4 Feb. 1911, WMW, box 10.

10. Joseph Grinnell to Annie Alexander, 14 May 1927, AA C-B1003, box 2.

11. Quoted in Walter Benjamin, "Little history of photography," in Walter Benjamin, *Selected Writings Volume 2 1927–1934*, ed. Michael W. Jennings, Howard Eiland, and Gary Smith (Cambridge: Harvard University Press, 1999), 507–30, p. 520.

12. Joseph Grinnell, "Conserve the collector," *Science* 41 (1915): 229–32, pp. 229 (quote), 232.

13. C. Hart Merriam to Witmer Stone, 5 Jan. 1924, ANS 678.

14. Joseph Grinnell to Wilfred Osgood, 30 Dec. 1921; also Osgood to Grinnell, 24 Dec. 1921; both in WHO, box 4.

15. Vernon Bailey to family, 10, 31 July 1887, VOB, box 1.

16. Grace Thompson Seton, diary, 6 July 1926, FMNH, Expeditions, box 11, no. 88.

17. Vernon Bailey to Alexander Ruthven, 20 Jan. 1910, AGR, box 51, f. 9.

18. Vernon Bailey to C. Hart Merriam, 1, 2 July 1887, both in VOB, box 13.

19. C. Hart Merriam to Vernon Bailey, 30 Dec. 1889; Bailey to family, 17, Jan. 1890 (quote); both in VOB, box 1.

20. Vernon Bailey to family, 28 Sept., 14 Dec. 1890, both in VOB, box 1.

21. Frank Chapman to William Brewster, 27 April 1890, quoted in Mark V. Barrow, Jr., "Birds and Boundaries: Community, Practice, and Conservation in North American Ornithology, 1865–1935," Ph.D. diss., Harvard University, 1992, p. 565.

22. Frank Chapman to William Brewster, 15 June 1890, quoted in Mark V. Barrow, Jr., *A Passion for Birds: American Ornithology after Audubon* (Princeton: Princeton University Press, 1998), 173.

23. Ulysses O. Cox to Henry F. Nachtrieb, 2 May 1901 (quote); also 12, 21 May 1900; all in HFN, box 4, f. 48.

24. Martin Rudwick has suggested that geologists experienced travel as a religious pilgrimage. Martin Rudwick, "Geological travel and theoretical innovation: the role of 'liminal' experience," *Social Studies of Science* 26 (1996): 143–59.

25. Joseph Grinnell to William E. Ritter, 8 June 1907, WER, box 10.

26. Sanborn, "Wilfred Hudson Osgood" (cit. n. 6), 104. Walter P. Taylor to Joseph Grinnell, 22 Dec. 1908, 20 Jan. 1909, both in JG-MVZ.

27. Harry Swarth to Joseph Grinnell, 29 Feb. 1910, 11 Oct. 1913, 11 Jan. 1916 (quotes); also Grinnell to Swarth, 30 July 1910; all in JG-MVZ.

28. Joseph Grinnell to Annie Alexander, 16 Dec. 1910 (quote), 13 Feb., 20 May 1910, all in AA C-B1003, box 1.

29. For example, Wilfred Osgood to Karl Schmidt, 5 July 1938, KPS, box 16, f. 2; Vernon Bailey to family, 16 Sept. 1888, 28 Sept. 1890, both in VOB, box 1.

30. Francis B. Sumner, *The Life History of An American Naturalist* (Lancaster, Pa.: Jaques Cattell Press, 1945), 61–64.

31. Joseph Grinnell to Annie Alexander, 11 June 1910, also 11 July 1908; Alexander to Grinnell, 1 May 1910; all in AA CU-120, box 1, f. 1.

32. Vernon Bailey to family, 14 April 1889, VOB, box 1.

33. Vernon Bailey to C. Hart Merriam, 1 July 1887 (quote), 2 July 1887, both in VOB, box 13.

34. Frank M. Chapman to Joel A. Allen, 6 March 1889, in *Frank M. Chapman in Florida: His Journals and Letters,* ed. Elizabeth S. Austin (Gainesville, University of Florida Press, 1967), 65–69; 20 March 1889, in ibid., 70–72 (quote, p. 71).

35. Walter P. Taylor to Joseph Grinnell, 2 June 1912; Grinnell to Taylor, 3 April, 5 June 1912; all in JG-MVZ.

36. Joseph Grinnell to Annie Alexander, 16 April 1908; also 11 July 1908; both in AA C-B1003, box 1.

37. Francis B. Sumner to William E. Ritter, 7 April 1914, FBS-SIO, box 5, f. 31.

38. Frances B. Sumner, "Some results of a twelve years' study of deer mice" [c. 1925], FBS-Fam, f. 25.

39. Edward A. Goldman to Wilfred Osgood, 27 March 1901, WHO, box 4.

40. Seton, diary, 8, 9 July 1926 (cit. n. 16).

41. Ulysses S. Cox to Henry F. Nachtrieb, 9 May 1901, HFN, box 4, f. 48.

42. Harry Swarth to Joseph Grinnell, 5 Sept. 1908 (quote); and 14 Aug. 1910; both in JG-MVZ.

43. Malcolm P. Anderson to Wilfred Osgood, 24 Dec. 1913, WHO, box 1.

44. Melbourne A. Carriker, Jr., "Report on the 1921–30 Peruvian expedition, n.d., ANS 900, box 1.

45. Frank M. Chapman to Joel A. Allen, 4 March 1889, in Austin, Chapman in Florida (cit. n. 35), 64–65.

46. Seton, diary, 9 Aug. 1927 (cit. n. 16).

47. Harry Swarth to Joseph Grinnell, 14 Aug. 1910, JG-MVZ.

48. Harold Hansen to Henry F. Nachtrieb, 14, 16 Aug. 1908, both in HFN, box 4, f. 48.

49. Thomas Large to Stephen A. Forbes, 28 Sept., 1, 4, 13 Oct. 1900, 23 May 1901, all in INHS 43/1/1, box 10. But see Large to Forbes, 1 June 1901, ibid.

50. D. Farrington, "Activities of the Field Museum," 1929, FMNH, Historical documents, box 1.

51. Alexander Ruthven, untitled history of the Michigan Museum [early 1930s], MMZ, box 1, ff. Histories.

52. Edith Clements, diary, 15 May 1918, FEC, box 114.

53. Joseph Grinnell to Annie Alexander, 29 Aug. 1912, AA C-B1003, box 1.

54. Frank S. Daggett to Joseph Grinnell, 24 May 1906, JG-UC, box 5.

55. Joseph Grinnell to Annie Alexander, 16 Oct. 1922, AA C-B1003, box 2.

56. Vernon Bailey to C. Hart Merriam, 20 June 1887, VOB, box 13.

57. Annie Alexander to Joseph Grinnell, 8 July 1911, AA CU-120, box 1, f. 2.

58. Vernon Bailey to C. Hart Merriam, 9 July 1887, VOB, box 13. Edward W. Nelson to Vernon Bailey, 22 June 1914, VOB, box 3.

59. Joseph Grinnell to Harry P. Swarth, 30 May 1912, JG-MVZ.

60. Walter P. Taylor to Joseph Grinnell, 17 July 1910 (quote); also Grinnell to Taylor, 10, 23 July 1910; all in JG-MVZ.

61. Ned Hollister to Wilfred Osgood, 29 March 1904, WHO, box 5.

62. Harry Swarth to Joseph Grinnell, 26 May 1912; Grinnell to Swarth, 29 May (quotes), 5 June 1912; all in JG-MVZ.

63. Harry Swarth to Joseph Grinnell, 1 June 1912, JG-MVZ.

64. For scientific work as investment, see Bruno Latour and Steven Woolgar, *Laboratory Life: The Construction of Scientific Facts* (London: Sage, 1979; new ed., Princeton: Princeton University Press, 1986), chap. 5.

65. Joseph Grinnell to Witmer Stone, 13 March 1903, ANS 684.

66. Vernon Bailey, "Hiram Bailey," n.d., VOB, box 11, f. Bailey biographical notes. Bailey to family, 10 Oct. 1887; 10 June, 16 Sept., 8 Oct. 1888; 28 Sept., 27 Oct., 22, 29 Dec. 1889; 24 Aug. 1890; all in VOB box 1.

67. John M. Coulter, "The future of systematic botany," *Botanical Gazette* 16 (1891): 243–54, p. 248.

68. Nathan Reingold, "Definitions and speculations: the professionalization of science in American in the nineteenth century." In *The Pursuit of Knowledge in the Early American Republic: American Scientific and Learned Societies from Colonial Times to the Civil War*, ed. Oleson, Alexandra C., and Sanborn C. Brown (Baltimore: Johns Hopkins University Press, 1976), 33–69.

69. These figures are for the natural history sciences, including paleontology but not ethnology, archaeology, geology, and minerology. Data are from *American Museum Annual Reports* 1871–1915, *American Men of Science*, and Clark Elliott, *Biographical Index to American Science, the Seventeenth Century to 1920* (Westport, Conn.: Greenwood Press, 1990).

70. Data are from *National Museum Annual Reports, American Men of Science*, and Elliott, *Index* (cit. n. 69). Paleontology is included but not ethnology, archaeology, and physical science. I omitted unpaid aides, custodians, and "honorary" curators except those whose primary appointment was in the Museum (most were employees of other agencies).

71. Data are from *Field Columbian Museum Annual Reports, American Men of Science*, and Elliott, *Index* (cit. n. 69).

72. Joseph Grinnell to Walter P. Taylor, 25 Nov. 1908; Grinnell to Wilfred Osgood, 27 June 1919; both in JG-MVZ.

73. Joseph Grinnell to C. Hart Merriam, 19 Dec. 1928, CHM, reel 58.

74. Joseph Grinnell to Witmer Stone, 15 Jan. 1912, ANS 684.

75. Harry Swarth to Joseph Grinnell, 20 July 1911, JG-MVZ.

76. Osgood to Grinnell, 24 Dec. 1921; Grinnell to Osgood, 30 Dec. 1921; both in WHO, box 4.

77. Herbert L. Stoddard to Wilfred Osgood, 8 Jan. 1924; Osgood to Stoddard, 9 Jan. 1924; both in WHO, box 9.

78. Harry Swarth to Joseph Grinnell, 21 Aug. 1913 (quote); also 9 Aug. 1913, 1 Jan., 28 Feb. 1914; Grinnell to Swarth, 13 Aug. 1913, 26 Feb. 1914; Grinnell to Wilfred Osgood, 19 Dec. 1921; all in JG-MVZ. Swarth to Grinnell, 16 Dec. 1915, JG-UC, box 18.

79. Susan Leigh Star, "Craft vs. commodity, mess vs. transcendence: how the right tool became the wrong one in the case of taxidermy and natural history," in *The Right Tools for the Job: At Work in Twentieth-Century Life Sciences*, ed. Adele E. Clarke and Joan H. Fujimura (Princeton: Princeton University Press, 1992), 257–86.

80. Stephen A. Forbes to Thomas Large, 24 April 1902; Large to Forbes, 3 Feb., 11 March 1902; all in INHS 43/1/2, letterpress, p. 665.

81. Thomas Large to Stephen A. Forbes, 16 July 1902, INHS 43/1/1, box 10.

82. Ibid., 27 March [1908], IHNS 43/1/1, box 10.

83. Steven Shapin, "The invisible technician," *American Scientist* 77 (1989): 554–63.

84. Adolph Hempel to Stephen A. Forbes, 17 Nov. 1898, INHS 43/1/1, box 6.

85. Robert E. Richardson to Stephen A. Forbes, 19 April 1908, 15 April 1909, both in INHS 43/1/1, box 10.

86. Malcolm Anderson to Joseph Grinnell, 28 Sept. 1902, 13 Jan. 1907; Wilfred Osgood to Charles B. Cory, 13 Dec. 1911; all in WHO, box 8, f. Osgood.

87. Malcolm Anderson to Wilfred Osgood, 13 Oct. 1911, WHO, box 1.

88. Wilfred Osgood to Malcolm Anderson, 6 Nov. 1911, WHO, box 1.

89. Ibid., 6 Nov. 1911, 26 Sept. 1912, 13 May 1914, all in WHO, box 1. Daniel G. Elliott to Malcolm P. Anderson, 17 Oct. 1913, WHO, box 12. Osgood to Charles B. Cory, 13 Dec 11, WHO, box 3. F. J. V. Skiff to Charles B. Cory, 31 Dec. 13, WHO, box 4, ff. Skiff.

90. Cherrie was hired by the Field Museum in 1894 to collect and to assist the curator in his managerial tasks. "He is a good collector and a rising young naturalist," Charles Cory observed, "but time will show how good an assistant he will make." Charles B. Cory to Frederic Skiff, 1 Sept. 1894, FMNH, Directors Papers, f. 13.

91. The voluminous correspondence between Wilfred Osgood and S. M. Klages affords an interesting glimpse into the world of freelance collectors. WHO, box 6.

92. Annie Alexander to Joseph Grinnell, 29 July 1915; Grinnell to Alexander, 3 Aug. 1915 (quote); both in AA CU-120, box 1, f. 3.

93. Harry Swarth to Joseph Grinnell, 23 May 1919, JG-MVZ.

94. Robert Ridgway to Anne Taylor, 26 June 1887, cited in Barrow, *Passion for Birds* (cit. n. 22), p. 3.

95. Working couples were not unusual in other sciences. See Pnina G. Abir-Am and Dorinda Outram, eds., *Uneasy Careers and Intimate Lives: Women in Science 1789–1979* (New Brunswick, N.J.: Rutgers University Press, 1987).

96. See Alex Soojung-Kim Pang, "Gender, culture, and astrophysical fieldwork: Elizabeth Campbell and the Lick Observatory-Crocker eclipse expeditions," *Osiris* 11 (1996): 17–43.

97. Malcolm Anderson to Wilfred Osgood, 13 May 1914, WHO, box 1.

98. Sterling, *Last of the Naturalists* (cit. n. 6), pp. 48, 71, 114–16.

99. Harriet Kofalk, *No Woman Tenderfoot: Florence Merriam Bailey, Pioneer Naturalist* (College Station: Texas A&M University Press, 1989), 89, 102, 107–108, 122, 126, 130, 159.

100. Austin, *Frank M. Chapman in Florida* (cit. n. 34), 140.

101. Joseph Grinnell to Annie Alexander, 2 Nov. 1907, AA C-B1003, box 1.

102. Carl L. Hubbs to Max M. Ellis, 14 May 1938, CLH, box 67, f. 49; Robert Miller to Hubbs, 29 July 1947, CLH, box 24, f. 77.

103. Barbara R. Stein, *On Her Own Terms: Annie Montague Alexander and the Rise of Science in the American West* (Berkeley: University of California Press, 2001), 17–18; also Marshall Field, Jr. to Stanley Field [May 1926], FMNH, Expeditions, box 11, no. 89.

104. Joseph Grinnell to Annie Alexander, 22 April 1908, AA C-B1003, box 1. Also Alexander to Grinnell, 25 April 1908, AA 67/121, quoted in Stein, ibid., 99–100.

105. Stein, ibid., 100–102, chap. 15.

106. Annie Alexander to Joseph Grinnell, 15 Jan. 1922, AA C-120, box 1, quoted in Stein, ibid., 92.

107. Traweek, *Beamtimes and Lifetimes* (cit. n. 2), chap. 3. But see Naomi Oreskes, "Objectivity or heroism? on the invisibility of women in science," *Osiris* 11 (1996): 87–116.

108. Marshall Field, Jr., to D. C. Davies, 21 April 1926; Davies to [Stanley?] Field, 24 April 1926; Evelyn Field to Marshall Field, Jr., 4 May 1926; all in FMNH, Expeditions, box 11 no. 89.

109. Paul Oehser, quoted in Kofalk, *No Woman Tenderfoot* (cit. n. 99), 112–13.

110. A. S. Byatt, "Morpho Eugenia," in *Angels and Insects: Two Novellas* (London: Chatto and Windus, 1992).

111. Harry Swarth to Joseph Grinnell, 28 Aug. 1911, JG-MVZ.

112. Margaret W. Rossiter, *Women Scientists in America: Struggles and Strategies to 1940* (Baltimore: Johns Hopkins University Press, 1982).

113. Henrika Kuklick and Robert E. Kohler, "Science in the field," *Osiris* 11 (1996): 1–16, pp. 10–13; and Kohler, *Landscapes and Labscapes: Exploring the Lab-Field Border in Biology* (Chicago: University of Chicago Press, 2002), 194–99.

114. Sterling, Last of the Naturalists (cit. n. 6), 39, 11–20. Charles C. Adams to Charles B. Davenport, 2 Sept. 1900, CBD-GEN.

115. Harry Swarth to Joseph Grinnell, 25 Oct. 1905, JG-UC, box 8.

116. Aven Nelson, "Popular ignorance concerning botany and botanists," *Plant World* 3 (1900): 33–36.

117. Edith Clements, diary 31 Aug. 1914, FEC, box 114.

118. Harry Swarth to Joseph Grinnell, 15 June 1909, JG-MVZ.

119. Joseph Grinnell to Annie Alexander, 1 Mar. 1923, AA C-B1003, box 2.

120. Thomas Large to Stephen A. Forbes, 9 July 1901, INHS 43/1/1, box 10. Charles Kofoid to Forbes, 21 July, 9 Aug. 1897, INHS 43/1/1, box 8. J. E. Hallinen to Forbes, 23 Oct. 1892, INHS 43/1/1, box 5.

121. For example, Joseph Grinnell to Annie Alexander, 26 May 1909, 14 Jan. 1910, both in AA C-B1003, box 1.

122. Charles A. Vogelsang to Charles Case, 15 March (quote), 3 April 1908, both in AGR, box 54, f. 1.

123. For example: Karen Halttunin, *Confidence Men and Painted Women: A Study of Middle-Class Culture in America, 1830–1870* (New Haven: Yale University Press, 1982).

124. Barrow, *Passion for Birds* (cit. n. 22), chap. 6.

125. Ernest L. Brown to Henry F. Nachtrieb, 12 Sept., 22 Oct., 8 Nov. 1891; 11 June 1892; 27 Nov., 31 Dec. 1893; 19 July 1902; all in HFN, box 4, ff. 47 and 48.

126. Edward A. Goldman to Wilfred Osgood, 5 April 1903, WHO, box 4. Roland M. Harper to Arthur Vestal, 31 May 1914, AGV, box 2.

127. Vernon Bailey to family, 8, 20 Feb. 1891, both in VOB, box 1. Bailey, "Into Death Valley 50 years ago," *Westways* (Dec. 1940): 8–11.
128. Seton, diary, 9 July 1926 (cit. n. 16).
129. Alexander Ruthven, untitled history of Michigan Museum [early 1930s], MMZ, box 1, f. histories 1.
130. Thomas Large to Stephen A. Forbes, 14 July 1901, INHS 43/1/1, box 10.
131. White, *Organic Machine* (cit. n. 1), x.
132. Paula Findlen, "Between carnival and lent: the scientific revolution at the margins of culture," *Configurations* 6 (1998): 243–67.

Chapter 6: Knowledge

1. On the practices and politics of classification in biomedicine, see Geoffrey C. Bowker and Susan Leigh Star, *Sorting Things Out: Classification and Its Consequences* (Cambridge: MIT Press, 1999).
2. Marshall Sahlins, *Islands of History* (Chicago: University of Chicago Press, 1985), chap. 1 (quote, p. 31).
3. Jürgen Haffer, "The history of species concepts and species limits in ornithology," *Bulletin of the British Ornithological Club* Centenary Supplement 112A (1992): 107–58; Ernst Mayr, "The species problem," in *The Species Problem*, ed. Mayr (Washington, D.C.: American Association for the Advancement of Science, 1957), 1–22; Mayr, "The species as a category, taxon, and population," in *Histoire du Concept d'Espèce dans les Sciences de la Vie*, ed. Jean-Louis Fisher and Jacques Roger (Paris: Fondation Singer-Polignac, 1987), 303–20.
4. Lisbet Koerner, *Linnaeus: Nature and Nation* (Cambridge: Harvard University Press, 1999), chap. 2.
5. Ernst Mayr, E. G. Linsley, R. L. Usinger, *Methods and Principles of Systematic Zoology* (New York: McGraw-Hill, 1953), 147.
6. Bruno Latour and Steve Woolgar, *Laboratory Life: The Construction of Scientific Facts* (2nd ed., Princeton: Princeton University Press, 1986), 87.
7. Robert Ridgway, "The Birds of North and Middle America: A Descriptive Catalogue . . . Part I," *Bulletin of the U.S. National Museum* 50 (1901), p. ix; Edward D. Cope, "A critical review of the characters and variations of the snakes of North America," *Proceedings of the U. S. National Museum* 14 (1891): 589–694, p. 589.

8. Mayr et al., *Methods and Principles*, (cit. n. 5), 63 (quote); also Ernst Mayr, *Systematics and the Origin of Species from the Viewpoint of a Zoologist* (New York: Columbia University Press, 1942).

9. Ernst Mayr, "Biological materials," National Research Council report, typscript, n.d. [1957], p. 13, KPS, box 12, f. 4.

10. Robert Ridgway, "Allen on individual and seasonal variation in the genus *Elainea*," *Auk* 7 (1890): 385–86.

11. Edgar Anderson, "Hybridization in American *Tradescantias*," *Annals of the Missouri Botanical Garden* 23 (1936): 511–25, p. 513.

12. Arnold E. Ortmann to Carl H. Eigenmann, 18 July 1903, CIW, f. Eigenmann.

13. Frank Chapman to Witmer Stone, 6 Aug. 1915, 16 Jan. 1922 (quote); 22 Aug. 1913, 30 July 1915; also 1, 9 May 1912 (on the worthlessness of "Bogata" collections); all in ANS 679.

14. Wilfred Osgood to Oldfield Thomas, 20 July 1911, WHO, box 9.

15. Edgar Anderson, *Plants, Man and Life* (Boston: Little Brown 1952; repr., Berkeley: University of California Press, 1969), 43–44.

16. Robert Ridgway, "Geographical, versus sexual, variation in *Oreortyx pictus*," *Auk* 11 (1894): 193–97.

17. Edgar Anderson, "The species problem in iris," *Annals of the Missouri Botanical Garden* 23 (1936): 457–509, p. 471.

18. Alfred C. Kinsey, "The Gall Wasp Genus Cynips," *Indiana University Studies* 16 (1929), pp. 20–23 (quote, pp. 21–22).

19. Kenneth S. Norris, "To Carl Leavitt Hubbs, a modern pioneer naturalist on the occasion of his eightieth year," *Copeia* (1974): 581–94.

20. Ernst Mayr to Edgar Anderson, 18 Jan. 1941, EM-HU, box 2.

21. Mayr et al., *Methods and Principles* (cit. n. 5), 106–107. This knack is akin to what James Scott has called "metis." James C. Scott, *Seeing Like a State: How Certain Schemes to Improve the Human Condition Have Failed* (New Haven: Yale University Press, 1998), chap. 9.

22. Carl H. Eigenmann to Charles B. Davenport, 7 April 1907, CBD-Gen.

23. Anderson, *Plants, Man and Life* (cit. n. 15), 40–41.

24. Mayr et al., "Biological materials," (cit. n. 9), p. 10.

25. Alexander Ruthven, untitled history of the Michigan Museum, no date [early 1930s], MMZ, box 1, f. histories 1.

26. Carl L. Hubbs, "The importance of race investigations on Pacific fishes," *Proceedings of the IVth Pacific Science Congress* vol. 3 (1929): 13–23. In a spoken version of his paper Hubbs allowed that he had per-

haps exaggerated: Hubbs, "An experimental attack on the species problem," 21 Nov. 1935, CLH, box 75, f. 101.

27. Mayr et al., *Methods and Principles* (cit. n. 5), p. 8 (quote); and Ernst Mayr, "Speciation phenomena in birds," *American Naturalist* 74 (1940): 249–78, pp. 249–50.

28. Ernst Mayr to Edgar Anderson, 28 Jan. 1941, EM-HU, box 2.

29. Robert E. Kohler, *Landscapes and Labscapes: Exploring the Lab-Field Border in Biology* (Chicago: University of Chicago Press, 2002), chaps. 7–8.

30. For example, Witmer Stone to Joseph Grinnell, 25 April, 1 Oct. 1902, 23 Feb. 1903, all in JG-UC, box 17.

31. Charles Baskerville, "The elements: verified and unverified," *Proceedings of the American Association for the Advancement of Science* 53 (1903–1904): 387–442; and V. Karpenko, "The discovery of supposed new elements: two centuries of errors," *Ambix* 27 (1980): 77–102.

32. Robert E. Kohler, *Lords of the Fly: Drosophila Genetics and the Experimental Life* (Chicago: University of Chicago Press, 1991), 48–49.

33. Mayr et al., *Methods and Principles* (cit. n. 5), 48–50, 59, 245; Mayr, "Speciation phenomena," (cit. n. 27), 257. Kinsey, "Gall Wasp Genus Cynips" (cit. n. 18), 18–19 (on the cycle of confidence and despair).

34. C. Hart Merriam, "Suggestions for a new method of discriminating between species and subspecies," *Science* 5 (1897): 753–58, p. 756; Gerrit S. Miller, Jr. and James A. G. Rehn, "Systematic results of the study of North American land mammals to the close of the year 1900," *Proceedings of the Boston Society for Natural History* 30 (1901): 1–352, p. 1; and Joel A. Allen, "Recent progress in the study of North American mammals," *Proceedings of the Linnean Society of New York* (1894): 17–45, pp. 18–19, and 44 (on American Museum). Allen's numbers differ from Miller and Rehn's.

35. David S. Jordan, "The origin of species through isolation," *Science* 22 (1905): 545–62, p. 562; Jordan to Alexander Ruthven, 4 Oct. 1927, AGR, box 51, f. 29.

36. Karl P. Schmidt, "Herpetology," in *A Century of Progress in the Natural Sciences 1853–1953* (San Francisco: California Academy of Sciences, 1955), 591–627; Alexander Ruthven to Hermon C. Bumpus, 19 Mar. 1906, AGR, box 53, f. 9.

37. Edward S. Ross, "Systematic entomology," ibid., 485–95, pp. 486–88.

38. Edgar Anderson, "The technique and use of mass collection in plant taxonomy," *Annals of the Missouri Botanical Garden* 28 (1941): 287–92.

39. Wilfred Osgood to Edward W. Nelson, 22 May 1919, WHO, box 7.

40. Wendell H. Camp to Ernst Mayr, 4 Nov. 1943, EM-HU, box 2.

41. Ernst Mayr to Edgar Anderson, 28 Jan. 1941, EM-HU, box 2.

42. Mark V. Barrow, Jr., *A Passion for Birds: American Ornithology after Audubon* (Princeton: Princeton University Press, 1998), 76.

43. Mayr, *Methods and Principles* (cit. n. 5), 30–32, 35, 243–44. Morphological differences between subspecies are in some cases more pronounced than differences between species.

44. On the concept and diagnosis of subspecies, see Mayr, *Methods and Principles* (cit. n. 5), 37–47, 188–97; and Barrow, *Passion for Birds* (cit. n. 42), 76–85.

45. For example, Joseph Grinnell, "The methods and uses of a research museum," *Popular Science Monthly* 77 (1910): 163–69.

46. Richard C. McGregor to Joseph Grinnell, 1 Sept. 1902, JG-UC, box 13.

47. Barrow, *Passion for Birds* (cit. n. 42), 88–101.

48. Joseph Grinnell to Witmer Stone, 16 Feb. 1903, ANS 684.

49. William A. Gosline, "Further thoughts on subspecies and trinomials," *Systematic Zoology* 3 (1954): 92–94, p. 92.

50. Haffer, "History of species concepts" (cit. n. 3), 146–48.

51. Frank M. Chapman, "Criteria for the determination of subspecies in systematic ornithology," *Auk* 41 (1924): 17–29, pp. 23, 18–19.

52. Ibid., pp. 19–25.

53. Mayr, "Speciation phenomena" (cit. n. 27), 256 and 261; also Mayr, *Systematics and the Origin of Species* (cit. n. 8), 166–67.

54. C. Hart Merriam, "Suggestions for a new method of discriminating between species and subspecies," *Science* 5 (1897): 753–58, pp. 753–54. In the end Merriam gave up on subspecies and named every form as a species, to his colleagues' dismay.

55. On the history of objectity: Lorraine Daston, "Objectivity and the escape from perspective," *Social Studies of Science* 22 (1992): 587–618; and Theodore M. Porter, *Trust in Numbers: The Pursuit of Objectivity in Science and Public Life* (Princeton: Princeton University Press, 1995).

56. Haffer, "History of species concepts" (cit. n. 3), 114–15, 118–31, 125 (quote); Leonhard Stejneger, "On the use of trinominals [*sic*] in Amer-

ican ornithology," *Proceedings of the U.S. National Museum* 7 (1884): 71–81, pp. 70–75.

57. Haffer, "History of species concepts: (cit. n. 3), 128–30.

58. E. F. Rivinus and E. M. Youssef, *Spencer Baird of the Smithsonian* (Washington, D.C.: Smithsonian Institution Press, 1992), esp. chap. 8; William A. Deiss, "Spencer F. Baird and his collectors," *Journal of the Society for the Bibliography of Natural History* 9 (1980): 635–45; and Daniel Goldstein, " 'Yours for science': the Smithsonian Institution's correspondents and the shape of scientific community in nineteenth-century America," *Isis* 85 (1994): 573–99.

59. Haffer, "History of species concepts" (cit. n. 3), 130–31.

60. Stejneger, "On the use of trinominals" (cit. n. 56), 76–77; also Joel A. Allen, "Stejneger on trinomials in American ornithology," *Auk* 1 (1884): 381–82.

61. From a manuscript quoted by Stejneger, "On the use of trinominals," ibid., 78–79.

62. Mary P. Winsor, *Reading the Shape of Nature: Comparative Zoology at the Agassiz Museum* (Chicago: University of Chicago Press, 1991), 88, 90–93, 96; Barrow, *Passion for Birds* (cit. n. 42), chap. 4; and Elliott Coues, "Progress of American ornithology," *American Naturalist* 5 (1871): 364–73, pp. 371–72.

63. Winsor, ibid., 76–80; Barrow, ibid., 76–83; Joel Asaph Allen, *Autobiographical Notes and a Bibliography of the Scientific Publications of Joel Asaph Allen* (New York: American Museum of Natural History, 1916), 18–23.

64. Barrow, *Passion for Birds* (cit. n. 42), 83–85; Coues, "Progress," (cit. n. 62), 371–72.

65. Coues, "Progress" (cit. n. 62), on pp. 371–72.

66. Joel A. Allen, "Are trinomials necessary?" *Auk* 1 (1884): 101–104 (quotes, pp. 104, 103); also Montague Chamberlain, "Are trinomials necessary?" ibid., 101–2. Ridgway, "Birds of North and Middle America" (cit. n. 7), x (quote); also Ridgway, "On the use of trinomials in zoological nomenclature," *Bulletin of the Nuttall Ornithological Club* 4 (1879): 129–34, pp. 129–31.

67. Stejneger, "On the use of trinominals" (cit. n. 56), 81, 78–79 (quotes).

68. Elliott Coues, "On the application of trinomial nomenclature to zoology," *Zoologist* 8 (1994): 241–47, pp. 242–43 (quote, p. 243). Also Ridgway, "On the use of trinomials" (cit. n. 66), 129–31; and William H. Dall, "Report of the Committee on Zoological Nomenclature," *Pro-*

ceedings of the American Association for the Advancement of Science 26 (1877): 7–56, pp. 31–33.

69. *The Code of Nomenclature and Check-List of North American Birds Adopted by the American Ornithologists' Union* (New York: American Ornithologists' Union, 1886), 30.

70. Ibid., 30–32; and Coues, "On the application of trinomial nomenclature" (cit. n. 68), 246 (quote). The fullest history of the AUO's adoption of trinomials is Barrow, *Passion for Birds* (cit. n. 42), 94–114.

71. Barrow, ibid., 96–101, esp. fig. 15 (p. 97). Also Robert W. Shufeld, "Progress in American ornithology, 1886–1895," *American Naturalist* 30 (1896): 357–72.

72. Data are taken from Karl P. Schmidt, *A Check List of North American Amphibians and Reptiles*, 6th ed. (Chicago: University of Chicago Press, for the American Society of Ichthyologists and Herpetologists, 1953). I have found no comparable source of data for mammal subspecies. A recent checklist of world birds gives dates for subspecies names. Edward C. Dickinson, ed., *The Howard and Moore Complete Checklist of the Birds of the World* (Princeton: Princeton University Press, 2003).

73. Barrow, *Passion for Birds* (cit. n. 42), 95.

74. Wilfred Osgood, *Revision of the Mice of the American Genus* Peromyscus (Bureau of Biological Survey, North American Fauna no. 28, Washington, D.C.: Government Printing Office, 1909), pp. 17–23.

75. Wilfred Osgood to Edward W. Nelson, 22 May 1919, WHO, box 7.

76. Joel A. Allen, "Zoological nomenclature," *Auk* 1 (1884): 338–53, pp. 338–39, 341, 349; Haffer, "History of species concepts" (cit. n. 3), 131; and Allen, "Are trinomials necessary?" (cit. n. 66), 104.

77. Coues, "On the application of trinomial nomenclature" (cit. n. 68), 243 (quote). Virtually identical language appears in *Code of Nomenclature* (cit. n. 69), 31.

78. Ridgway, "On the use of trinomials" (cit. n. 66), on pp. 131.

79. Ridgway, "Birds of North and Middle America" (cit. n. 7), x.

80. Miller and Rehn, "Systematic results" (cit. n. 34), 1 (quote).

81. *Code of Nomenclature* (cit. n. 69), 31.

82. Allen, "On trinomial nomenclature," Zoologist 7 (1883): 97–100, pp. 99–100.

83. J. Gordon Edwards, "A new approach to infraspecific categories," *Systematic Zoology* 3 (1954): 1–20, pp. 1–2 (quote, 1).

84. Ernst Mayr, "Notes on nomenclature and classification," *Systematic Zoology* 3 (1954): 86–89.

85. Edward O. Wilson and William L. Brown, "The subspecies concept and taxonomic application," *Systematic Zoology* 2 (1953): 97–111. Clines were popularized in the late 1930s by Julian Huxley: for example, Julian S. Huxley, "Clines: an auxiliary method in taxonomy," *Bijdragen tot de Dierkunde* 27 (1939): 491–520.

86. Wilson and Brown, ibid., 107–109.

87. Charles G. Sibley, "The contribution of avian taxonomy," *Systematic Zoology* 3 (1954): 105–110, p. 105.

88. William H. Burt, "The subspecies category in mammals," *Systematic Zoology* 23 (1954): 99–104, p. 99.

89. Sibley, "Contribution of avian taxonomy" (cit. n. 87), p. 105; Haffer, "History of species concepts" (cit. n. 3), 147–48.

90. Data are compiled from Schmidt, *Check List* (cit. n. 72).

91. Theodore H. Hubbell, "The naming of geographical variant populations: or what is all the shooting about?" *Systematic Zoology* 3 (1954): 113–21, p. 114.

92. Edwards, "New approach" (cit. n. 83), 3–4 (quotes, p. 4).

93. William A. Gosline, "Further thoughts on subspecies and trinomials," *Systematic Zoology* 3 (1954): 92–94, p. 92.

94. Wilson and Brown, "Subspecies concept" (cit. n. 85), 103–104; and Edwards, "New approach" (cit. n. 83), 10–12.

95. Mayr, "Notes on nomenclature" (cit. n. 84), 87. Mayr opposed subspecies splitting and favored broadly inclusive, polytypic species.

96. For a similar case in a laboratory science, see Kohler, *Lords of the Fly* (cit. n. 32), 53–62.

97. David Lack, "The taxonomy of the robin *Erithacus rubecula* (Linn)," *Bulletin of the British Ornithologists' Club* 66 (1946): 55–65, pp. 62–63.

98. For example, ibid., 63.

99. Ernst Mayr to Charles G. Sibley, 16 Nov. 1953, quoted in Sibley, "Contribution of avian taxonomy" (cit. n. 87), 107–108.

100. For example, Edwards, "New approach" (cit. n. 83), 10–12; and Mayr, *Methods and Principles* (cit. n. 5), 193–97.

101. Wilson and Brown, "Subspecies concept" (cit. n. 85), 109; Hubbell, "Naming of geographical variant populations" (cit. n. 91), 116, 120.

102. On early efforts to standardize taxonomic rules: Gordon McOuat, "Species, rules, and meaning: the politics of language and the ends of definitions in nineteenth-century natural history," *Studies in the History and Philosophy of Science* 27 (1996): 473–519; idem, "Cataloguing power: delineating 'competent naturalists' and the meaning of species

in the British Museum," *British Journal for the History of Science* 34 (2001): 1–28; and idem, "The politics of 'natural kinds': practices of classification in the age of reform," in *Spaces of Classification*, ed. Ursula Klein (Berlin: Max Planck Institut für Wissenschaftsgeschichte, preprint no. 240, 2003), 97–114.

103. Landmarks of this shift in systematics are: Julian Huxley, *The New Systematics* (Oxford: Oxford University Press, 1940); Huxley, *Evolution and the Modern Synthesis* (New York: Harper, 1942); Mayr, *Systematics and the Origin of Species* (cit. n. 8); and A. J. Cain, *Animal Species and Their Evolution* (London: Hutchinson, 1954).

104. Wendell H. Camp to Ernst Mayr, 4 Nov. 1943; also Mayr to Camp, 1 Nov. 1943; both in EM-HU, box 2. The article referred to is Wendell H. Camp and Charles L. Gilly, "The structure and origin of species," *Brittonia* 4 (1943): 323–85.

105. Anderson, "Hybridization" (cit. n. 11), 512–15.

Chapter 7: Envoi

1. Ernst Mayr, "Biological materials," National Research Council report [1957], p. 15, KPS, box 12, f. 4.

2. For example: Ernst Mayr and William B. Provine, eds., *The Evolutionary Synthesis: Perspectives on the Unification of Biology* (Cambridge: Harvard University Press, 1980); Vassiliki Betty Smocovitis, *Unifying Biology: The Evolutionary Synthesis and Evolutionary Biology* (Princeton: Princeton University Press, 1996); and Joseph A. Cain, "Managing Synthesis: Community Infrastructure in the Synthesis Period of American Evolutionary Studies," Ph.D. diss., University of Minnesota, 1996.

3. For example: Harold K. Steen, ed., *Forest and Wildlife Science in America: A History* (Asheville: Forest History Society, 1999); Thomas R. Dunlap, *Saving America's Wildlife* (Princeton: Princeton University Press, 1988); Christian C. Young, "Defining the range: the development of carrying capacity in management practice," *Journal of the History of Biology* 31 (1998): 61–83; Ronald C. Tobey, *Saving the Prairies: The Life Cycle of the Founding School of American Plant Ecology, 1895–1955* (Berkeley: University of California Press, 1981); and Curt Meine, *Aldo Leopold: His Life and Work* (Madison: University of Wisconsin Press, 1988).

4. Donald Worster, *Nature's Economy: A History of Ecological Ideas* (New York: Cambridge University Press, 1977; 2nd ed. 1994), part 6.

Also Peter Bowler, *The Norton History of the Environmental Sciences* (New York: Norton, 1993), chap. 11.

5. Herbert L. Stoddard, *Memoirs of a Naturalist* (Norman: University of Oklahoma Press, 1969); Joseph Grinnell to Wilfred Osgood, 19 Dec. 1921 (on Dixon), JG-MVZ.

6. A. Hunter Dupree, *Science in the Federal Government* (Cambridge: Harvard University Press, 1957), 157–61, 182–83.

7. Jenks Cameron, *The Bureau of Biological Survey: Its History, Activities, and Organization* (Baltimore: Johns Hopkins University Press, 1929), 36–42 (increasing economic emphasis), and 42–147 (predator control); Dunlap, *Saving America's Wildlife* (cit. n. 3).

8. Osgood resigned from the Survey in 1910 and in 1915 declined an offer to succeed Merriam as its director. Edward W. Nelson to Wilfred Osgood, 18 July, 29 Aug. 1915; Osgood to Nelson, 20 Aug. 1915; all in WHO, box 7.

9. Rolland C. Allen, "Annual report of the director to the Board of Scientific Advisors to the Michigan Geological and Biological Survey," Dec. 1913, AGR, box 54, f. 6.

10. Charles A. Kofoid to William E. Ritter, 7 Sept., 9 Oct., 7 Nov. 1912, all in WER, box 13.

11. Charles Cadwalader to Melbourne A. Carriker, Jr., 16 Aug. 1933, Carriker Papers, box 1, f. D., ANS 900.

12. These generalizations are based on the Museum's annual reports.

13. Mark Barrow, *A Passion for Birds: American Ornithology after Audubon* (Princeton: Princeton University Press, 1998), 206–207, citing Witmer Stone, "The ornithology of today and tomorrow," in *The Fiftieth Anniversary of the Nuttal Ornithological Club* (Cambridge: Nuttall Ornithological Club, 1924), 7–25, p. 8.

14. Witmer Stone, "American ornithological literature 1883–1923," in *Fifty Years' Progress of American Ornithology 1883–1933* (Lancaster: American Ornithologists' Union, 1933), 29–49, p. 45.

15. Edward W. Nelson, "Memorandum for field naturalists of the Biological Survey," 12 June 1917, WHO, box 7.

16. Robin W. Doughty, *Feather Fashion and Bird Preservation: A Study in Nature Protection* (Berkeley: University of California Press, 1975); and Barrow, *Passion for Birds* (cit. n. 13), chaps. 6, 7.

17. C. Hart Merriam to Witmer Stone, 5 Jan 1924, ANS 678.

18. Joseph Grinnell to Susie L. Dyer, no date, quoted in Sally G. Kohlstedt, The Nature Study Movement: Science in Public Schooling, 1890–1932 (unpublished ms.), chap. 10.

19. Richard C. McGregor to Joseph Grinnell, 20 July 1903, JG-UC, box 13.

20. Joseph Grinnell to Walter K. Fisher, 7 June 1912, JG-MVZ. On the American Ornithologists' Union's response to bird-protection movements, see Barrow, *Passion for Birds* (cit. n. 13), pp. 141–46, and chaps. 5, 6.

21. Henry F. Nachtrieb to Governor A. E. Rice, 1 May 1908; Nachtrieb to Knute Nelson, 22 Jan. 1909; Nachtrieb to Governor John A. Johnson, 24 Aug. 1908; all in HFN box 1.

22. Alexander Ruthven to George Shiras, 27 Oct. 1912, AGR, box 52, f. 8; Ruthven to Bryant Walker, 13 June 1912, AGR, box 52, f. 18. Ruthven to William Oates (Game Warden), 13 May 1912; Oates to Ruthven, 14 May 1912; both in AGR, box 54, f. 5.

23. Joseph Grinnell to Annie Alexander, 19 April 1912, AA C-B1003, box 1; 10 March 1921 (quote), 20 Jan. 1927, both in AA C-B1003, box 2.

24. Mayr, "Biological materials," (cit. n. 1), p. 15.

25. Ernst Mayr, *Principles of Systematic Zoology* (New York: McGraw-Hill, 1969), 105, 183–85.

26. George H. Lauff and David Reichle, "Experimental ecological reserves," *Ecological Society of America Bulletin* 60 (1979): 4–11; and Joel B. Hagen, "Problems in the institutionalization of tropical biology: the case of the Barro Colorado Island biological laboratory," *History and Philosophy of the Life Sciences* 12 (1990): 225–47.

27. *American Museum of Natural History Annual Report* 67 (1935): 32, and ibid., 68 (1936): 38; A. E. Parr, "Towards new horizons," ibid., 78 (1946–1947): 9–24, pp. 20–21; ibid., 80 (1948–1949): [unpaginated]; Parr, "Purposes and progress: report of the director," ibid., 82 (1950–1951): 7–36, pp. 9–12.

28. See n. 1; also Leo F. Laporte, *George Gaylord Simpson: Paleontologist and Evolutionist* (New York: Columbia University Press, 2000).

29. John M. Kennedy, "Philanthropy and Science in New York City: The American Museum of Natural History, 1868–1968," Ph.D. diss., Yale University, 1968, 239–48; also Theodosius Dobzhansky to Robert C. Murphy, 7 March 1947, Ernst Mayr Papers BD65m, American Philosophical Society Library, Philadelphia, Pa.

30. For example: Dona Brown, *Inventing New England: Regional Tourism in the Nineteenth Century* (Washington, D.C.: Smithsonian Institution Press, 1995), 201–18; Hal K. Rothman, *Devil's Bargains: Tourism in the Twentieth-Century American West* (Lawrence: University of Kansas Press, 1998); and Paul S. Sutter, *Driven Wild: How the Fight Against*

Automobiles Launched the Modern Wilderness Movement (Seattle: University of Washington Press, 2002).

31. For example: Philip V. Scarpino, *Great River: An Environmental History of the Upper Mississippi, 1890–1950* (Columbia: University of Missouri Press, 1985); and Daniel W. Schneider, "Enclosing the floodplain: resource conflict on the Illinois River, 1880–1920," *Environmental History* 1:2 (1996): 70–96.

32. For example: William U. Chandler, *The Myth of TVA: Conservation and Development in the Tennessee Valley, 1933–1983* (Cambridge: Ballinger, 1984); Norris Hundley, Jr., *The Great Thirst: Californians and Water, 1770s–1990s* (Berkeley: University of California Press, 1992); and Richard White, *Organic Machine: The Remaking of the Columbia River* (New York: Hill and Wang, 1995).

33. For example: Sutter, *Driven Wild* (cit. n. 30), chap. 2; and Adam Rome, *The Bulldozer in the Countryside: Suburban Sprawl and the Rise of American Environmentalism* (New York: Cambridge University Press, 2001).

34. Robert Ridgeway to Witmer Stone, 27 Feb. 1920, ANS 681.

35. Daniel Goldstein, "Midwestern Naturalists: Academies of Science in the Mississippi Valley, 1850–1900," Ph.D. diss., Yale University, 1989; and Elizabeth B. Keeney, *The Botanizers: Amateur Scientists in Nineteenth-Century America* (Chapel Hill: University of North Carolina Press, 1992), chap. 2. Behind this decline in natural history societies is the waning authority of town culture in the 1920s, an important subject that has been largely ignored. But see Page Smith, *As a City upon a Hill: The Town in American History* (New York: Knopf, 1966), chap. 13.

36. Cindy Aron, *Working at Play: A History of Vacations in the United States* (New York: Oxford University Press, 1999), chaps. 7–9.

37. For example: Jennifer Price, *Flight Maps: Adventures with Nature in Modern America* (New York: Basic Books, 1999), chaps. 3–5; Susan G. Davis, *Spectacular Nature: Corporate Culture and Sea World Experience* (Berkeley: University of California Press, 1997); Gregg Mitman, *Reel Nature: America's Romance with Wildlife on Film* (Cambridge: Harvard University Press, 1999).

38. Data are from Charles G. Sibley and Burt L. Monroe, Jr., *Distribution and Taxonomy of Birds of the World* (New Haven: Yale University Press, 1990).

39. Data are from Karl P. Schmidt, *A Check List of North American Amphibians and Reptiles*, 6th ed. (Chicago: University of Chicago Press, 1953).

40. Data are from James H. Honacki, Kenneth E. Kinman, and James W. Koeppl, *Mammal Species of the World: A Taxonomic and Geographic Reference* (Lawrence, Kans.: Association of Systematics Collections, and Allen Press, 1982).

41. Vannever Bush, *Science the Endless Frontier: A Report to the President on a Program for Postwar Scientific Research* (Washington, D.C.: Government Printing Office, 1945: repr., National Science Foundation, 1960).

42. Dorinda Outram. "New spaces in natural history," in *Cultures of Natural History*, ed. Nicholas Jardine, James A. Secord, and E. C. Sparry (Cambridge: Cambridge University Press, 1996), 249–65.

43. William C. Sturtevant, "Museums as anthropological data banks," *Southern Anthropological Society Proceedings* 7 (1973): 40–55; and Chris Gosden and Chantal Knowles, *Collecting Colonialism: Material Culture and Colonial Exchange* (Oxford: Berg, 2001).

44. Edward O. Wilson, "The current state of biological diversity," in *Biodiversity*, ed. Wilson and Frances M. Peter (Washington, D.C.: National Academy Press, 1988), 3–20.

45. R. W. Taylor, "Descriptive taxonomy: past, present, and future," in *Australian Systematic Entomology: A Bicentenary Perspective*, ed. E. Highley and R. W. Taylor (Melbourne: Commonwealth Scientific and Industrial Research organization, 1983), 93–134, pp. 100–104.

46. Edward O. Wilson, "The biological diversity crisis: a challenge to science," *Issues in Science and Technology* 2 (1985): 20–29; R. M. May, "How many species are there on earth?" *Science* 241 (1988): 1441–49.

Selected Bibliography

Abir-Am, Pnina G., and Dorinda Outram, editors. *Uneasy Careers and Intimate Lives: Women in Science 1789–1979*. New Brunswick: Rutgers University Press, 1987.

Allen, David E. *The Naturalist in Britain: A Social History.* London: Allen Lane, 1976; reprinted, Princeton University Press, 1994.

Altherr, Thomas L. "The American hunter-naturalist and the development of the code of sportsmanship." *Journal of Sport History* 5 (1978): 7–22.

Anderson, Edgar. *Plants, Man and Life*. Cambridge: Cambridge University Press, 1952; reprinted Berkeley: University of California Press 1969.

Aron, Cindy. *Working at Play: A History of Vacations in the United States*. New York: Oxford University Press, 1999.

Austin, Elizabeth S., editor. *Frank M. Chapman in Florida: His Journals and Letters*. Gainesville, University of Florida Press, 1967.

Bailey, Peter. *Leisure and Class in Victorian England: Rational Recreation and the Contest for Control, 1830–1885*. London: Routledge & Kegan Paul, 1978.

Barrow, Mark V., Jr. *A Passion for Birds: American Ornithology after Audubon*. Princeton: Princeton University Press, 1998.

Bartlett, Richard A. *Great Surveys of the American West*. Norman: University of Oklahoma Press, 1962.

Biagioli, Mario. *Galileo Courtier: The Practice of Science in the Age of Absolutism*. Chicago: University of Chicago Press, 1993.

Blackman, Betsy. "Going to the mountains: a social history." In *Resorts of the Catskills*. New York: St. Martin's Press, 1979, 71–99.

Bluestone, Daniel M. "From promenade to park: the gregarious origins of Brooklyn's park movement." *American Quarterly* 39 (1987): 529–50.

Blumin, Stuart M. "Hypothesis of middle-class formation in nineteenth-century America: a critique and some proposals." *American Historical Review* 90 (1985): 299–338.

Bowker, Geoffrey C., and Susan Leigh Star. *Sorting Things Out: Classification and Its Consequences*. Cambridge: MIT Press, 1999.

Brown, Dona. *Inventing New England: Regional Tourism in the Nineteenth Century.* Washington: Smithsonian Institution Press, 1995.

Browne, Janet. "Biogeography and empire." In *Cultures of Natural History,* edited by Jardine et al. (1996), 305–21.

Bulkley, Peter B. "Identifying the White Mountain tourist, 1853–1854: origin, occupation, and wealth as a definition of the early hotel trade." *Historical New Hampshire* 35 (1980): 106–62.

Camerini, Jane. "Wallace in the field." *Osiris* 11 (1996): 44–65.

Cameron, Jenks. *The Bureau of Biological Survey: Its History, Activities, and Organization.* Baltimore: Johns Hopkins University Press, 1929.

Cannon, Susan Faye. "Humboldtian science." In *Science in Culture: The Early Victorian Period.* New York: Science History Publications, 1978, 73–110.

Chapman, Frank M. *Autobiography of a Bird Lover.* New York: Appleton-Century, 1933.

Cole, Douglas J. *Captured Heritage: The Scramble for Northwest Coast Artifacts.* Seattle: University of Washington Press, 1985.

Croker, Robert A. *Stephen Forbes and the Rise of American Ecology.* Washington: Smithsonian Institution Press, 2001.

Cronon, William. "Revisiting the vanishing frontier: the legacy of Frederick Jackson Turner." *Western Historical Quarterly* 18 (1987): 157–76.

———. *Nature's Metropolis: Chicago and the Great West.* New York: Norton, 1991.

Daston, Lorraine. "Objectivity and the escape from perspective." *Social Studies of Science* 22 (1992): 587–618.

———. "Fear and loathing of the imagination in science." *Daedalus* 127:1 (1998): 73–85.

Daston, Lorraine, and Peter Galison. "The image of objectivity." *Representations* 40 (1992): 81–128.

Deiss, William A. "Spencer F. Baird and his collectors." *Journal of the Society for the Bibliography of Natural History* 9 (1980): 635–45.

Dickinson, Edward C., editor. *The Howard and Moore Complete Checklist of the Birds of the World.* Princeton: Princeton University Press, 2003.

Doughty, Robin W. *Feather Fashions and Bird Preservation: A Study in Nature Protection.* Berkeley: University of California Press, 1975.

Dunlap, Thomas R. *Saving America's Wildlife.* Princeton: Princeton University Press, 1988.

Dupree, A. Hunter. *Science in the Federal Government: A History of Policies and Activities to 1940*. Cambridge: Harvard University Press, 1957.

Earle, Carville. "Place your bets: rates of frontier expansion in American history, 1650–1890." In *Cultural Encounters with the Environment: Enduring and Evolving Geographic Themes*, edited by Alexander B. Murphy and Douglas L. Johnson. Lanham, Md.: Rowman and Littlefield, 2000, 79–105.

Elliott, Clark. *Biographical Index to American Science, the Seventeenth Century to 1920*. Westport, Conn.: Greenwood Press, 1990.

Findlen, Paula. *Possessing Nature: Museums, Collecting, and Scientific Culture in Early Modern Italy*. Berkeley: University of California Press, 1994.

———. "Between carnival and lent: the scientific revolution at the margins of culture." *Configurations* 6 (1998): 243–67.

Frost, Darrel R. "Amphibian Species of the World: An Online Reference." American Museum of Natural History, 1998 and ongoing. http://research.amnh.org/herpetology/amphibia/index.html

Gelber, Steven M. "Working at playing: the culture of the work place and the rise of baseball." *Journal of Social History* 16:4 (June 1983): 3–20.

Goetzmann, William H. *Army Exploration in the American West, 1803–1863*. New Haven: Yale University Press, 1959.

———. *Exploration and Empire: The Explorer and the Scientist in the Winning of the American West*. New York: Knopf, 1966.

———. "Exploration and the culture of science: the long good-bye of the twentieth century." In *Making America: The Society and Culture of the United States*, edited by Luther S. Leudtke. Chapel Hill: University of North Carolina Press, 1992, 413–31.

Goetzmann, William H., and Kay Sloan. *Looking Far North: The Harriman Expedition to Alaska, 1899*. New York: Viking, 1982.

Goldstein, Daniel. "Midwestern Naturalists: Academies of Science in the Mississippi Valley, 1850–1900." Ph.D. dissertation, Yale University, 1989.

———. " 'Yours for science': the Smithsonian Institution's correspondents and the shape of scientific community in nineteenth-century America." *Isis* 85 (1994): 573–99.

Golinski, Jan. *Making Natural Knowledge: Constructivism and the History of Science*. New York: Cambridge University Press, 1998.

Grove, Richard H. *Green Imperialism: Colonial Expansion, Tropical Island Edens, and the Origin of Environmentalism, 1600–1860*. New York: Cambridge University Press, 1995.

Grover, Kathryn, editor. *Hard at Play: Leisure in America, 1840–1940*. Amherst: University of Massachusetts Press, 1992.

Haffer, Jürgen. "The history of species concepts and species limits in ornithology." *Bulletin of the British Ornithological Club* Centenary Supplement 112A (1992): 107–58.

Hagen, Joel B. "Experimentalists and naturalists in twentieth-century botany: experimental taxonomy, 1920–1950." *Journal of the History of Biology* 17 (1984): 249–70.

Halttunen, Karen. *Confidence Men and Painted Women: A Study of Middle-Class Culture in America, 1830–1870*. New Haven: Yale University Press, 1982.

Harraway, Donna. "Teddy bear patriarchy: taxidermy in the Garden of Eden, New York City, 1908–30." In *Primate Visions: Gender, Race, and Nature in the World of Modern Science*. New York: Routledge, 1989, 26–58.

Harris, Neil. "On vacation." In *Resorts of the Catskills*. New York: St. Martin's Press, 1979, 101–108.

Hays, Robert G. *State Science in Illinois: The Scientific Surveys, 1850–1978*. Carbondale: Southern Illinois University Press, 1980.

te Heesen, Anke, and Emma C. Spary, editors. *Sammeln als Wissen: Das Sammeln und seine wissenschaftsgeschichtliche Bedeutung*. Göttingen: Wallenstein Verlag, 2001.

Hevly, Bruce. "The heroic science of glacier motion." *Osiris* 11 (1996): 66–86.

Higham, John. "The reorientation of American culture in the 1890s." In *Writing American History: Essays on Modern Scholarship*. Bloomington: University of Indiana Press, 1970, 73–102.

Honacki, James H., Kenneth E. Kinman, and James W. Koeppl. *Mammal Species of the World: A Taxonomic and Geographic Reference*. Lawrence: Association of Systematics Collections, and Allen Press, 1982.

Huth, Hans. *Nature and the American: Three Centuries of Changing Attitudes*. Berkeley: University of California Press, 1957.

Iliffe, Rob. "Science and voyages of discovery." In *The Cambridge History of Science*, vol. 4, *Eighteenth-Century Science*, edited by Roy Porter. Cambridge: Cambridge University Press, 2003, 618–45.

Jacoby, Karl. *Crimes Against Nature: Squatters, Poachers, Thieves, and the Hidden History of American Conservation*. Berkeley: University of California Press, 2001.

Jardine, Nicholas, James A. Secord, and E. C. Sparry, editors. *Cultures of Natural History*. Cambridge: Cambridge University Press, 1996.

Keeney, Elizabeth B. *The Botanizers: Amateur Scientists in Nineteenth-Century America*. Chapel Hill: University of North Carolina Press, 1992.

Klein, Kerwin L. "Frontier products: tourism, consumerism, and the Southwestern public lands, 1890–1990." *Pacific History Review* 62 (1993): 39–71.

Knell, Simon J. *The Culture of English Geology, 1815–1851*. Aldershot, U.K.: Ashgate, 2000.

Koelsch, William A. "Antebellum Harvard students and the recreational exploration of the New England landscape." *Journal of Historical Geography* 8 (1982): 362–72.

Koerner, Lisbet. *Linnaeus: Nature and Nation*. Cambridge: Harvard University Press, 1999.

Kofalk, Harriet. *No Woman Tenderfoot: Florence Merriam Bailey, Pioneer Naturalist*. College Station: Texas A&M University Press, 1989.

Kohler, Robert E. *Partners in Science: Foundations and Natural Scientists, 1900–1945*. Chicago: University of Chicago Press, 1991.

———. "Moral economy, material culture, and community in *Drosophila* genetics." In *The Science Studies Reader*, edited by Mario Biagioli. London: Routledge, 1999, 243–57.

———. "Place and practice in field biology." *History of Science* 40 (2002): 189–210.

———. *Landscapes and Labscapes: Exploring the Lab-Field Border in Biology*. Chicago: University of Chicago Press, 2002.

Kohlstedt, Sally G. "Henry A. Ward: the merchant naturalist and American museum development." *Journal of the Society for the Bibliography of Natural History* 9 (1980): 647–61.

———. "Collectors, cabinets, and summer camp: natural history in the public life of nineteenth century Worcester." *Museum Studies Journal* 2 (1985): 10–23.

———. "International exchange and national style: a view of natural history museums in the United States, 1850–1900." In *Scientific Colonialism*, edited by Reingold and Rothenberg (1987), 167–90.

Kuklick, Henrika, and Robert E. Kohler, editors. "Science in the Field." *Osiris* 11 (1996).

Larsen, Anne L. "Not Since Noah: The English Scientific Zoologists and the Craft of Collecting, 1800–1840." Ph.D. dissertation, Princeton University, 1993.

Latour, Bruno, and Steven Woolgar. *Laboratory Life: The Construction of Scientific Facts.* Second edition, Princeton: Princeton University Press, 1986.

Löfgren, Orvar. *On Holiday: A History of Vacationing.* Berkeley: University of California Press, 1999.

Lowenthal, David. "Geography, experience, and imagination: towards a geographical epistemology." *Annals of the Association of American Geographers* 51 (1961): 241–60.

Lutts, Ralph H. *The Nature Fakers: Wildlife Science and Sentiment.* Golden, Colo.: Fulcrum, 1990.

McCook, Stuart. " 'It may be truth, but it is not evidence': Paul du Chaillu and the legitimation of evidence in the field science." *Osiris* 11 (1996): 177–200.

McCorvie, Mary R., and Christopher L. Lant. "Drainage district formation and the loss of midwestern wetlands, 1850–1930." *Agricultural History* 67:4 (Fall 1993): 13–39.

McOuat, Gordon. "Species, rules, and meaning: the politics of language and the ends of definitions in nineteenth-century natural history." *Studies in the History and Philosophy of Science* 27 (1996): 473–519.

———. "Cataloguing power: delineating 'competent naturalists' and the meaning of species in the British Museum." *British Journal for the History of Science* 34 (2001): 1–28.

———. "The politics of 'natural kinds': practices of classification in the age of reform." In *Spaces of Classification*, edited by Ursula Klein. Berlin: Max Plank Institut für Wissenschaftsgeschichte, preprint no. 240, 2003, 97–114.

Maienschein, Jane. "Pattern and process in early studies of Arizona's San Francisco Peaks." *BioScience* 44 (1994): 479–85.

Manning, Thomas G. *Government in Science: The U.S. Geological Survey 1867–1894.* Lexington: University of Kentucky Press, 1967.

Marx, Leo. *The Machine in the Garden: Technology and the Pastoral Ideal in America.* New York: Oxford University Press, 1964.

Mayr, Ernst. *Systematics and the Origin of Species from the Viewpoint of a Zoologist.* New York: Columbia University Press, 1942.

———. *Principles of Systematic Zoology.* New York: McGraw-Hill, 1969.

Mayr, Ernst, E. G. Linsley, and R. L. Usinger. *Methods and Principles of Systematic Zoology*. New York: McGraw-Hill, 1953.

Meindle, Christopher F. "Past perceptions of the great American wetland: Florida's Everglades during the early twentieth century." *Environmental History* 5 (2000): 378–95.

Meinig, Donald W., editor. *The Interpretation of Ordinary Landscapes: Geographical Essays*. New York: Oxford University Press, 1979.

———. *The Shaping of America: A Geographical Perspective on 500 Years of History. Volume 3. Transcontinental America 1850–1915*. New Haven: Yale University Press, 1998.

Merrill, Frederick J. H. "Natural History Museums of the United States and Canada." *New York State Museum Bulletin* 62. Albany: University of the State of New York, 1903.

Miller, David P., and P. Reill, editors. *Visions of Empire: Voyages, Botany, and Representations of Nature*. Cambridge: Cambridge University Press, 1996.

Mitman, Gregg. "Cinematic nature: Hollywood technology, popular culture, and the American Museum of Natural History." *Isis* 84 (1993): 637–61.

———. *Reel Nature: America's Romance with Wildlife on Film*. Cambridge: Harvard University Press, 1999.

Nash, Roderick. *Wilderness and the American Mind*. New Haven: Yale University Press, 1967.

Nyhart, Lynn K. "Science, art, and authenticity in natural history displays." In *Models: The Third Dimension of Science*. Edited by Soroya de Chadarevian and Nick Hopwood. Stanford: Stanford University Press, 2004, 307–35.

Oreskes, Naomi. "Objectivity or heroism? on the invisibility of women in science." *Osiris* 11 (1996): 87–116.

Osborne, Michael A. *Nature, the Exotic, and the Science of French Colonialism*. Bloomington: Indiana University Press, 1994.

Outram, Dorinda. "New spaces in natural history." In *Cultures of Natural History*, edited by Jardine et al. (1996), 249–65.

Pang, Alex Soojung-Kim. "The social event of the season: solar eclipse expeditions and Victorian culture." *Isis* 84 (1993): 252–77.

———. "Gender, culture, and astrophysical fieldwork: Elizabeth Campbell and the Lick Observatory–Crocker eclipse expeditions." *Osiris* 11 (1996): 17–43.

———. *Empire of the Sun: Victorian Solar Eclipse Expeditions*. Stanford: Stanford University Press, 2002.

Peters, James L. "Collections of birds in the United States and Canada: study collections." In *Fifty Years' Progress of American Ornithology 1883–1933*. Lancaster, Pa.: American Ornithologists' Union, 1933, 131–41.

Pomeroy, Earl. *In Search of the Golden West: The Tourist in Western America*. Lincoln: University of Nebraska Press, 1957.

Porter, Theodore M. "Objectivity as standardization: the rhetoric of impersonality in measurement, statistics, and cost-benefit analysis." *Annals of Scholarship* 9 (1992): 19–59.

———. *Trust in Numbers: The Pursuit of Objectivity in Science and Public Life*. Princeton: Princeton University Press, 1995.

Price, Jennifer. *Flight Maps: Adventures with Nature in Modern America*. New York: Basic Books, 1999.

Prince, Hugh. *Wetlands of the American Midwest: A Historical Geography of Changing Attitudes*. Chicago: University of Chicago Press, 1997.

Reiger, John F. *American Sportsmen and the Origins of Conservation*. New York: Winchester Press, 1975.

Reingold, Nathan. "Definitions and speculations: the professionalization of science in American in the nineteenth century." In *The Pursuit of Knowledge in the Early American Republic: American Scientific and Learned Societies from Colonial Times to the Civil War*, edited by Alexandra C. Oleson and Sanborn C. Brown. Baltimore: Johns Hopkins University Press, 1976, 33–69.

Reingold, Nathan, and Marc Rothenberg, editors. *Scientific Colonialism: A Cross-Cultural Comparison*. Washington, D.C.: Smithsonian Institution Press, 1987.

Renehan, Edward J., Jr. *John Burroughs: An American Naturalist*. Post Mills, Vt.: Chelsea Green, 1992.

Riffenburgh, Beau. *The Myth of the Explorer: The Press, Sensationalism, and Geographical Discovery*. New York: Oxford University Press, 1994.

Rivinus, E. F., and E. M. Youssef. *Spencer Baird of the Smithsonian*. Washington, D.C.: Smithsonian Institution Press, 1992.

Robbins, David. "Sport, hegemony and the middle class: the Victorian mountaineers." *Theory, Culture and Society* 4 (1987): 579–601.

Rodgers, Daniel T. *The Work Ethic in Industrial America 1850–1920*. Chicago: University of Chicago Press, 1974.

Rome, Adam. *The Bulldozer in the Countryside: Suburban Sprawl and the Rise of American Environmentalism*. New York: Cambridge University Press, 2001.

Ross, Edward S. "Systematic entomology." In *A Century of Progress in the Natural Sciences, 1853–1953*. San Francisco: California Academy of Sciences, 1955, 485–95.

Rothman, Hal K. *Devil's Bargains: Tourism in the Twentieth-Century American West*. Lawrence, Kans.: University of Kansas Press, 1998.

Rudwick, Martin. "Encounters with Adam, or at least the hyaenas: nineteenth-century visual representations of the deep past." In *History, Humanity, and Evolution*, edited by James Moore. New York: Cambridge University Press, 1989, 231–52.

———. "Geological travel and theoretical innovation: the role of 'liminal' experience." *Social Studies of Science* 26 (1996): 143–59.

Runte, Alfred. *National Parks: The American Experience*. Lincoln: University of Nebraska Press, 1979; revised edition, 1987.

Sahlins, Marshall. *Islands of History*. Chicago: University of Chicago Press, 1985.

Schlereth, Thomas J. "Chautauqua: a middle landscape of the middle class." *Old Northwest* 12 (1986): 265–78.

Schmidt, Karl P. *A Check List of North American Amphibians and Reptiles*. Sixth edition, Chicago: University of Chicago Press, for the American Society of Ichthyologists and Herpetologists, 1953.

———."Herpetology." In *A Century of Progress in the Natural Sciences 1853–1953*. San Francisco: California Academy of Sciences, 1955, 591–627.

Schmitt, Peter J. *Back to Nature: The Arcadian Myth in Urban America*. New York: Oxford University Press, 1969; reprinted, Baltimore: Johns Hopkins University Press, 1990.

Schneider, Daniel W. "Enclosing the floodplain: resource conflict on the Illinois River, 1880–1920." *Environmental History* 1:2 (1996): 70–96.

Schneider, Paul. *The Adirondacks: A History of America's First Wilderness*. Boston: Henry Holt, 1997.

Scott, James C. *Seeing Like a State: How Certain Schemes to Improve the Human Condition Have Failed*. New Haven: Yale University Press, 1998.

Sellers, Richard W. *Preserving Nature in the National Parks: A History*. New Haven: Yale University Press, 1997.

Shapin, Steven. "The house of experiment." *Isis* 79 (1988): 373–404.

Shapin, Steven. "Who was Robert Hooke?" In *Robert Hooke: New Studies*, edited by Michael Hunter and Simon Schaffer. Woodbridge, U.K.: Boydell Press, 1989, 253–86.

———. " 'A scholar and a gentleman': the problematic identity of the scientific practitioner in early modern England." *History of Science* 29 (1991): 279–327.

Sheets-Pyenson, Susan. "How to 'grow' a natural history museum: the building of colonial collections, 1850–1900." *Archives of Natural History* 15 (1887): 121–47.

Sibley, Charles G., and Burt L. Monroe, Jr. *Distribution and Taxonomy of Birds of the World*. New Haven: Yale University Press, 1990.

Sorrenson, Richard. "The ship as a scientific instrument in the eighteenth century." *Osiris* 11 (1996): 221–36.

Spirn, Anne W. "Constructing nature: the legacy of Frederick Law Olmsted." In *Uncommon Ground: Toward Reinventing Nature*, ed. William Cronon. New York: W. W. Norton, 1995, 91–113.

Stafford, Robert A. *Scientist of Empire: Sir Roderick Murchison, Scientific Exploration and Victorian Imperialism*. Cambridge: Cambridge University Press, 1989.

Star, Susan Leigh. "Craft vs. commodity, mess vs. transcendence: how the right tool became the wrong one in the case of taxidermy and natural history." In *The Right Tools for the Job: At Work in Twentieth-Century Life Sciences*, edited by Adele E. Clarke and Joan H. Fujimura. Princeton: Princeton University Press, 1992, 257–86.

Star, Susan Leigh,and James Griesemer. "Institutional ecology: 'translations' and boundary objects: amateurs and professionals in Berkeley's Museum of Vertebrate Zoology, 1907–39." *Social Studies of Science* 19 (1989): 387–420.

Steen, Harold K., editor. *Forest and Wildlife Science in America: A History*. Asheville, N.C.: Forest History Society, 1999.

Stein, Barbara R. *On Her Own Terms: Annie Montague Alexander and the Rise of Science in the American West*. Berkeley: University of California Press, 2002.

Sterling, Keir. *Last of the Naturalists: The Career of C. Hart Merriam*. New York: Arno, 1977.

Stewart, Larry. "Global pillage: science, commerce, and empire." In *The Cambridge History of Science, vol. 4, Eighteenth-Century Science*, edited by Roy Porter. Cambridge: Cambridge University Press, 2003, 825–44.

Stilgoe, John. "The wildering of rural New England 1850 to 1950." *New England and Saint Lawrence Valley Geographical Society Proceedings* 10 (1980): 1–6.

———. *Borderland: Origins of the American Suburb, 1820–1939*. New Haven: Yale University Press, 1988.

Stone, Witmer. "American ornithological literature 1883–1923." In *Fifty Years' Progress of American Ornithology 1883–1933*. Lancaster: American Ornithologists' Union, 1933, 29–41.

Strauss, David. "Toward a consumer culture: 'Adirondack Murray' and the wilderness vacation." *American Quarterly* 39 (1987): 270–86.

Sturtevant, William C. "Museums as anthropological data banks." *Proceedings of the Southern Anthropological Society* 7 (1973): 40–55.

Sutter, Paul S. *Driven Wild: How the Fight Against Automobiles Launched the Modern Wilderness Movement*. Seattle: University of Washington Press, 2002.

Terrie, Philip G. *Contested Terrain: A New History of Nature and People in the Adirondacks*. Syracuse: Syracuse University Press, 1997.

Tobey, Ronald C. *Saving the Prairie: The Life Cycle of the Founding School of American Plant Ecology, 1895–1955*. Berkeley: University of California Press, 1981.

Tobin, Gary A. "The bicycle boom of the 1890s: the development of private transportation and the birth of the modern tourist." *Journal of Popular Culture* 7 (1974): 838–49.

Traweek, Sharon. *Beamtimes and Lifetimes: The World of High Energy Physicists*. Cambridge: Harvard University Press, 1988.

Trefethen, James B. *An American Crusade for Wildlife*. New York: Winchester Press, 1975.

Turner, Paul V. *Campus: An American Planning Tradition*. Cambridge: MIT Press, 1984.

Uminowicz, Glenn A. "Sport in a middle-class utopia: Asbury Park, New Jersey, 1871–1895." *Journal of Sport History* 11 (1984): 51–73.

Warren, Louis. *The Hunter's Game: Poachers and Conservationists in Twentieth-Century America*. New Haven: Yale University Press, 1997.

Waterman, Laura, and Guy Waterman. *Forest and Crag: A History of Hiking, Trail Blazing, and Adventure in the Northeast Mountains*. Boston: Appalachian Mountain Club, 1989.

White, Richard. *Land Use, Environment, and Social Change: The Shaping of Island County Washington*. Seattle: University of Washington Press, 1980.

White, Richard. *"It's Your Misfortune and None of My Own": A History of the American West.* Norman: University of Oklahoma Press, 1991.

———. *The Organic Machine: The Remaking of the Columbia River.* New York: Hill and Wang, 1995.

Wiggins, David K. "Work, leisure, and sport in America: the British traveller's image, 1830–1860." *Canadian Journal of the History of Sport* 13 (1982): 28–60.

Williams, Michael. *Americans and Their Forests: A Historical Geography.* New York: Cambridge University Press, 1989.

Wilson, Edward O. "The current state of biological diversity." In *Biodiversity,* edited by Wilson. Washington, D.C.: National Academy Press, 1988, 3–18.

Winsor, Mary P. *Reading the Shape of Nature: Comparative Zoology at the Agassiz Museum.* Chicago: University of Chicago Press, 1991.

Wonders, Karen. *Habitat Dioramas: Illusions of Wilderness in Museums of Natural History.* Stockholm: Almqvist and Wiksell, 1993.

Worster, Donald. *Nature's Economy: A History of Ecological Ideas.* New York: Cambridge University Press, 1977; second edition, 1994.

Zetzel, Susanna S. "The garden in the machine: the construction of nature in Olmsted's Central Park." *Prospects* 14 (1989): 291–339.

Zunz, Olivier. *Making America Corporate, 1870–1920.* Chicago: University of Chicago Press, 1990.

Index